Digital Signal Processing

Digital Signal Processing

Wataru Mayeda

PTR Prentice Hall
Englewood Cliffs, New Jersey 07632

Library of Congress Cataloging-in-Publication Data

Mayeda, Wataru,
[Dijitaru shingō shori no kiso. English]
Digital signal processing / Wataru Mayeda.
p. cm.
Translation of: Dijitaru shingō shori no kiso.
Includes bibliographical references and index.
ISBN 0-13-211301-5
1. Signal processing—Digital techniques. I. Title.
TK5102.5.M2313 1993
621.382′2—dc20 92-4350
CIP

Editorial/production supervision
and interior design: Laura A. Huber
Acquisitions editor: Karen Gettman
Cover design: Wanda Lubelska Design
Prepress buyer: Mary Elizabeth McCartney
Manufacturing buyer: Susan Brunke

Original Japanese edition
Fundamentals of Digital Signal Processing by Wataru Mayeda
First published in 1985 by Ohmsha, Ltd.

The publisher offers discounts on this book when ordered in bulk quantities. For more information, write:
Special Sales/Professional Marketing
Prentice Hall
Professional Technical Reference Division
Englewood Cliffs, New Jersey 07632

Printed in the United States of America

10 9 8 7 6 5 4 3 2 1

ISBN 0-13-211301-5

Prentice-Hall International (UK) Limited, *London*
Prentice-Hall of Australia Pty. Limited, *Sydney*
Prentice-Hall Canada Inc., *Toronto*
Prentice-Hall Hispanoamericana, S.A., *Mexico*
Prentice-Hall of India Private Limited, *New Delhi*
Prentice-Hall of Japan, Inc., *Tokyo*
Simon & Schuster Asia, Pte. Ltd., *Singapore*
Editora Prentice-Hall do Brasil, Ltda., *Rio de Janeiro*

Contents

Preface

Recent advancement of Large-Scale Integrated (LSI) and Very-Large-Scale Integrated (VLSI) technology has made it feasible for digital systems to replace analog systems even though the former are structurally more complicated and sophisticated. For example, compact disk (CD) players have become so popular that many manufacturers have decided not to produce any more long-playing (LP) systems. Because of these changes, it is currently very difficult to obtain a new needle for playing LP records.

Why are digital signals replacing analog signals? Let us take a tape recorder as an example. The following are difficulties in recording and playing back an analog signal by an ordinal analog tape recorder: (1) The mechanism to move a tape should maintain exactly the same speed when a tape is used for either recording or playing back a high-fidelity analog signal. Maintenance of the same speed for the entire length of a tape, however, is almost impossible with an economical mechanism. (2) When a tape is moving for recording or playing back a high-fidelity analog signal, the air gap between the tape and the recording head should remain unchanged, which is impossible. (3) Magnetic material on a tape should be homogeneous for the entire length of the tape to record an analog signal precisely, which is unlikely for almost all available cassette tapes.

If we use digital signals instead of analog signals, however, it is only necessary to record and read 1 and 0 to resolve the preceding problems. Although there are new problems associated with digital signals, rapid advancement in VLSI manufacturing techniques has developed a complicated and sophisticated circuit that can be placed in a chip, making reasonably priced products available.

One machine, a DAT player, has perfect duplicating capability because it uses digital signals. This perfect performance will have an unfair impact on the business of recorded cassette tapes, however. Hence, manufacturers of DAT players must modify the machines not to possess a perfect duplicating capability to put the DAT players on the market.

The applications of digital signal processing are expanding to areas such as sonar, radar, acoustics, speech communication, data communication, picture communication, biomedical engineering, audio engineering, video engineering, and many others. Also, new techniques and theories of digital signal processing are rapidly being developed to widen areas of using digital signals. Soon digital signal processing will no longer be a field of only specialists but will be given as a general culture course.

In this book, we assume that readers are familiar with or understand analog signals. Before introducing a process or a transformation on digital signals, whenever there is a similar case in analog signals, we will show the analog case briefly so that reader can understand the process more easily. Also, this book aims to show simple and clear mathematical development for basic theories of digital signal processing including the Hilbert transformation rather than examples in several areas of applications of digital signals.

In Chapter 2, solving linear difference equations by employing an almost identical method for solving linear differential equations is discussed. In Chapter 3, Fourier transformation and convolution of analog signals are briefly reviewed before introducing discrete function Fourier transformation and convolution of discrete signals. In Chapter 4, solving differential equations by using Laplace transformations is reviewed before solving difference equations by using Z-transformations. In Chapter 5, signal flow graphs and Mason's formula for obtaining a solution from a signal flow graph for algebraic functions are informally shown; then the reader will learn how to use signal flow graphs to solve difference equations and to obtain discrete transfer functions of discrete systems.

In Chapter 6, a discrete function Fourier transformation is introduced. Because the result of this discrete function Fourier transformation is not a discrete function, it should be sampled to use the result as a discrete function. By modifying this transformation, however, we can obtain a transformation known as a discrete Fourier transformation (DFT), the result of which is a discrete function. This transformation yields an effective transformation for discrete signals, which is known as FFT and is explained extensively in Chapter 7.

Several digital filters associated with analog filters are introduced in Chapter 8. As we have familiarity with complex frequencies employed to solve problems in physics and electrical engineering, we can form complex discrete signals by the use of the Hilbert transformation, which is explained in Chapter 9. Finally, use of FFT to transform two-dimensional discrete signals is explained carefully so you can see that processing two-dimensional discrete signals is as easy as that of one-dimensional discrete signals.

There are answers for approximately half of the problems at the end of chapters. Also, there are references and an appendix at the end of this book showing some helpful properties of complex numbers and techniques for solving simultaneous difference equations by signal flow graphs.

After completing this book, readers can turn to highly technical books on digital filters such as *Digital Signal Processing* by A. V. Oppenheim and R. W. Schafer, and *Theory and Application of Digital Signal Processing* by L. R. Rabiner and B. Gold. Readers who are interested in mathematical developments on transformations should read the classic book *The Fast Fourier Transformation* by E. O. Brigham.

I acknowledge many researchers who gave me suggestions on this book. Also, I thank my wife Chieko Mayeda, who solved the problems and prepared the answer section in the last part of this book.

Digital Signal Processing

1
Digital Signals and Analog Signals

1.1 INTRODUCTION

It has been said that digital communication started in 1835 when Morse code was employed for distance communication. Morse code uses a sequence of two basic signals called dot "·" and dash "-" to indicate characters. These two basic signals are clearly distinguishable, and it is almost impossible to mistake "·" for "-" or "-" for "·". Hence, we can say that Morse code is a very reliable communication method. Furthermore, the characters that appear often are represented by simpler combinations of basic signals than those that seldom appear in regular communications. For example, the characters *e* and *t* are represented by "·" and "-". Conversely, characters *q* and *z* are represented by "--·-" and "--··". Because of this, Morse code is an efficient communication method. It is very surprising that such a reliable and efficient communication method was developed more than 150 years ago.

An analog signal is a signal that has a value at every instant of time, such as music from a stereo system. A triangular wave and a square wave as shown in Fig. 1.1 are analog signals. Previously, I said that Morse code is a digital signal. When an SOS is transmitted by Morse code, however, the voltage at the input of the transmitter will be one, as shown in Fig. 1.2. The

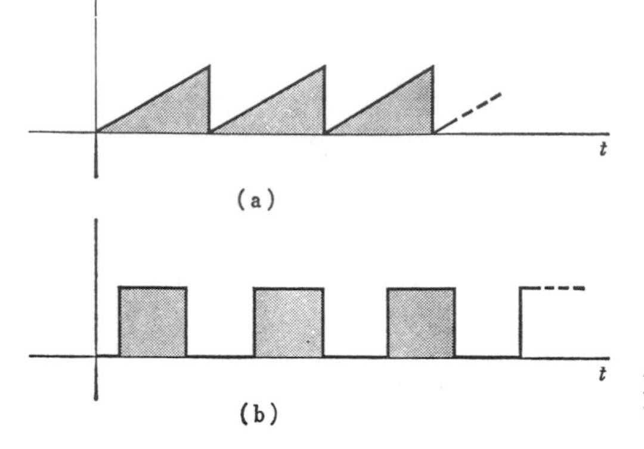

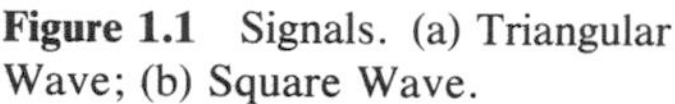
Figure 1.1 Signals. (a) Triangular Wave; (b) Square Wave.

only difference between the shape in Fig. 1.2 and that in Fig. 1.1(b) is the width of the square waves. This means the signal indicating SOS from the transmitter is an analog signal. Then what is a digital signal? To understand the difference between analog signals and digital signals, we need to know about a special signal known as a discrete signal.

Consider an analog signal shown in Fig. 1.3(a). By taking Δt as an interval, we can divide a time axis as shown in Fig. 1.3(b), where $t_0, t_1, \cdots$ are points on the time axis such that $t_{i+1} - t_1 = \Delta t$ for $i = 0, 1, 2, \cdots$. Let $x(t_i)$ be the value of $x(t)$ at t_i. These points $t_0, t_1, \cdots, t_i$ are called sampling points, and obtaining a set of $x(t_0), x(t_1), \cdots, x(t_i), \cdots$ is called sampling. The result obtained by this sampling as shown in Fig. 1.3(c) is a discrete signal. By representing t_i and $x(t_i)$ with finite n bit binary numbers, we have a set of binary signals $x_d(m_0), x_d(m_1), \cdots$ which is a digital signal. The process of changing a discrete signal to a digital signal is known as quantization. In other words, analog signals, discrete signals, and digital signals can be stated as follows:

1. An analog signal is one defined at every point on an axis that represents an independent variable.
2. A discrete signal is one defined only at each sampled point on an axis that represents an independent variable.
3. A digital signal is one obtained by quantizing a discrete signal with a finite n bit binary number.

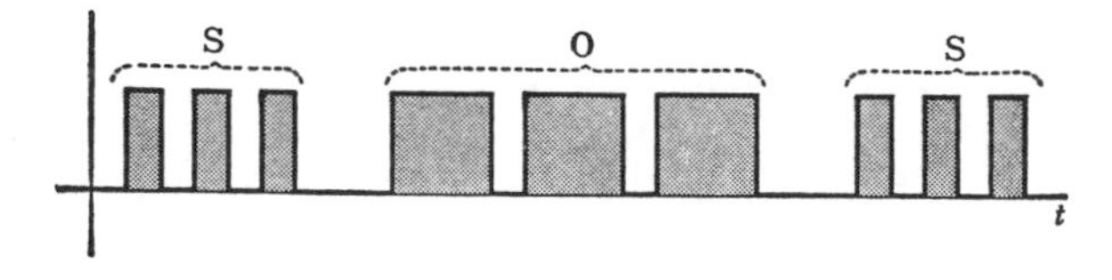

Figure 1.2 Morse Code SOS

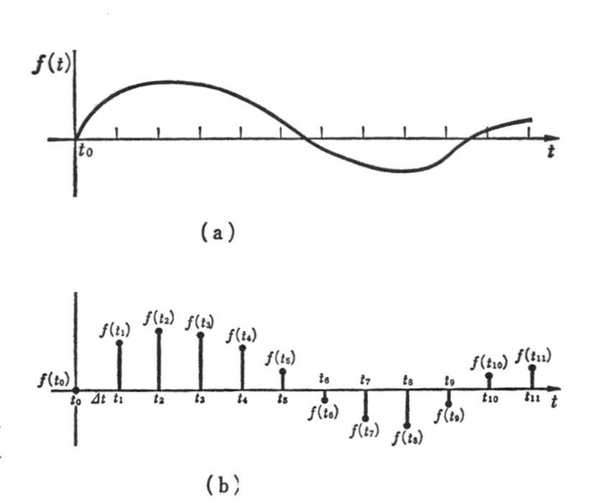

Figure 1.3 Analog and Discrete Signals. (a) Analog Signal; (b) Discrete Signal.

1.2 DIGITAL SIGNAL

A digital signal can be considered as a sequence of binary numbers. Is there any relationship between two adjacent sets of binary numbers in a digital signal? Consider a digital signal entering a computer in a bank that is a set of data, each of which is the savings balance of a customer. Because there is almost no relationship between savings of customers, two adjacent sets of binary numbers indicating savings of two customers will have no relationships. Another example is when a digital signal $\{x_d(n)\}$ entering a computer is a set of prices of stocks in a stock market. Suppose $x_d(i)$ is a present stock price of a company i and $x_d(i + 1)$ is a present stock price of stock of another company $i + 1$. Adjacent $x_d(i)$ and $x_d(i + 1)$ may have some relationship if these companies are in a similar business or somehow related to each other. If company i and company $i + 1$ are completely different businesses and have no relationship, however, then adjacent $x_d(i)$ and $x_d(i + 1)$ are independent numbers. For these kinds of digital signals, a major interest is to maintain each $x_d(i)$ in $\{x_d(n)\}$ without having any error (noise) during storing and transferring these digital signals. Hence, a major problem of digital signal process for these digital signals is to check whether there exists errors and if so how to correct errors to obtain an original digital signal.

Consider a digital signal that is a sequence of coded letters expressing information. Because a set of random letters will very seldom form a meaningful word, and a set of random words will not, in general, make a meaningful expression, there may be a relationship between adjacent $x_d(i)$ and $x_d(i + 1)$. Such a relationship, however, may not be sufficient to fix $x_d(i + 1)$ from knowing previous $x_d(k)$ for $k < i + 1$. Hence, these digital signals are very similar to those previously shown such that either no error (noise) will be produced during storing and transferring of digital signals, or some kind of protection on a digital signal will be placed.

1.3 QUANTIZATION

Recent advancement of VLSI technology has made it possible to produce an economical instrument for essential analog systems by the use of digital signal processing technology such as CD systems. From an analog signal $x(t)$ such as music, a discrete signal $\{x(k\Delta t)\}$ is obtained by sampling, which was explained in Section 1.1. Then each $x(k\Delta t)$ in this discrete signal $\{x(k\Delta t)\}$ is changed into a binary form $x_d(n)$ to obtain a digital signal $\{x_d(n)\}$ by quantization. This digital signal $\{x_d(n)\}$ is stored in a CD by the use of small dips called pits on a spiral line like a groove in an LP record.

When a CD is played by a CD player, a digital signal on the CD is read by a laser beam. This digital signal is processed and is changed back to an analog signal to reproduce music. Unlike an LP record, which must employ a needle to read an analog signal on an LP record, a CD player will not produce a needle's fricative sound, the surface of a CD will not deface by frequently playing it, and CD systems have many other advantages. CD systems produce new problems, however.

One problem is that a little stain such as a fingerprint, a little dust, or a tiny scratch can produce errors in digital signals, which may produce a large change in the corresponding analog signal. Because an analog signal will be produced regardless of existence of errors in a digital signal, errors must be either corrected or modified so that a reasonable analog signal can be reproduced. This requires not an error detection but an error correction. Furthermore, if errors in $x_d(n)$ are such that they cannot be corrected, $x_d(n)$ must be modified so that the corresponding analog signal will not produce unpleasant sound. Hence, like those digital signals in the previous section, the digital signal on a CD uses an error-correcting code and a data-modifying scheme. Because I do not cover these topics in this book, readers who are interested in these subjects should look elsewhere for a book on coding theory.

The second problem is that an error is produced when a discrete signal is changed to a digital signal, which is known as quantization noise. For example, a decimal number 0.4 will be $0.011001100110011\cdots$. Hence, if it is required to have a finite number of bits in each $x_d(n)$ in a digital signal $\{x_d(n)\}$, such as 16, the binary representation of a decimal number 0.4 will be either 0.01100110011001 or 0.01100110011010. In either case, there will be an error.

It is clear that more bits for each $x_d(n)$ can reduce the size of an error. Because the maximum of total binary numbers that can be stored in a CD surface is fixed, more bits for each $x_d(n)$ means less total $x_d(n)$ in a digital signal $\{x_d(n)\}$, which corresponds to reduction of the maximum playing time for each CD. Furthermore, more bits for each $x_d(n)$ requires more complicated digital circuits in CD systems. Hence, this problem is one of economic feasibility of CD systems, which is not discussed in this book. The last problem involves reproduction, which is studied in the next section.

1.4 DISCRETE SIGNAL CONCEPT

In addition to CD systems, there are many physical systems that use digital signals produced from analog signals. As is mentioned in Section 1.1, digital signals are obtained by quantizing discrete signals that result from analog signals by sampling. Consider a system shown in Fig. 1.4.

The top part of Fig. 1.4 shows that an analog signal $x(t)$ is entering at an input terminal of an analog system $h(t)$, and an analog signal $y(t)$ is produced at an output terminal of the system. An example of this is that $x(t)$ will be music picked up by a microphone. This analog signal $x(t)$ is entered at an input terminal of an amplifier $h(t)$, which may make some adjustment on some frequency components to obtain a specified sound, and a resultant analog signal $y(t)$ is produced at its output terminal. At the output terminal a speaker system may be connected so that an analog signal $y(t)$ is transferred to an acoustic sound.

When a digital system is employed instead of an analog system, an input signal $x(t)$ is transferred by an A/D converter to a digital signal $\{x(n)\}$ as shown in the left-hand side of the bottom part of Fig. 1.4. This digital signal $\{x(n)\}$ is entered at an input terminal of a digital system $\{h(n)\}$, which may process the digital signal $\{x(n)\}$ to a desirable signal digitally, and a resultant digital signal $\{y(n)\}$ will be produced at its output terminal. Then this digital signal $\{y(n)\}$ is transferred by a D/A converter to an analog signal, which should be the same signal as $y(t)$ as shown in the right-hand side of the bottom part of Fig. 1.4.

A practical example of this is a Pulse Code Modulation (PCM) system, which is shown in Fig. 1.5. When an analog signal is entered into a PCM system, it is transformed to a digital signal by an A/D converter. This digital signal is coded, and then modulated and transmitted. When the modulated

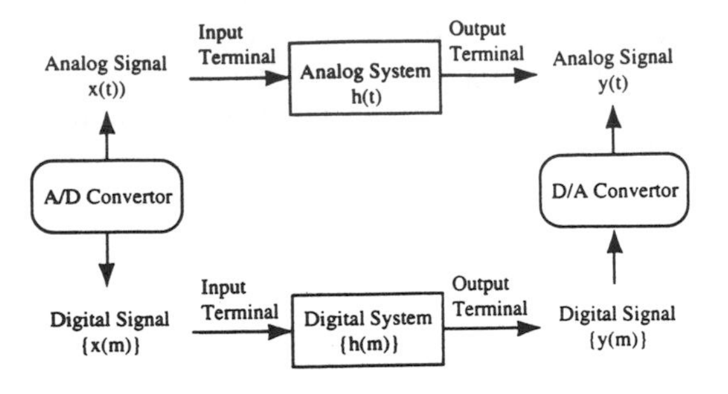

Figure 1.4 Analog System and Digital System

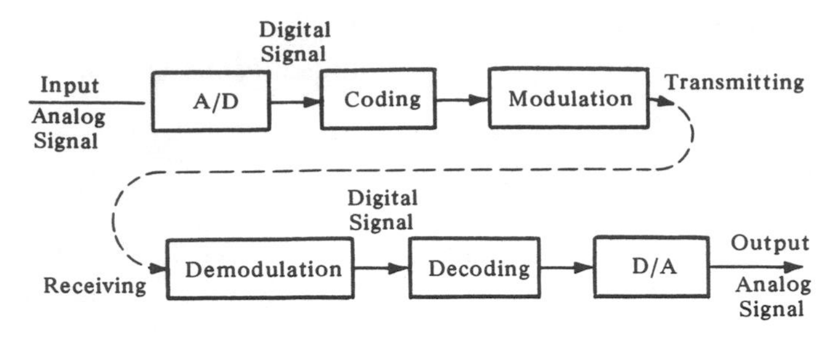

Figure 1.5 Pulse Code Modulation System

digital signal is received, it is demodulated and decoded. Then it will change back to an analog signal by a D/A converter.

It is clear that when a digital signal $\{x(n)\}$ obtained from an analog signal $x(t)$ has errors, it is impossible to obtain from the output signal $\{y(n)\}$ of a digital system an analog signal that is the same signal as analog signal $y(t)$ at the output terminal of the corresponding analog system. There will be a quantization noise (error) as shown in Section 1.3 when each $x(n)$ in a digital signal $\{x(n)\}$ is expressed by a finite number of bits. Hence, investigating a kind of an analog signal that can be produced from a digital signal that was obtained by sampling and quantizing an analog signal should be done without having any errors in the digital signal. That is, a digital signal should have no error by quantization when a relationship between a digital signal and an analog signal is studied.

When a digital signal obtained from a discrete signal has no quantization noise, the digital signal and the discrete signal are the same. Hence, in this book, I use discrete instead of digital signals for studying relationships with analog signals, and relationships between an input and output signal of digital systems.

In Section 1.1, we have learned that a discrete signal is obtained from an analog signal by sampling. If a sampling interval is infinitesimally small, then an analog signal and the resultant discrete signal are intuitively very closely related; we can almost be certain that the analog signal can be recovered from this discrete signal. Conversely, it is very difficult to figure out how to obtain an original analog signal back from a digital signal obtained by using a very large sampling interval. Hence, there should be a maximum sampling interval that can be employed to obtain a discrete signal from an analog signal so that the analog signal can be recovered from the resultant discrete signal.

1.5 SUMMARY

1. An analog signal is a function whose independent variable is continuous.

2. A discrete signal can be obtained by sampling an analog signal.
3. A digital signal can be obtained by quantizing a discrete signal. The independent variable must be quantized also.
4. A/D transformation is employed to transform an analog signal into a digital signal.
5. D/A transformation is employed to transform a digital signal into an analog signal.
6. Quantization noise is error produced by changing a number to a finite n bit binary number.

2
Characteristics of Discrete Signals

2.1 DISCRETE SIGNAL

Suppose $x(t)$ is an analog signal and $\{x(nt_0)\}$ is a discrete signal obtained by sampling $x(t)$ at $t = 0, t_0, 2t_0, 3t_0, \cdots$. Is it possible to reproduce the given $x(t)$ from $\{x(nt_0)\}$? Fig. 2.1(a) shows an analog signal $x_1(t)$, which is sampled at $0, t_0, 2t_0, 3t_0, \ldots$ where $x_1(kt_0)$ indicates the value at kt_0. Fig. 2.1(b) shows another analog signal $x_2(t)$, which is also sampled at $0, t_0, 2t_0, 3t_0, \ldots$ where $x_2(kt_0)$ is the value at kt_0. Suppose $x_1(kt_0) = x_2(kt_0)$ for $k = 0, 1, 2, 3, \ldots$. In other words, these analog signals $x_1(t)$ and $x_2(t)$ have exactly the same discrete signal when they are sampled at $0, t_0, 2t_0, 3t_0, \ldots$. When we reproduce an analog signal from a discrete signal $\{x_1(nt_0)\}$, which is the same as $\{x_2(nt_0)\}$, are we getting $x_1(t)$ or $x_2(t)$? Before we can answer this question, we must study properties of discrete signals.

For convenience, we will modify the independent variable in an analog signal so that the discrete signal of the analog signal $x(t)$ will be $\{x(n)\}$. Also, we assume that sampling points will be extended into the negative side of the axis so that for any given analog signal $x(t)$ sampling points are $\ldots, -3, -2, -1, 0, 1, 2, 3, \ldots$. Hence, the discrete signal of an analog signal $x(t)$ will be $\{\ldots, x(-3), x(-2), x(-1), x(0), x(1), x(2), x(3), \ldots\}$. Let $\{x_1(n)\}$ and $\{x_2(n)\}$ be discrete signals. Then the sum of $\{x_1(n)\}$ and $\{x_2(n)\}$ is

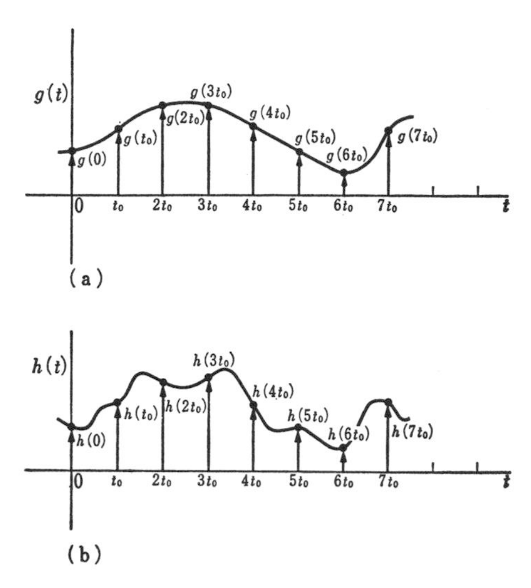

Figure 2.1 Analog Signals. (a) Sampling Analog Signal $x_1(t)$; (b) Sampling Analog Signal $x_2(t)$.

$$\{x_1(n)\} = \{x_1(n) + x_2(n)\} \tag{2.1}$$

This is reasonable because if we add corresponding analog signals $x_1(t)$ and $x_2(t)$, and then sample the resultant analog signal at . . . , $-3, -2, -1, 0, 1, 2, 3,$ . . . , we can see that at each sampling point k the value will be $x_1(k) + x_2(k)$.

Multiplying a constant c to a discrete signal $\{x(n)\}$ will produce a discrete signal as

$$c\{x(n)\} = \{cx(n)\} \tag{2.2}$$

For example, suppose $x_1(t) = \sin^2 t/5$, and $x_2(t) = \cos^2 t/5$. Corresponding discrete signals are

$$\{x_1(n)\} = \{\cdots, x_1(0) = 0, x_1(1) = 0.039469, \\ x_1(2) = 0.151647, x_1(3) = 0.318821, \cdots\}$$

and

$$\{x_2(n)\} = \{\cdots, x_2(0) = 1, x_2(1) = 0.960531, \\ x_2(2) = 0.848353, x_2(3) = 0.681179, \cdots\}$$

The sum of discrete signals $\{x_1(n)\}$ and $\{x_2(n)\}$ will be

$$\{x_1(n)\} + \{x_2(n)\} = \{\cdots, x_1(0) + x_2(0) = 1, x_1(1) + x_2(1) = 1, \\ x_1(2) + x_2(2) = 1, x_1(3) + x_2(3) = 1, \cdots\}$$

The product of a constant $c = -1$ and discrete signal $\{x_1(n)\}$ will be

$$(-1) \times \{x_1(n)\} = \{\cdots, -x_1(0) = 0, -x_1(1) = -0.039469, \\ -x_1(2) = -0.151647, -x_1(3) = -0.318821, \cdots\}$$

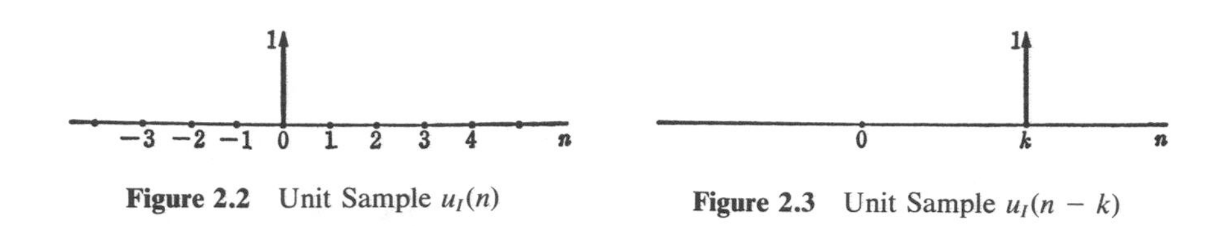

Figure 2.2 Unit Sample $u_I(n)$

Figure 2.3 Unit Sample $u_I(n - k)$

There are two important discrete signals for analysis of discrete systems. One of these is known as a unit sample, and the other is a unit step. A unit sample symbolized by $u_I(n)$ is defined as

$$u_I(n) = \begin{cases} 1 & n = 0 \\ 0 & n \neq 0 \end{cases} \tag{2.3}$$

Fig. 2.2 shows a unit sample which has 1 only at $t = 0$. Notice that a unit sample is a discrete signal. Hence, it should be indicated by $\{u_I(n)\}$. When a sequence of terms in the parentheses can be expressed by either few equations or symbols that will show a single value at every sampling point and undefine all other points, however, then we will use these equations or symbols without the parentheses for the representation of a discrete signal. Because a unit sample $u_I(n)$ defined by Equation 2.3 clearly satisfies the preceding restrictions, we will use the symbol $u_I(n)$ to indicate such a discrete signal.

Suppose the location of 1 is shifted so that 1 is only at $t = k$ as shown in Fig. 2.3. We call this discrete signal a unit sample also. We employ a symbol $u_I(n - k)$ to indicate the location of 1, however. In other words, unit sample $u_I(n - k)$ is defined as

$$u_I(n - k) = \begin{cases} 1, & n = k \\ 0, & n \neq k \end{cases} \tag{2.4}$$

By using unit samples, a discrete signal $\{x(n)\}$ can be expressed as

$$\{x(n)\} = \left\{ \sum_{r=-\infty}^{\infty} x(r)u_I(n - r) \right\} \tag{2.5}$$

For example, discrete signal x_1 in Fig. 2.4(a) is

$$\{x_1(n)\} = \{\cdots 0, x_1(-3) = 1, x_1(-2) = 0, x_1(-1) = 0, x_1(0) = 0, x_1(1) = 0, \\ x_1(2) = 3, x_1(3) = 0, x_1(4) = 0, x_1(5) = 5, x_1(6) = 0, 0, \cdots\}$$

If we use unit samples, the preceding sequence can be expressed as

$$\{x_1(n)\} = \{u_I(n + 3) + 3u_I(n - 2) + 5u_I(n - 5)\}$$

This $x_1(n) = u_I(n + 3) + 3u_I(n - 2) + 5u_I(n - 5)$ gives the value at every sampling point. For sampling point -3, $n = -3$ gives $x_1(-3) = 1$. For sampling point 2, $n = 2$ gives $x_1(2) = 3$, and so on. Hence,

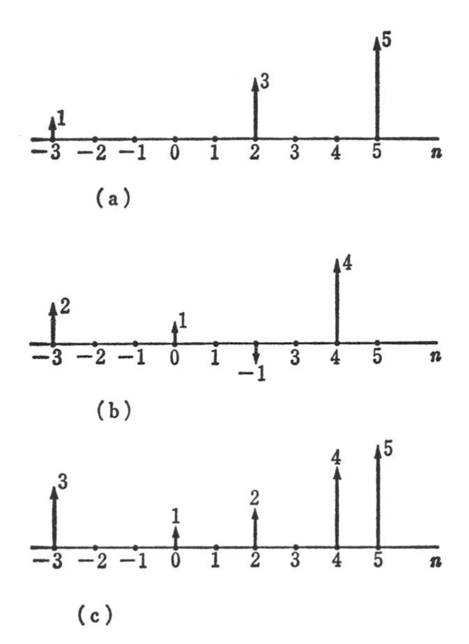

Figure 2.4 Discrete Signals. (a) Discrete Signal $x_1(n)$; (b) Discrete Signal $x_2(n)$; (c) Discrete Signal $x_1(n) + x_2(n)$.

$$x_1(n) = u_I(n + 3) + 3u_I(n - 2) + 5u_I(n - 5)$$

indicates a discrete signal in Fig. 2.4(a). Similarly, a discrete signal $\{x_2(n)\}$ shown in Fig. 2.4(b) can be expressed as

$$x_2(n) = 2u_I(n + 3) + u_I(n) - u_I(n - 2) + 4u_I(n - 4)$$

The sum of these two discrete signals $x_1(n)$ and $x_2(n)$ is

$$x_1(n) + x_2(n) = 3u_I(n + 3) + u_I(n) + 2u_I(n - 2) + 4u_I(n - 4) + 5u_I(n - 5)$$

which is shown in Fig. 2.4(c). This indicates that a representation of discrete signals by equations is not ambiguous.

From the preceding discussion, we have learned that a discrete signal can be expressed as a sum of unit samples. Also, by delaying k units, $u_I(n)$ becomes a unit sample $u_I(n - k)$. Hence, changing n by $n - k$ of a discrete signal corresponds to delaying the discrete signal by k units. In other words, if we need to delay a discrete signal $x(n)$ by k units, it is only necessary to change n by $n - k$ as $x(n - k)$. For example, delaying a discrete signal $x_1(n)$ in Fig. 2.4(a) by 3 units can be accomplished by changing n to $n - 3$ as

$$x_1(n - 3) = u_I(n) + 3u_I(n - 5) + 5u_I(n - 8)$$

which is shown in Fig. 2.5. Notice that every signal in $x_1(n)$ (Fig. 2.4[a]) is moved to the right by 3 units to produce $x_1(n - 3)$ (Fig. 2.5). If a discrete signal $x_1(n)$ is delayed by k units, the resultant signal becomes identical with the discrete signal $x_2(n)$. Then $x_1(n)$ and $x_2(n)$ will satisfy

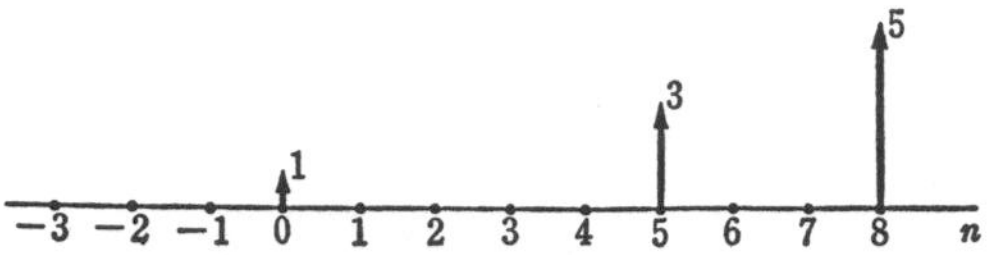

Figure 2.5 Discrete Signal $x_1(n - 3)$

$$x_1(n - k) = x_2(n) \tag{2.6}$$

Any sequence $\{x_1(n)\}$ can be expressed by an equation using unit samples, which has been shown previously. Hence, the symbol $x_1(n)$ can be employed for expressing a discrete signal. Similarly, the symbol $x_2(n - k)$ can be a symbol for the discrete signal, which is the result of a delay in k units of a discrete signal $x_2(n)$.

Another important discrete signal is called a unit step symbolized by $u_s(n)$, which is defined as

$$u_s(n) = \begin{cases} 1, & n \geq 0 \\ 0, & n < 0 \end{cases} \tag{2.7}$$

Fig. 2.6 shows a unit step. Because a unit step is a discrete signal, it can be expressed by unit samples as

$$u_s(n) = \sum_{k=0}^{\infty} u_I(n - k) \tag{2.8}$$

2.2 DIFFERENCE EQUATION

A system shown in Fig. 2.7 indicates that entering a signal x at its input terminal produces a signal y at its output terminal. Signal x is known as an "input signal," an "input function," or simply an "input." Signal y is called an "output signal," an "output function," or simply an "output." When x and y are both discrete signals, the system is called a "discrete system" or simply a "system."

Difference equations are commonly employed to specify such a discrete system. Hence, we will discuss some properties and a method of solving difference equations.

A linear constant coefficient difference equation is of the form

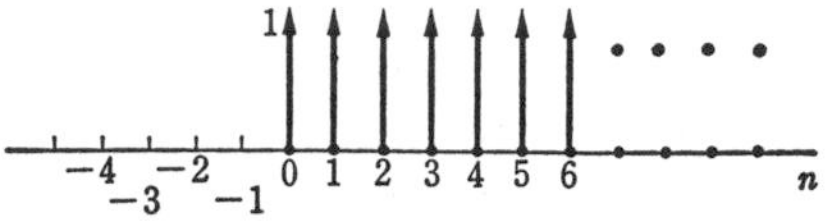

Figure 2.6 Unit Step $u_s(n)$

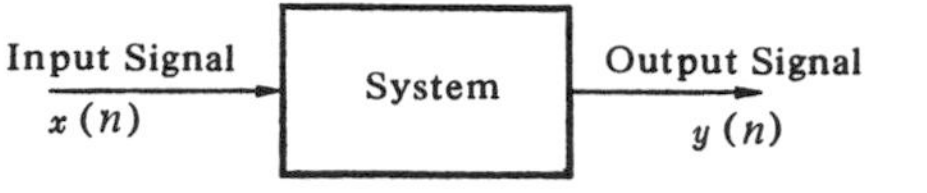

Figure 2.7 Input and Output Signals

$$\sum_{p=0}^{N} a_p y(n-p) = \sum_{q=0}^{N} b_q x(n-q) \tag{2.9}$$

where a_p and b_q are constants. The right-hand side of the preceding equation $\sum_{q=0}^{N} b_q x(n-q)$ is called an input function. For example, in

$$y(n) - 3y(n-1) = 2n \tag{2.10}$$

the right-hand side $2n$ is an input function. A solution of a difference equation (2.9) can be obtained by a method that is almost identical with a method of obtaining solutions of linear differential equations. That is, we assume that a solution consists of two parts, one of which is a general solution $y_g(n)$, the other is a particular solution $y_p(n)$, and the sum of $y_g(n) + y_p(n)$ is the solution.

The steps to obtain a solution $y(n) = y_g(n) + y_p(n)$ are as follows:

1. To obtain a general solution $y_g(n)$, we set the right-hand side of Equation 2.9 to 0 as

$$\sum_{p=0}^{N} a_p y(n-p) = 0 \tag{2.11}$$

Then we assume that general solution $y_g(n)$ is of the form $K\alpha^n$. From the equation obtained by substituting this $y_g(n) = K\alpha^n$ in Equation 2.11, we can obtain the values of α. For example, to solve Equation 2.10, we first set the right-hand side of equation to 0 as

$$y(n) - 3y(n-1) = 0$$

Next, substitute general solution $y_g(n) = K\alpha^n$ into the equation, which gives

$$K\alpha^n - 3K\alpha^{n-1} = 0$$

This can be expressed as

$$K\alpha^{n-1}(\alpha - 3) = 0$$

Hence, the value of α is

$$\alpha = 3$$

Thus, the general solution $y_g(n)$ is

$$y_g(n) = K3^n$$

2. To obtain a particular solution, we assume $y_p(n)$ to have the same form as the input function. Substituting $y_p(n)$ into Equation 2.9, we obtain the values of all constants in $y_p(n)$. For example, the input function of Equation 2.10 is $2n$. Hence, we assume a particular solution is of the form

$$y_p(n) = Cn + D$$

Substituting this into Equation 2.10, we obtain

$$Cn + D - 3[C(n - 1) + D] = 2n$$

which may be simplified to

$$-2Cn + 3C - 2D = 2n$$

This equation must be satisfied for all values of n. Hence, the following two equations must be satisfied:

$$-2C = 2$$

$$3C - 2D = 0$$

From these equations, we can obtain the values of constants C and D as

$$C = -1$$

$$D = -\frac{3}{2}$$

Hence, particular solution $y_p(n)$ is

$$y_p(n) = -n - \frac{3}{2}$$

3. To obtain a solution $y(n)$, we add the general solution $y_g(n)$ and the particular solution $y_p(n)$ obtained previously. This solution contains constants K, which must be determined by the use of initial conditions. For example, the solution of $y(n) = y_g(n) + y_p(n)$ of Equation 2.10, which we have found so far, is

$$y(n) = K3^n - n - \frac{3}{2} \tag{2.12}$$

Suppose that a given initial condition for Equation 2.10 is

$$y(0) = 0$$

Then substituting n in Equation 2.12 by 0, we have

$$y(0) = 0 = K - \frac{3}{2}$$

Hence, the value of K is

$$K = \frac{3}{2}$$

Thus solution $y(n)$ of Equation 2.10 with $y(0) = 0$ is

$$y(n) = \frac{3^{n+1}}{2} - n - \frac{3}{2}$$

Consider the following difference equations:

$$\begin{cases} y(n) - 2y(n-1) - 3y(n-2) = x(n) \\ x(n) = \begin{cases} 4^n, & n \geq 0 \\ 0, & n < 0 \end{cases} \\ y(-1) = 0,\ y(-2) = 0 \end{cases} \tag{2.13}$$

where the second equation specifies the input function, and the third equation indicates the initial conditions. To obtain solution $y(n) = y_g(n) + y_p(n)$, first we calculate the general solution $y_g(n)$ by assuming $y_g(n) = K\alpha^n$ and substitute this into a difference equation in Equation 2.16 with the right-hand side being 0 as

$$K\alpha^n - 2(K\alpha^{n-1}) - 3(K\alpha^{n-2}) = 0$$

This can be expressed as

$$K\alpha^{n-2}(\alpha^2 - 2\alpha - 3) = 0$$

Hence, we have two values of α, which are $\alpha = 3$ and $\alpha = -1$. Thus, general solution $y_g(n)$ is

$$y_g(n) = K_1 3^n + K_2(-1)^n$$

Because the input function is 4^n, we assume a particular solution $y_p(n)$ to be

$$y_p(n) = A4^n + B$$

Substituting this into Equation 2.13, we have

$$A4^{n-2}(4^2 - 2 \times 4 - 3) - 4B = 4^n$$

This equation must be satisfied for all values of $n > 0$. Hence, constants A and B must be

$$A = 16/5 \qquad \text{and} \qquad B = 0$$

Thus, particular solution $y_p(n)$ is

$$y_p(n) = \frac{16}{5} \times 4^n$$

The solution $y(n)$, which is the sum of general solution $y_g(n)$ and particular solution $y_p(n)$, becomes

$$y(n) = K_1 3^n + K_2(-1)^n + \frac{4^{n+2}}{5} \tag{2.14}$$

In this solution, two constants K_1 and K_2 must be determined using given initial conditions $y(-1) = 0$ and $y(-2) = 0$. It should be noted that Equation 2.14 contains particular solution $y_p(n)$, which is obtained by assuming that input function $x(n)$ is 4^n. However, $x(n) = 4$ only when $n \geq 0$. That is, this particular solution is valid only when $n \geq 0$. Thus, we cannot use given initial conditions that are at $n = -1$ and $n = -2$ for $y(n)$ in Equation 2.14 to fix coefficients K_1 and K_2. To obtain new initial conditions, setting $n = 0$ in Equation 2.13, we have

$$y(0) - 2y(-1) - 3y(-2) = 4^0$$

Because $y(-1) = 0$ and $y(-2) = 0$ are given as initial conditions, insertion of these values into the preceding equation gives

$$y(0) = 1 \tag{2.15}$$

Next, setting $n = 1$ in Equation 2.13, we have

$$y(1) - 2y(0) - 3y(-1) = 4$$

Substituting the initial conditions $y(-1) = 0$ and $y(0) = 1$ (Equation 2.15) in the preceding equation, we obtain

$$y(1) = 6$$

These $y(0) = 1$ and $y(1) = 6$ are new initial conditions for $n \geq 0$. Hence, we can use these conditions to determine K_1 and K_2 in Equation 2.14. First, setting $n = 0$ in Equation 2.14 gives

$$y(0) = K_1 3^0 + K_2(-1)^0 + \frac{4^2}{5}$$

Because $y(0) = 1$, we obtain

$$K_1 + K_2 + \frac{4^2}{5} = 1 \tag{2.16}$$

Then setting $n = 1$ in Equation 2.14 gives

$$y(1) = K_1 3^1 + K_2(-1)^1 + \frac{4^3}{5}$$

Because $y(1) = 6$, we obtain

$$3K_1 - K_2 + \frac{4^3}{5} = 6 \tag{2.17}$$

These two equations (Equations 2.16 and 2.17) give the values of K_1 and K_2 as

$$K_1 = -\frac{9}{4}$$

$$K_2 = \frac{1}{20}$$

By inserting these K_1 and K_2 into Equation 2.14, we obtain the solution $y(n)$ as

$$y(n) = -\frac{3^{n+2}}{4} + \frac{(-1)^n}{20} + \frac{4^{n+2}}{5}$$

for $n \geq 0$.

2.3 UNIT SAMPLE RESPONSE

A digital system is said to satisfy initial rest conditions if its input signal $x(n) = 0$ for $n < n_0$ fixes its output signal $y(n) = 0$ for $n < n_0$. In other words, if a system whose output signal $y(n)$ will always be 0 until the nonzero signal is entered into an input terminal, then the system is said to satisfy initial rest conditions. When a deference equation is employed to express an input-output relationship of a system, then the system satisfying initial rest conditions means that all initial conditions are properly fixed. For example, if a difference equation

$$y(n) - 2y(n-1) - 3y(n-2) = x(n)$$

is the input-output relationship of a system satisfying initial rest conditions, and

$$x(n) = \begin{cases} 4^n, & n \geq 0 \\ 0, & n < 0 \end{cases}$$

is given as an input signal (function), then $x(n) = 0$ for $n < 0$ fixes its initial conditions $y(n) = 0$ for all $n < 0$.

When an input signal of a discrete system is a unit sample, then its output signal is called a unit sample response* and is represented by the symbol $h(n)$. For example, if an input-output relationship of a discrete system is given by

$$y(n) - 3y(n-1) = x(n)$$

then the unit sample response of this system can be calculated by the following steps:

* It is also known as a ''transfer function.''

First, we set input signal $x(n)$ to be a unit sample as

$$y(n) - 3y(n - 1) = u_I(n) \tag{2.18}$$

To obtain a general solution $y_g(n)$, we substitute $y_g(n) = K\alpha^n$ into Equation 2.18 with its right-hand side being set to 0 as

$$K\alpha^n - 3K\alpha^{n-1} = 0$$

From this equation, we can obtain

$$\alpha = 3$$

Hence, a general solution $y_g(n)$ will be

$$y_g(n) = K3^n$$

Normally, we should obtain a particular solution $y_p(n)$ next. Because the input signal is a unit sample $u_I(n)$, however, we have 0 at the right-hand side of Equation 2.18 except when $n = 0$, at which the equation becomes

$$y(0) - 3y(-1) = 1$$

Suppose a given system satisfies initial rest conditions, then $y(-1) = 0$. Thus, the preceding equation becomes

$$y(0) = 1$$

By considering this equation as an initial condition, deference equation representing a given system is

$$y(n) - 3y(n - 1) = 0 \tag{2.19}$$

whose right hand side is 0. Hence, a solution $y(n)$ consists only of a general solution. In other words, a solution of Equation 2.18 is

$$y(n) = K3^n \tag{2.20}$$

with an initial condition $y(0) = 1$. Setting $n = 0$ in Equation 2.20, we have

$$y(0) = K3^0 = 1$$

Hence, K will be

$$K = 1$$

and solution $y(n)$ becomes

$$y(n) = 3^n$$

Thus, the unit sample response $h(n)$ of a given system is

$$h(n) = 3^n$$

We can obtain a unit sample response by a different method as follows:

From initial conditions $y(-1) = 0$ and $y(0) = 1$, and Equation 2.19, we can obtain

$$\left.\begin{array}{l} y(-1) = 0 \\ y(0) \quad = 1 \end{array}\right\} \text{ Initial Conditions}$$

$$\left.\begin{array}{ll} y(1) - 3y(0) = 0 & \Rightarrow y(1) = 3 \\ y(2) - 3y(1) = 0 & \Rightarrow y(2) = 3^2 \\ y(3) - 3y(2) = 0 & \Rightarrow y(3) = 3^3 \\ \quad\vdots & \quad\vdots \\ y(n) - 3y(n-1) = 0 & \Rightarrow y(n) = 3^n \end{array}\right\} \text{ By Equation 2.19}$$

which clearly show that the unit sample response $h(n)$ is 3^n.

2.4 LINEAR SHIFT INVARIANT SYSTEM

To classify a system, we employ the symbol $\mathcal{T}(x(n))$ to indicate the output signal $y(n)$ of a system when an input signal is $x(n)$. Hence,

$$y(n) = \mathcal{T}(x(n))$$

indicates not only an output signal but also an input signal $x(n)$, which produces the output signal. Suppose an input signal is $\alpha_1 x_1(n) + \alpha_2 x_2(n)$. If the equation

$$y(n) = \mathcal{T}(\alpha_1 x_1(n) + \alpha_2 x_2(n)) = \alpha_1 \mathcal{T}(x_1(n)) + \alpha_2 \mathcal{T}(x_2(n))$$

is satisfied, then the system is said to be linear or is called a linear system.

It has already been discussed that the input signal $x(n)$ can be expressed as a function of unit samples as

$$x(n) = \sum_{k=-\infty}^{\infty} x(k) u_I(n - k)$$

Hence, if a system is linear, output signal $y(n)$ can always be expressed as

$$y(n) = \mathcal{T}(x(n)) = \sum_{k=-\infty}^{\infty} \mathcal{T}(x(k) u_I(n - k))$$

Each term $x(k)$ of the preceding discrete signal is a constant. Hence, the preceding equation becomes

$$y(n) = \sum_{k=-\infty}^{\infty} x(k) \mathcal{T}(u_I(n - k)) \tag{2.21}$$

We also know that the output signal $y(n)$ is a unit sample response when input signal $x(n)$ is a unit sample. Hence, Equation 2.21 can be written as

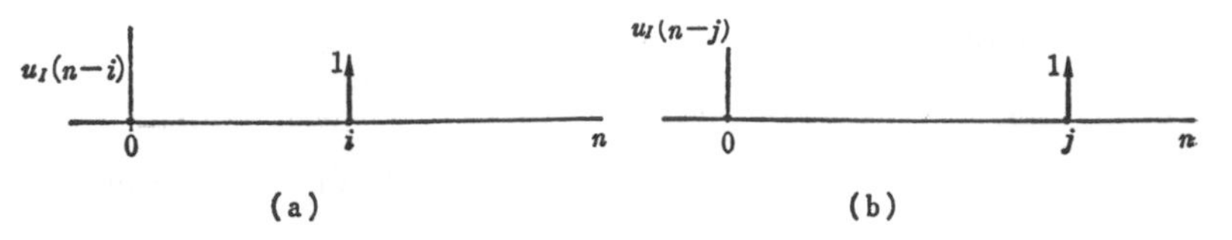

Figure 2.8 Unit Samples $u_I(n - i)$ and $u_I(n - j)$. (a) $u_I(n - i)$; (b) $u_I(n - j)$.

$$y(n) = \sum_{k=-\infty}^{\infty} x(k)h_k(n) \tag{2.22}$$

Do these unit sample responses . . . , $h_1(n)$, $h_2(n)$, . . . , $h_i(n)$, . . . , $h_j(n)$, . . . , $h_m(n)$, . . . in Equation 2.22 have any relationship with each other? A unit sample response $h_i(n)$ is the output signal when a unit sample $u_I(n - i)$ is entered. Similarly, a unit sample response $h_j(n)$ is the output signal when a unit sample $u_I(n - j)$ is entered. The difference between $u_I(n - i)$ and $u_I(n - j)$ is the time when a unit sample appears as shown in Fig. 2.8. In other words, $u_I(n - i)$ is a discrete signal that can be obtained by delaying $i - j$ units of $u_I(n - j)$.

A system is said to be "shift invariant" or is a "shift invariant system" if its output signal delays k units when its input signal delays k units. In other words, let $x(n)$ be an input signal and $y(n)$ be the corresponding output signal of a system. Then the system is shift invariant if the output signal becomes $y(n - k)$ when its input signal $x(n)$ is changed to $x(n - k)$.

A unit sample response $h(n)$ is an output signal. For a shift invariant system, its output signal $h(n)$ becomes $h(n - j)$ when its input signal $u_I(n)$ is changed to $u_I(n - j)$. Hence, for a shift invariant system, Equation 2.22 can be rewritten as

$$y(n) = \sum_{k=-\infty}^{\infty} x(k)h(n - k) \tag{2.23}$$

The right-hand side of this equation is usually expressed with an asterisk (*) as

$$y(n) = x(n) * h(n) \tag{2.24}$$

and is called a "convolution of $x(n)$ and $h(n)$." In other words, if a system is linear shift invariant, then the output signal $y(n)$ is a convolution of the input signal $x(n)$ and the unit sample response $h(n)$.

By changing $n - k$ to r in Equation 2.23, k becomes $n - r$, and Equation 2.23 becomes

$$y(n) = \sum_{r=-\infty}^{\infty} x(n - r)h(r) \tag{2.25}$$

Hence, $x(n) * h(n)$ can be expressed as

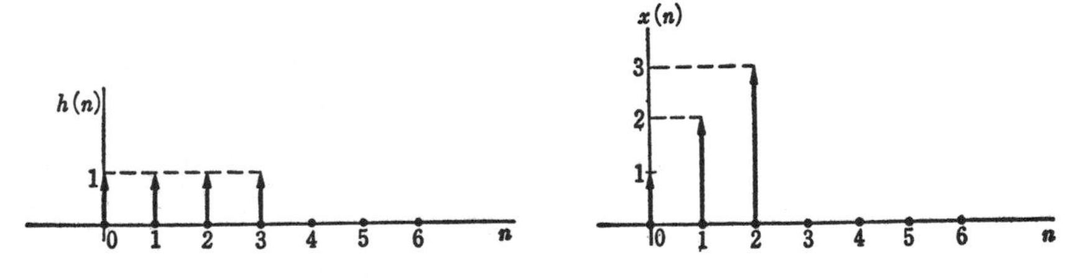

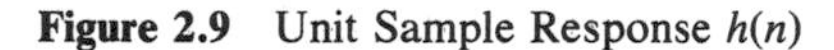
Figure 2.9 Unit Sample Response $h(n)$

Figure 2.10 Input Signal $x(n)$

$$x(n) * h(n) = \sum_{k=-\infty}^{\infty} x(k)h(n-k) = \sum_{k=-\infty}^{\infty} x(n-k)h(k) \tag{2.26}$$

Here, we will use an example to investigate a property of a linear shift invariant system. Suppose the unit sample response of a linear shift invariant system is one shown in Fig. 2.9, that is, we assume that the output signal of the system is $h(n)$ in Fig. 2.9 when a unit sample $u_I(n)$ is applied. Then what will be the output signal when an input signal $x(n)$ shown in Fig. 2.10 is applied to an input of the system?

Using unit samples, this input signal $x(n)$ can be expressed as

$$x(n) = u_I(n) + 2u_I(n-1) + 3u_I(n-2)$$

Hence, the input signal $x(n)$ may be considered as the sum of $u_I(n)$, $2u_I(n-1)$, and $3u_I(n-2)$. Thus, the output signal $y(n)$ is the sum of $h(n)$, $2h(n-1)$, and $3h(n-2)$, or

$$y(n) = h(n) + 2h(n-1) + 3h(n-2) \tag{2.27}$$

The output signal $h(n)$ for input signal $u_I(n)$ is shown in Fig. 2.11(a). The output signal $2h(n-1)$ caused by input signal $2u_I(n-1)$ is one in Fig. 2.11(b). Also, Fig. 2.11(c) is the output signal $3h(n-2)$ corresponding to input signal $3u_I(n-2)$.

The output signal $y(n)$ is the sum of these output signals in Fig. 2.11(a), (b), and (c), which is shown in Fig. 2.11(d). Equation 2.27 also shows that for each value of n, e.g., when $n = a$, $y(a)$ can be obtained by convolution. For example, from Equation 2.27, $y(2)$ is equal to

$$y(2) = h(2) + 2h(1) + 3h(0) = \sum_{k=0}^{2} x(k)h(2-k)$$

To understand convolution, let us study properties of $h(n-k)$. When $n = 0$, $h(n-k)$ becomes $h(-k)$, which is equal to $h(n)$ with n being set to $-k$. This means that the figure $h(-k)$ can be obtained from the figure of $h(n)$ by interchanging the plus side and the minus side of the n axis. For example,

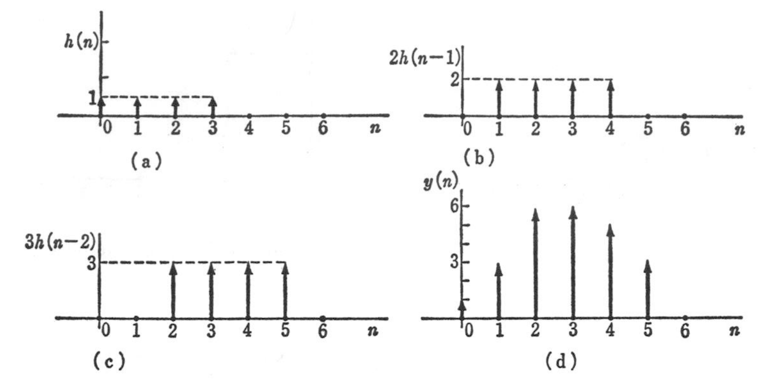

Figure 2.11 Output Signal $y(n)$. (a) $h(n)$; (b) $2h(n-1)$; (c) $3h(n-2)$; (d) $y(n)$.

interchanging the plus side and the minus side of the n axis in Fig. 2.9(a) gives $h(-k)$ as shown in Fig. 2.12(a).

For $n = \alpha$, $h(\alpha - k)$ can be considered as follows: Assume that $h(-k)$ in Fig. 2.12(a) is $h(n)$. Then assume that delaying α units gives $h(n - \alpha)$. Hence, $h(\alpha - k)$ can be obtained by delaying α units of $h(-k)$ in Fig. 2.12(a) as shown in Fig. 2.12(b). Hence, the figure of $h(1 - k)$ can be obtained by shifting the figure $h(-k)$ in Fig. 2.12(a) to the right by 1 unit. The figure of $h(2 - k)$ can be obtained by shifting the figure of $h(1 - k)$ to the right by 1 unit, and so on.

Now the meaning of $y(n) = \sum_{k=-\infty}^{\infty} x(k)h(n-k)$ is easily seen. For $n = 0$, $\sum_{k=-\infty}^{\infty} x(k)h(-k)$ means multiplying $x(k)$ in Fig. 2.13(a) and $h(-k)$ in Fig. 2.13(b) at each k as shown in Fig. 2.13(c). Then we add all nonzero values together to obtain $y(0)$. In Fig. 2.13(c), there is only one nonzero value at $k = 0$ which is 1. Hence, $y(0) = 1$. For $n = 1$, $h(1 - k)$ in Fig. 2.14(b) is obtained from $h(-k)$ in Fig. 2.13(b) by shifting all nonzero value to the right by 1 unit. Then, we multiply $x(k)$ in Fig. 2.14(a) and $h(1 - k)$ in

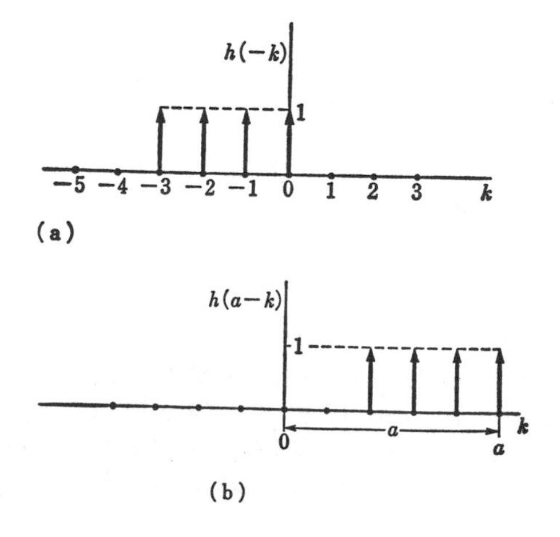

Figure 2.12 $h(-k)$ and $h(\alpha - k)$. (a) $h(-k)$ obtained from $h(n)$ in Fig. 2.9; (b) $h(\alpha - k)$.

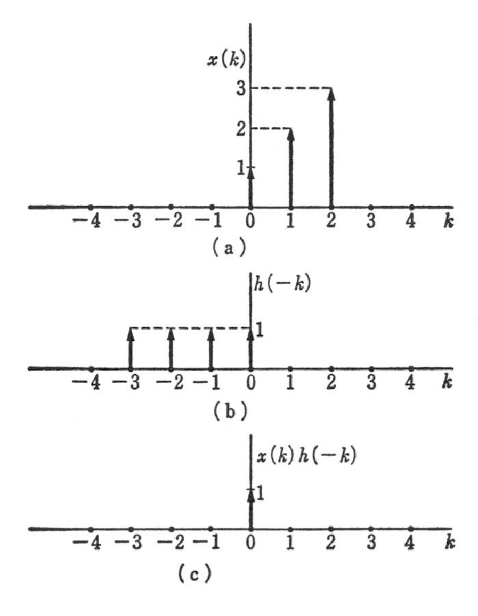

Figure 2.13 $x(k)h(-k)$. (a) $x(k)$; (b) $h(-k)$; (c) $x(k)h(-k)$.

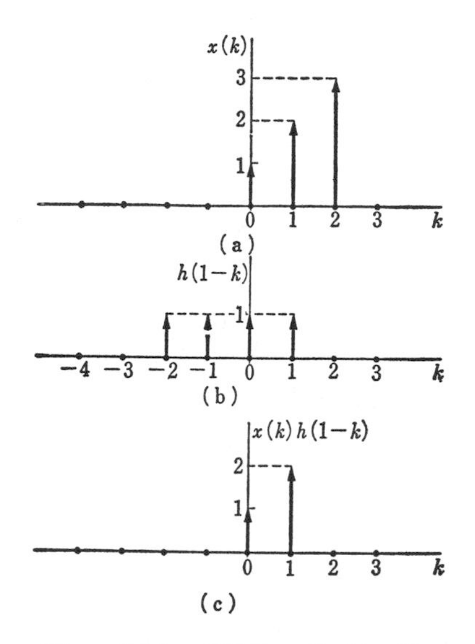

Figure 2.14 $x(k)h(1-k)$. (a) $x(k)$; (b) $h(1-k)$; (c) $x(k)h(1-k)$.

Fig. 2.14(b) at each k as shown in Fig. 2.14(c). Because $y(1) = \sum_{k=-\infty}^{\infty} x(k)h(1-k)$, we add all nonzero values in Fig. 2.14(c) together to obtain $y(1) = 1 + 2 = 3$.

For $n = 2$, we obtain $h(2-k)$ in Fig. 2.15(b) by shifting $h(1-k)$ in Fig. 2.14(b) to the right by 1 unit. Multiplying $x(k)$ in Fig. 2.15(a) and $h(2-k)$ at each k produces the result shown in Fig. 2.15(c). Adding nonzero values in Fig. 2.15(c), i.e., $1 + 2 + 3 = 6$, we obtain $y(2) = 6$. For $y(3)$, multiplying $x(k)$ in Fig. 2.16(a) and $h(3-k)$ in Fig. 2.16(b), which is obtained by shifting $h(2-k)$ in Fig. 2.15(b) to the right by 1 unit, the result shown in Fig. 2.16(c) is obtained. Then adding nonzero values in the result, we derive the value $y(3) = 6$. Similarly, we shift $h(3-k)$ in Fig. 2.16(b) to the right by 1 unit to obtain $h(4-k)$ in Fig. 2.17(b). Multiplying $x(k)$ in Fig. 2.17(a) and $h(4-k)$, the result is shown in Fig. 2.17(c), where there are two nonzero values which are 2 and 3. Then adding nonzero values together in the result gives $y(4) = 5$. Again we shift $h(4-k)$ in Fig. 2.17(b) to the right by 1 unit to obtain $h(5-k)$ as shown in Fig. 2.18(b). Then multiplying $x(k)$ in Fig. 2.18(a) and $h(5-k)$ produces the result as shown in Fig. 2.18(c) in which there is only one nonzero value having a value of 3. Hence, $y(5) = 3$. Again we shift $h(5-k)$ in Fig. 2.18(b) to the right by 1 unit to obtain $h(6-k)$ as shown in Fig. 2.19(b). Then multiplying $x(k)$ in Fig. 2.19(a) and $h(6-k)$, we have a result as shown in Fig. 2.19(c), where there are no nonzero values. Hence, $y(6) = 0$. Because any further shifting of $h(6-k)$ to the right and multiplying it with $x(k)$ will not produce a result having

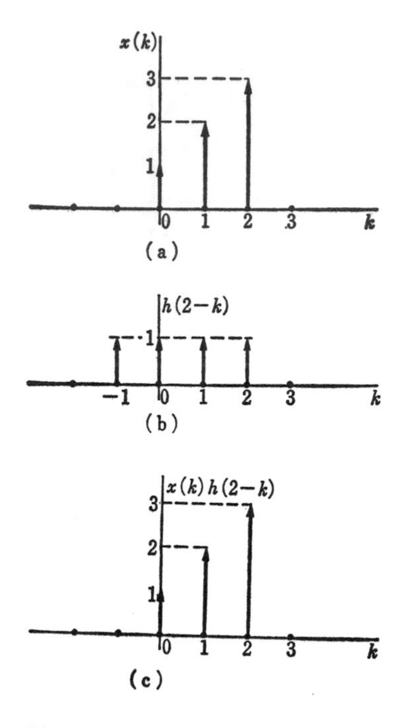

Figure 2.15 $x(k)h(2-k)$. (a) $x(k)$; (b) $h(2-k)$; (c) $x(k)h(2-k)$.

Figure 2.16 $x(k)h(3-k)$. (a) $x(k)$; (b) $h(3-k)$; (c) $x(k)h(3-k)$.

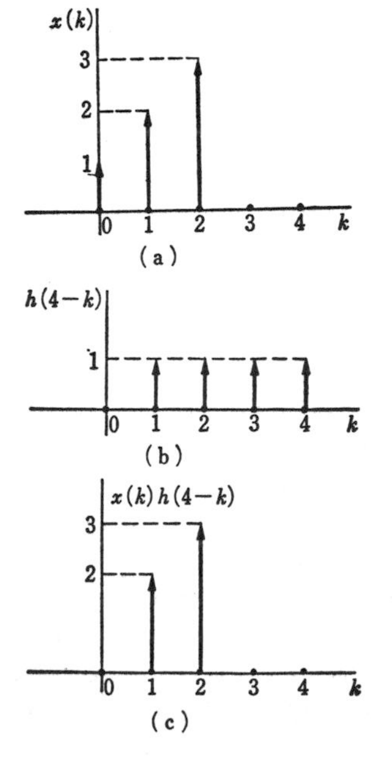

Figure 2.17 $x(k)h(4-k)$. (a) $x(k)$; (b) $h(4-k)$; (c) $x(k)h(4-k)$.

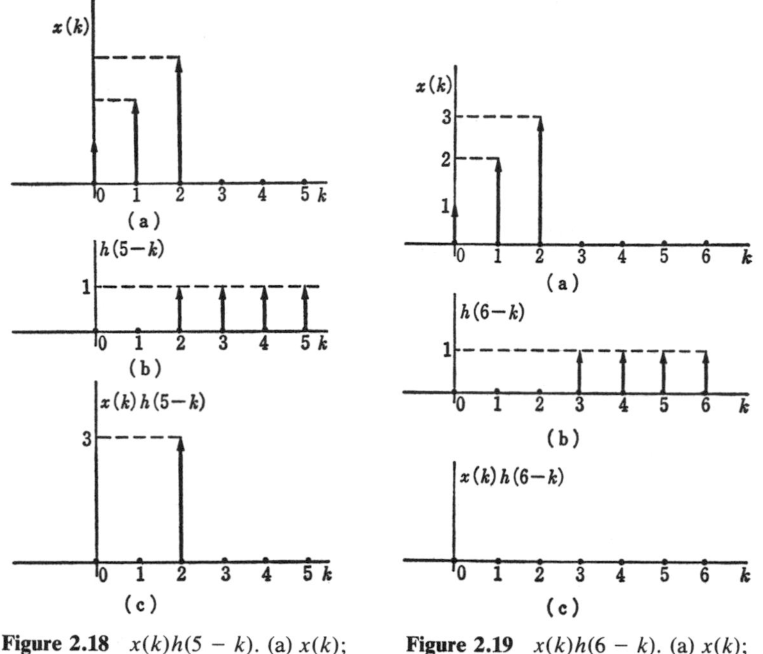

Figure 2.18 $x(k)h(5-k)$. (a) $x(k)$; (b) $h(5-k)$; (c) $x(k)h(5-k)$.

Figure 2.19 $x(k)h(6-k)$. (a) $x(k)$; (b) $h(6-k)$; (c) $x(k)h(6-k)$.

nonzero values, we can see that $y(n) = 0$ for all $n > 6$. We did not calculate with any negative value of n. This is because $y(n) = 0$ for all $n < 0$ is clear from Fig. 2.13. Notice that shifting $h(-k)$ to the left by 1 unit gives $h(-1-k)$. The preceding process is the convolution of $x(n)$ and $h(n)$ given by Equation 2.26.

2.5 SUMMARY

1. A unit sample $u_I(n)$ is defined as

$$u_I(n) = \begin{cases} 1, & n = 0 \\ 0, & n \neq 0 \end{cases}$$

2. A unit step $u_s(n)$ is defined as

$$u_s(n) = \begin{cases} 1, & n \geq 0 \\ 0, & n < 0 \end{cases}$$

3. Unit sample response $h(n)$ of a system is defined as its output signal $y(n)$ when its input signal is a unit sample.
4. A system satisfying initial rest conditions is a system whose output signal $y(n) = 0$ for all $n < n_0$ when input signal $x(n) = 0$ for all $n < n_0$.

5. A linear system satisfies

$$\mathcal{T}(\alpha_1 x_1(n) + \alpha_2 x_2(n)) = \alpha_1 \mathcal{T}(x_1(n)) + \alpha_2 \mathcal{T}(x_2(n))$$

where α_1 and α_2 are constants and symbol $\mathcal{T}(x(n))$ indicates the output signal when an input signal is $x(n)$.

6. A shift invariant system has the property that its output signal $y(n)$ becomes $y(n - k)$ when its input signal $x(n)$ is changed to $x(n - k)$.

7. Convolution of $x(n)$ and $h(n)$ is defined as

$$x(n) * h(n) = \sum_{k=-\infty}^{\infty} x(k)h(n - k) = \sum_{k=-\infty}^{\infty} x(n - k)h(k).$$

2.6 PROBLEMS

1. Solve the following difference equations
   a. $y(n) + 5y(n - 1) = 0, \quad y(-1) = 1$
   b. $y(n) + y(n - 1) - 6y(n - 2) = 4^n, \quad y(-1) = 0, y(-2) = 0$
   c. $y(n) + y(n - 1) - 2y(n - 2) = 3^n, \quad y(0) = 0, y(1) = 0$
   d. $y(n) - 3y(n - 1) + 2y(n - 2) = 5^n, \quad y(-1) = -2, y(-2) = 0$
   e. $y(n) + 4y(n - 2) = 2, \quad y(-1) = 4, y(-2) = 0.$
2. Calculate the unit sample response $h(n)$ of systems specified by the following equations under the assumption that these systems satisfy initial rest conditions.
   a. $y(n) + 5y(n - 1) = x(n)$
   b. $y(n) + y(n - 1) - 6y(n - 2) = x(n)$
   c. $y(n) + y(n - 1) - 2y(n - 2) = x(n)$
   d. $y(n) - 3y(n - 1) + 2y(n - 2) = x(n)$
   e. $y(n) + 4y(n - 2) = x(n)$
3. Calculate the output $y(n)$ of each of linear shift invariant systems whose unit sample responses are shown in Fig. 2.20 when an input $x(n)$ is

$$x(n) = u_I(n) + 2u_I(n - 1) + 4u_I(n - 2)$$

4. Calculate the output signal $y(n)$ of each linear shift invariant system whose unit sample responses are specified in Fig. 2.21 when an input signal $x(n)$ is the same as the system unit sample response.
5. Calculate the output signal $y(n)$ of a linear shift invariant system whose unit sample response is

$$h(n) = \sum_{k=1}^{6} k u_I(n - k),$$

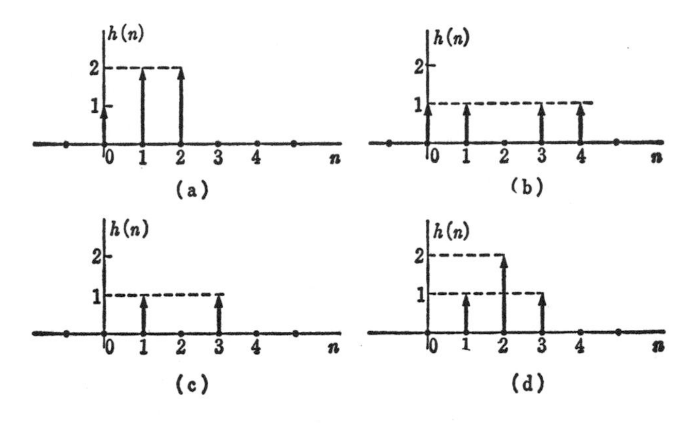

Figure 2.20 Unit Sample Responses

for each of the following input signals:

a. $x(n) = \sum_{k=1}^{3} u_I(n - k)$

b. $x(n) = \sum_{k=1}^{2} u_I(n - 2k)$

c. $x(n) = \sum_{k=1}^{6} u_I(n - k)$

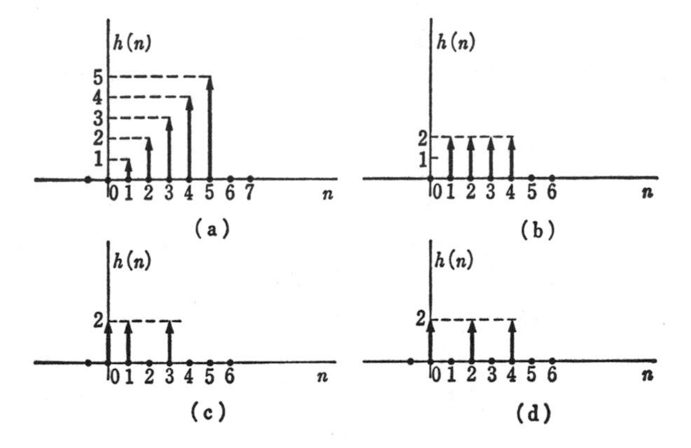

Figure 2.21 Unit Sample Responses

3
Digital Signal and Fourier Transformation

3.1 FOURIER SERIES

Showing an output signal y for a given input signal x is one way of indicating properties of a system. A ratio $y(j\omega)/x(j\omega)$ between a resultant output signal $y(j\omega)$ with respect to the particular input signal $x(j\omega) = e^{j\omega}$ is a typical way of expressing characteristics of electrical networks and systems, however. This ratio is also employed for indicating properties of discrete systems. We will study the Fourier series, which is a basic mathematical tool necessary for discussing the ratio between a resultant output signal and a given input signal.

If a given function $x(t)$ satisfies

$$x(t + T) = x(t) \tag{3.1}$$

with a real number $T > 0$, then $x(t)$ is called a "periodic function," and the smallest positive number T satisfying Equation 3.1 is called a "period."

Expressing a periodic function $x(t)$ as

$$x(t) = \frac{1}{2} a_0 + \sum_{k=1}^{\infty} [a_k \cos (2\pi k f_0 t) + b_k \sin (2\pi k f_0 t)] \tag{3.2}$$

which is known as a "Fourier series." To determine constants a_k and b_k in

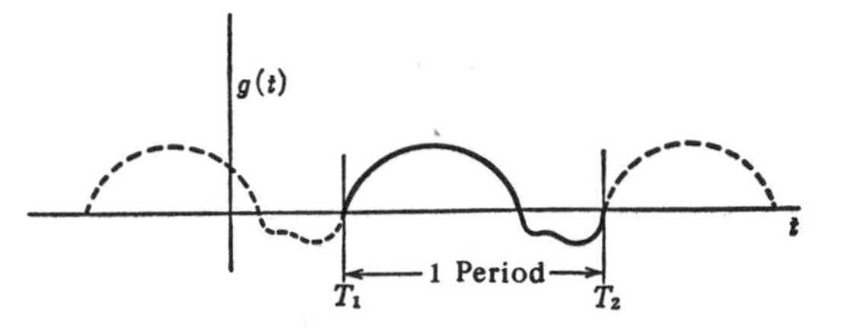

Figure 3.1 One Period of Periodic Function $x(t)$

the right-hand side of Equation 3.2, we will assume that one period of $x(t)$ is from T_1 to T_2 as shown in Fig. 3.1. We call $f_0 = 1/(T_2 - T_1)$ as the "fundamental frequency." Multiplying $\cos(2\pi m f_0 t)$ to both sides of Equation 3.2 and integrating it from T_2 to T_1 where m is a positive integer, we have

$$\int_{T_1}^{T_2} x(t)\cos(2\pi m f_0 t)\,dt = \frac{1}{2}a_0\int_{T_1}^{T_2}\cos(2\pi m f_0 t)\,dt + \sum_{k=1}^{\infty} a_k \int_{T_1}^{T_2}\cos(2\pi k f_0 t)\cos(2\pi m f_0 t)\,dt + \sum_{k=1}^{\infty} b_k \int_{T_1}^{T_2}\sin(2\pi k f_0 t)\cos(2\pi m f_0 t)\,dt \tag{3.3}$$

The second term in the right-hand side of Equation 3.3 can be expressed as

$$a_k\int_{T_1}^{T_2}\cos(2\pi k f_0 t)\cos(2\pi m f_0 t)\,dt = \frac{a_k}{2}\int_{T_1}^{T_2}\cos[2\pi(k+m)f_0 t]\,dt + \frac{a_k}{2}\int_{T_1}^{T_2}\cos[2\pi(k-m)f_0 t]\,dt \tag{3.4}$$

Because both k and m are positive integers, $\cos[2\pi(k+m)f_0 t]$ can be written as

$$\cos[2\pi(k+m)f_0 t] = \cos(2\pi p f_0 t), \qquad p = 1, 2, \cdots$$

By changing t to $t + 1/f_0$, the preceding equation becomes

$$\cos\left[2\pi p f_0\left(t + \frac{1}{f_0}\right)\right] = \cos[2\pi p f_0 t + 2\pi p] = \cos(2\pi p f_0 t)$$

which indicates that $\cos(2\pi p f_0 t)$ is a periodic function since $T = T_2 - T_1$ is a period of $\cos(2\pi p f_0 t)$, the first term in the right-hand side of Equation 3.4 is to integrate a periodic function from T_1 to T_2, which consists of exactly p periods. As shown in Fig. 3.2, the positive area and the negative area under the curve of $\cos(2\pi p f_0 t)$ in one period are the same. Hence, the first term in the right-hand side of Equation 3.4 is 0. Similarly, the second term in the

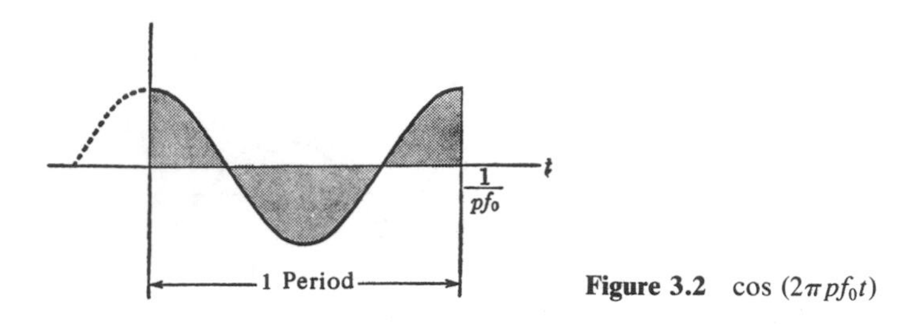

Figure 3.2 $\cos(2\pi p f_0 t)$

right-hand side of Equation 3.4 is 0 when $k - m \neq 0$. When $k = m$, this second term becomes

$$\frac{a_k}{2}\int_{T_1}^{T_2} \cos\,[2\pi(k - m)f_0 t]\; dt = \frac{a_k}{2}\int_{T_1}^{T_2} dt = \frac{a_k}{2}\,T$$

Hence, Equation 3.4 becomes

$$a_k \int_{T_1}^{T_2} \cos\,(2\pi k f_0 t)\cos\,(2\pi m f_0 t)\; dt = \begin{cases} \dfrac{a_k}{2}\,T, & k = m \\ 0, & \text{Otherwise} \end{cases} \tag{3.5}$$

The last term in the right-hand side of Equation 3.3 can be expressed as

$$b_k \int_{T_1}^{T_2} \sin\,(2\pi k f_0 t)\cos\,(2\pi m f_0 t)\; dt$$

$$= \frac{b_0}{2}\int_{T_1}^{T_2} \sin\,[2\pi(k + m)f_0 t]\; dt + \frac{b_k}{2}\int_{T_1}^{T_2} \sin\,[2\pi(k - m)f_0 t]\; dt$$

The difference between the previous case and this case is that the function to be integrated is a sine rather than a cosine. Hence, even when $k = m$, the integration will result 0. That is,

$$b_k \int_{T_1}^{T_2} \sin\,(2\pi k f_0 t)\cos\,(2\pi m f_0 t)\; dt = 0 \tag{3.6}$$

Finally, the first term of the right-hand side of Equation 3.3, which is

$$\frac{a_0}{2}\int_{T_1}^{T_2} \cos\,(2\pi m f_0 t)\; dt$$

is also 0 except when $m = 0$ or

$$\frac{a_0}{2}\int_{T_1}^{T_2} \cos\,(2\pi m f_0 t)\; dt = \begin{cases} \dfrac{a_0}{2}\,T, & m = 0 \\ 0, & \text{Otherwise} \end{cases} \tag{3.7}$$

Thus by Equations 3.5, 3.6, and 3.7, Equation 3.3 becomes

$$\int_{T_1}^{T_2} x(t) \cos (2\pi m f_0 t)\, dt = \frac{a_m}{2} T \tag{3.8}$$

Changing a dummy variable m to k, we have

$$a_k = \frac{2}{T} \int_{T_1}^{T_2} x(t) \cos (2\pi k f_0 t)\, dt \tag{3.9}$$

This is a formula to calculate a constant a_k.

By multiplying $\sin (2\pi m f_0 t)$ to both sides of Equation 3.2 and integrate it from T_2 to T_1, we can obtain a formula for a constant b_k as

$$b_k = \frac{2}{T} \int_{T_1}^{T_2} x(t) \sin (2\pi k f_0 t)\, dt \tag{3.10}$$

Equations 3.9 and 3.10 are known as "Fourier coefficients." Recall that $T = T_2 - T_1$ is a period and $f_0 = 1/T$ is the fundamental frequency of $x(t)$. As an example, the Fourier series expression of a function $x(t)$ shown in Fig. 3.3 can be obtained by the following steps: By Equation 3.9, a_k is

$$\begin{aligned} a_k &= \frac{2}{T} \int_{T_1}^{T_2} x(t) \cos (2\pi k f_0 t)\, dt = \frac{2A}{T} \int_{-\tau/2}^{\tau/2} \cos (2\pi k f_0 t)\, dt \\ &= \frac{2A}{T} \frac{\sin (2\pi k f_0 t)}{2\pi k f_0} \bigg|_{-\tau/2}^{\tau/2} = \frac{2A\tau}{T} \frac{\sin (\pi k f_0 \tau)}{\pi k f_0 \tau} \end{aligned} \tag{3.11}$$

By setting $k = 0$, we have $a_0 = 2A\tau/T$. With $k = 1$, we have $a_1 = (2A\tau/T) \sin (\pi f_0 \tau)/\pi f_0 \tau$ and so on.

Another Fourier coefficient b_k is (by Equation 3.10)

$$b_k = \frac{2}{T} \int_{T_1}^{T_2} x(t) \sin (2\pi k f_0 t)\, dt = \frac{2A}{T} \int_{-\tau/2}^{\tau/2} \sin (2\pi k f_0 t)\, dt = 0$$

Hence, the Fourier series of $x(t)$ in Fig. 3.3 can be obtained by substituting a_k in Equation 3.11 into Equation 3.2 as

$$x(t) = \frac{A\tau}{T} + \sum_{k=1}^{\infty} \left(\frac{2A\tau}{T} \frac{\sin (\pi k f_0 \tau)}{\pi k f_0 \tau} \right) \cos (2\pi k f_0 t)$$

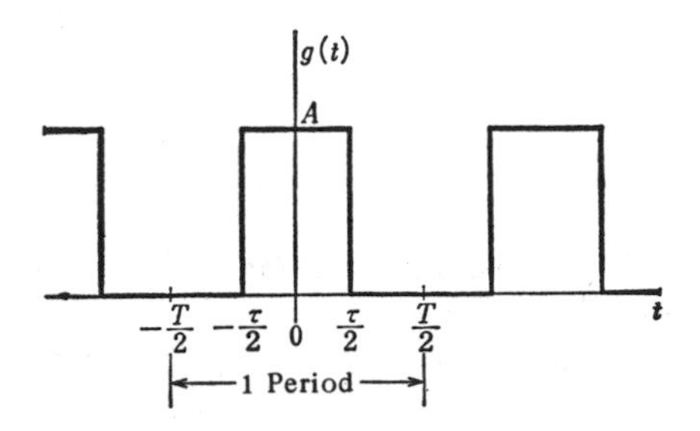

Figure 3.3 Periodic Function $x(t)$

By using Equation a-29, $\cos(2\pi k f_0 t)$ in Equation 3.2 can be expressed as

$$\cos(2\pi k f_0 t) = \frac{e^{j2\pi k f_0 t} + e^{-j2\pi k f_0 t}}{2}$$

Also by Equation a-31, $\sin(2\pi k f_0 t)$ can be expressed as

$$\sin(2\pi k f_0 t) = \frac{e^{j2\pi k f_0 t} - e^{-j2\pi k f_0 t}}{2j}$$

Using these, the Fourier series in Equation 3.2 can be changed as

$$x(t) = \sum_{k=-\infty}^{\infty} C_k e^{j2\pi k f_0 t} \tag{3.12}$$

which is called the "Fourier exponential series" of a function $x(t)$. The coefficient C_k is

$$C_k = \frac{1}{T}\int_{T_1}^{T_2} x(t) e^{-j2\pi k f_0 t}\, dt \tag{3.13}$$

Showing C_k at frequency kf_0 for all k, we can obtain a graph called the "frequency spectrum." As an example, consider a function $x(t)$ shown in Fig. 3.3. Using Equation 3.13, C_k is

$$C_k = \frac{1}{T}\int_{-\tau/2}^{\tau/2} A e^{-j2\pi k f_0 t}\, dt = \frac{A\tau}{T}\frac{\sin(\pi k f_0 \tau)}{\pi k f_0 \tau} \tag{3.14}$$

Substituting this into Equation 3.12, we have

$$x(t) = \sum_{k=-\infty}^{\infty} \frac{A\tau}{T}\frac{\sin(\pi k f_0 \tau)}{\pi k f_0 \tau} e^{j2\pi k f_0 t} \tag{3.15}$$

which is the Fourier exponential series of $x(t)$. By changing $e^{j2\pi k f_0 t}$ to trigonometric functions, the above expression becomes

$$x(t) = \sum_{k=-\infty}^{\infty} \frac{A\tau}{T}\frac{\sin(\pi k f_0 \tau)}{\pi k f_0 \tau}[\cos(2\pi k f_0 t) + j\sin(2\pi k f_0 t)]$$

$$= \sum_{k=-\infty}^{-1} \frac{A\tau}{T}\frac{\sin(\pi k f_0 \tau)}{\pi k f_0 \tau}[\cos(2\pi k f_0 t) + j\sin(2\pi k f_0 t)] + \frac{A\tau}{T}$$

$$+ \sum_{k=1}^{\infty} \frac{A\tau}{T}\frac{\sin(\pi k f_0 \tau)}{\pi k f_0 \tau}[\cos(2\pi k f_0 t) + j\sin(2\pi k f_0 t)]$$

Combining these summations together properly, the equation becomes the Fourier series of $x(t)$.

By setting $\tau/T = 0.2$, Equation 3.14 becomes

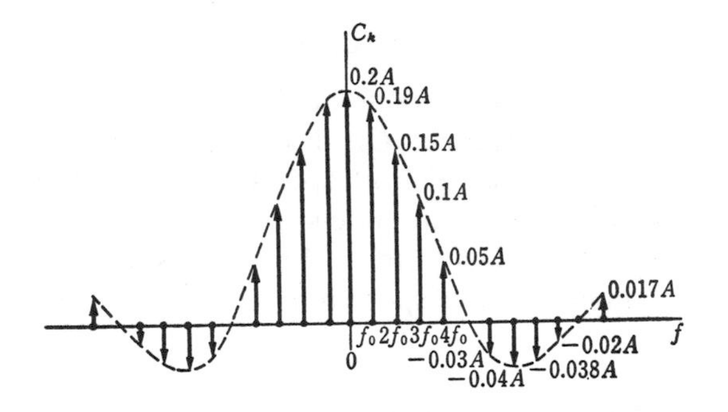

Figure 3.4 Frequency Spectrums

$$C_k = 0.2A \frac{\sin(0.2\pi k)}{0.2\pi k}$$

Let us obtain the frequency spectrum from this C_k. For $k = 0, 1, 2, \ldots, C_0$, $C_1, C_2, \ldots$ are

$$\begin{aligned}
k &= 0, \quad C_0 = 0.2A \\
k &= 1, \quad C_1 = 0.19A \\
k &= 2, \quad C_2 = 0.15A \\
k &= 3, \quad C_3 = 0.1A \\
&\vdots \qquad\quad \vdots
\end{aligned}$$

It can be seen that the value of C_k for $k = -1, -2, -3, \ldots$ and that for $k = 1, 2, 3, \ldots$ are the same. Hence, the value of the frequency spectrum at $f = 0$ is C_0, the value at $f = \pm f_0$ is C_1, the value at $f = \pm 2f_0$ is C_2, and so on. We can obtain the frequency spectrum as shown in Fig. 3.4.

Equation 3.12 shows that a periodic function $x(t)$ can be expressed as

$$x(t) = \sum_{k=-\infty}^{\infty} C_k e^{j2\pi k f_0 t} = \cdots + C_{-1} e^{-j2\pi f_0 t} + C_0 + C_1 e^{j2\pi f_0 t} + C_2 e^{j2\pi(2f_0)t} + \cdots$$

This indicates that $x(t)$ is constructed by $C_k e^{j2\pi k f_0 t}$ called a "frequency component." In other words, the kind of components by which $x(t)$ is formed is indicated by the frequency spectrum of $x(t)$.

When a coefficient C_k is a complex number, the frequency spectrum cannot be shown as one in Fig. 3.4. However, by expressing C_k as

$$C_k = |C_k| e^{j\phi_k}$$

each term in the right-hand side of Equation 3.12 becomes

$$C_k e^{j2\pi k f_0 t} = |C_k| e^{j(2\pi k f_0 t + \phi_k)}$$

This indicates that the maximum value of $C_k e^{j2\pi k f_0 t}$ is $|C_k|$, and the minimum value of $C_k e^{j2\pi k f_0 t}$ is $-|C_k|$. Hence, $|C_k|$ shows the amplitude of $C_k e^{j2\pi k f_0 t}$ and ϕ_k shows the argument (phase) of $C_k e^{j2\pi k f_0 t}$. Thus, a graph obtained by indicating $|C_k|$ at kf_0 is called the "amplitude spectrum," and a graph showing ϕ_k at kf_0 is called the "phase spectrum." When C_k is a complex number, an amplitude spectrum and a phase spectrum are necessary to indicate C_k.

3.2 FOURIER TRANSFORMATION

The coefficient C_k of a Fourier exponential series in Equation 3.13 indicates the size of each frequency component of a periodic function $x(t)$. Hence, expressing $x(t)$ by such a series is very useful for showing properties of $x(t)$. However, $x(t)$ cannot be expressed by a Fourier exponential series if $x(t)$ is not a periodic function. A modification of a Fourier exponential series so that it can be used for a nonperiodic function is a "Fourier transformation."

We have studied that a Fourier exponential series of a periodic function $x(t)$ is

$$x(t) = \sum_{k=-\infty}^{\infty} C_k e^{j2\pi k f_0 t} \tag{3.16}$$

The coefficient C_k in this equation is given by

$$C_k = \frac{1}{T} \int_{T_1}^{T_2} x(t) e^{-j2\pi k f_0 t} \, dt \tag{3.17}$$

where T is the period of $x(t)$, $T_2 - T_1 = T$ and f_0 is the fundamental frequency of $x(t)$, which is equal to

$$f_0 = \frac{1}{T} \tag{3.18}$$

Consider a situation in which the period T of $x(t)$ becomes larger and larger. When $T \rightarrow \infty$, $x(t)$ will become a nonperiodic function. Hence, if Equation 3.17 gives a finite nonzero value for $T \rightarrow \infty$, we can obtain a coefficient C_k for a nonperiodic function. Unfortunately, there is $1/T$ in the right-hand side of Equation 3.17, which makes $C_k \rightarrow 0$ when $T \rightarrow \infty$. This indicates that if we remove $1/T$ from the right-hand side of Equation 3.17, we may be able to use this series to express a nonperiodic function. $T \rightarrow \infty$ can be accomplished by making $T_1 \rightarrow -\infty$ and $T_2 \rightarrow \infty$. From Equation 3.18, $f_0 \rightarrow 0$ as $T \rightarrow \infty$. Hence, by replacing kf_0 in Equation 3.17 by a variable f, we can obtain an equation for a coefficient C_k for a nonperiodic function $x(t)$ as

$$C_k = \int_{-\infty}^{\infty} x(t)e^{-j2\pi ft}\,dt \tag{3.19}$$

Because we have removed $1/T$ from Equation 3.17, we must put $1/T$ in Equation 3.16 as

$$x(t) = \frac{1}{T}\sum_{k=-\infty}^{\infty} C_k e^{j2\pi kf_0t}$$

This equation becomes

$$x(t) = \int_{-\infty}^{\infty} C_k e^{j2\pi kft}\,df \tag{3.20}$$

as $T \to \infty$. Instead of using the symbol C_k, the symbol $X(j\omega)$ or the symbol $\mathcal{F}(x(t))$ for indicating the Fourier transformation of a function $x(t)$ are commonly employed. Similarly, the inverse Fourier transformation is symbolized as $\mathcal{F}^{-1}[X(j\omega)]$. That is, expressing Equations 3.19 and 3.20 as

$$\mathcal{F}[x(t)] = X(j\omega) = \int_{-\infty}^{\infty} x(t)e^{-j\omega t}\,dt \tag{3.21}$$

and

$$\mathcal{F}^{-1}[X(j\omega)] = x(t) = \int_{-\infty}^{\infty} X(j\omega)e^{j\omega t}\,df \tag{3.22}$$

are the "Fourier transformation" and the "inverse Fourier transformation" of a function $x(t)$ where the symbol ω is known as angular frequeny which is equal to $2\pi f$.

As an example, let us obtain the Fourier transformation of a function $x(t)$ given by

$$x(t) = \begin{cases} A, & |t| < T_0 \\ 0, & |t| > T_0 \end{cases}$$

which is shown in Fig. 3.5(a). By Equation 3.21,

$$X(j\omega) = \int_{-T_0}^{T_0} Ae^{-j\omega t}\,dt = 2AT_0\frac{\sin \omega T_0}{\omega T_0}$$

is the Fourier transformation of $x(t)$, which is shown in Fig. 3.5(b).

One of the reasons that the Fourier transformation is useful for analysis of electrical networks is the existence of a so-called impulse function. An impulse function $\delta(t - t_0)$ is defined as

$$\left.\begin{aligned} &\text{(I)} \quad \int_{-\infty}^{\infty} \delta(t - t_0)\,dt = 1 \\ &\text{(II)} \quad \delta(t - t_0) = 0, \quad \text{where } t \neq t_0 \end{aligned}\right\} \tag{3.23}$$

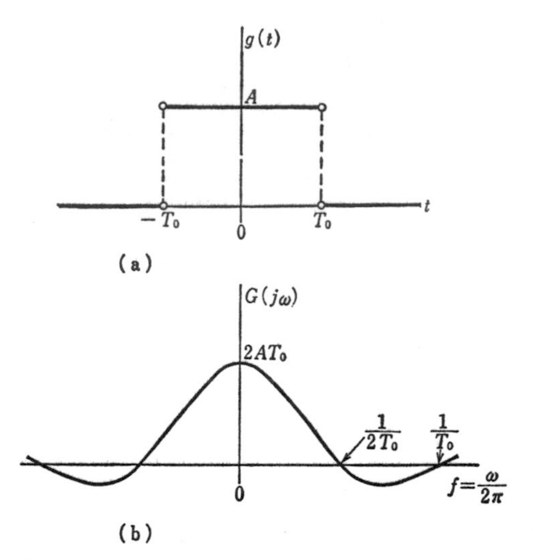

Figure 3.5 $x(t)$ and $X(j\omega)$. (a) $x(t)$; (b) $X(j\omega)$.

Some well-known impulse functions are as follows:

$$\delta(t) = \lim_{a\to\infty} \int_{-a}^{a} e^{\pm j\omega t}\, df \tag{3.24}$$

$$\delta(t) = \lim_{a\to\infty} \int_{-a}^{a} \cos \omega t df, \qquad \text{where } \omega = 2\pi f \tag{3.25}$$

$$\delta(t) = \lim_{a\to\infty} \frac{\sin(2\pi a t)}{\pi t} \tag{3.26}$$

Also several books use a function $x(t)$ shown in Fig. 3.6 as an example of an impulse function $\delta(t - t_0)$ because it satisfies Equation 3.23.

Let us investigate properties of impulse functions. Because $\delta(t - t_0)$ is 0 except $t = t_0$ by Equation 3.23, product $\phi(t)\delta(t - t_0)$ of a function $\phi(t)$, and an impulse function $\delta(t - t_0)$ and product $\phi(t_0)\delta(t - t_0)$ are the same.

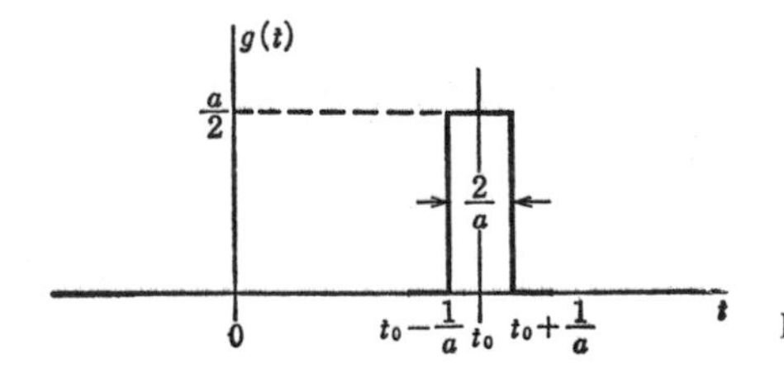

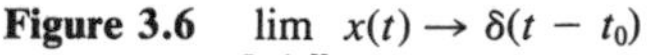

Figure 3.6 $\lim_{a\to\infty} x(t) \to \delta(t - t_0)$

Hence,

$$\int_{-\infty}^{\infty} \Phi(t)\delta(t - t_0)\, dt = \Phi(t_0) \tag{3.27}$$

which is a very useful property.

When a function $x(t)$ is

$$x(t) = A\delta(t)$$

the Fourier transformation of $x(t)$ is (by Equation 3.21)

$$X(j\omega) = \int_{-\infty}^{\infty} A\delta(t)e^{-j\omega t}\, dt = Ae^0 = A$$

Also, the inverse Fourier transformation of this result is (by Equations 3.22 and 3.24)

$$\mathscr{F}^{-1}[A] = A\delta(t) \tag{3.28}$$

By replacing at by τ in $\int_{-\infty}^{\infty} \delta(at)\phi(t)\, dt$, we have

$$\int_{-\infty}^{\infty} \delta(at)\phi(t)\, dt = \frac{1}{a}\int_{-\infty}^{\infty} \delta(\tau)\phi\left(\frac{\tau}{a}\right) d\tau = \frac{1}{a}\,\phi(0)$$

Hence, a relationship between $\delta(at)$ and $\delta(t)$ is

$$\delta(at) = \frac{1}{a}\,\delta(t) \tag{3.29}$$

which is a very important relationship.

Let us obtain the Fourier transformation of a function $e^{j2\pi f_0 t}$. By Equation 3.21,

$$\mathscr{F}[e^{j2\pi f_0 t}] = \int_{-\infty}^{\infty} e^{j2\pi f_0 t}e^{-j\omega t}\, dt = \lim_{a\to\infty}\int_{-a}^{a} e^{-j2\pi(f-f_0)t}\, dt \qquad \text{where } \omega = 2\pi f$$

We can see that this is identical with Equation 3.24 if we interchange f and t. Hence, we have

$$\mathscr{F}[e^{j2\pi f_0 t}] = \delta(f - f_0) \tag{3.30}$$

Using this result, we can obtain the Fourier transformation of cos $(2\pi f_0 t)$ as

$$\mathscr{F}[\cos(2\pi f_0 t)] = \int_{-\infty}^{\infty} \frac{e^{j2\pi f_0 t} + e^{-j2\pi f_0 t}}{2}\, e^{-j\omega t}\, dt = \frac{1}{2}\,[\delta(f - f_0) + \delta(f + f_0)]$$

which is shown in Fig. 3.7.

The Fourier transformation of an impulse series $\Sigma\, \delta(t - nT)$, which is very useful when we study the relationship between a digital signal and an analog signal, is

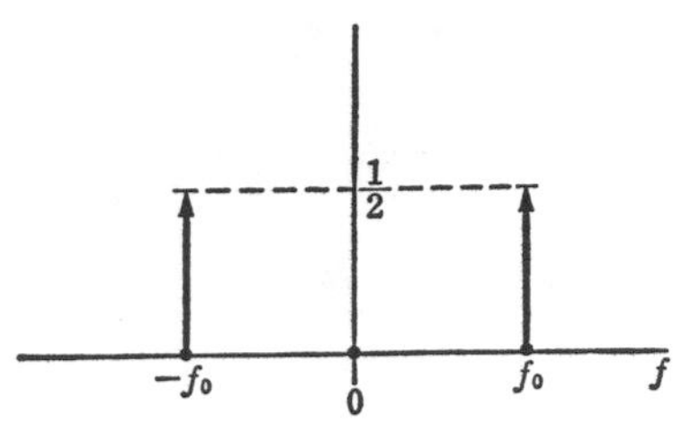

Figure 3.7 Fourier transformation of $\cos(2\pi f_0 t)$

$$\mathcal{F}\left[\sum_{n=-\infty}^{\infty} \delta(t - nT)\right] = \frac{1}{T}\sum_{n=-\infty}^{\infty} \delta\left(f - \frac{n}{T}\right) \tag{3.31}$$

To prove that this is correct, we will show that the inverse Fourier transformation of the right-hand side of Equation 3.31 becomes $\sum_{n=-\infty}^{\infty} \delta(t - nT)$. By Equations 3.22 and 3.23, the inverse Fourier transformation of the right-hand side of Equation 3.31 is

$$\mathcal{F}^{-1}\left[\frac{1}{T}\sum_{n=-\infty}^{\infty} \delta\left(f - \frac{n}{T}\right)\right] = \int_{-\infty}^{\infty} \frac{1}{T}\sum_{n=-\infty}^{\infty} \delta\left(f - \frac{n}{T}\right) e^{j2\pi ft}\, df = \frac{1}{T}\sum_{n=-\infty}^{\infty} e^{j(2\pi nt/T)}$$

which can be written as

$$\begin{aligned}\frac{1}{T}\sum_{n=-\infty}^{\infty} e^{j(2\pi nt/T)} &= \lim_{N\to\infty} \frac{1}{T}\sum_{n=-N}^{N} e^{j(2\pi nt/T)} = \lim_{N\to\infty} \frac{1}{T}\,\frac{e^{-j(2\pi Nt/T)} - e^{j(2\pi(N+1)t/T)}}{1 - e^{j(2\pi t/T)}} \\ &= \lim_{N\to\infty} \frac{1}{T}\,\frac{\sin\left[\left(N + \frac{1}{2}\right)\frac{2\pi t}{T}\right]}{\sin\frac{\pi t}{T}}\end{aligned} \tag{3.32}$$

The right-hand side of this equation is a periodic function with a period T. Hence if the right-hand side becomes $\delta(t)$ between $-T/2$ and $T/2$, then the right-hand side is $\sum_{n=-\infty}^{\infty} \delta(t - nT)$ in the t axis. By letting $t/T = \tau$, the right-hand side of Equation 3.32 becomes

$$\begin{aligned}&\lim_{N\to\infty} \frac{1}{T}\,\frac{\sin\left[2\pi\left(N + \frac{1}{2}\right)\tau\right]}{\sin 2\pi\left(\frac{\tau}{2}\right)} \\ &\qquad = \lim_{N\to\infty} \frac{2\pi}{T}\left(\frac{\sin\left[2\pi\left(N + \frac{1}{2}\right)\tau\right]}{\pi\tau}\right)\left(\frac{\frac{\tau}{2}}{\sin 2\pi\left(\frac{\tau}{2}\right)}\right)\end{aligned} \tag{3.33}$$

$(\tau/2)/\sin 2\pi\, \tau/2$ in the right-hand side is finite between $-T/2$ and $T/2$, and is $1/2\pi$ at $\tau = 0$. Also by Equation 3.26,

$$\lim_{N\to\infty} \left(\frac{2\pi}{T}\right) \frac{\sin\left[2\pi\left(N + \frac{1}{2}\right)\tau\right]}{\pi\tau} = \frac{2\pi}{T}\,\delta(\tau)$$

By Equation 3.29, the right-hand side of this equation is $2\pi\delta(t)$. Hence, Equation 3.33 is

$$\lim_{N\to\infty} \frac{1}{T} \frac{\sin\left[\left(N + \frac{1}{2}\right)\frac{2\pi t}{T}\right]}{\sin \frac{\pi t}{T}} = \delta(t)$$

between $-T/2$ and $T/2$, which proves that Equation 3.31 is correct.

3.3 CONVOLUTION

An analog function $y(t)$ being a convolution of the analog function $x(t)$ and $h(t)$ means that $y(t)$ is equal to

$$y(t) = \int_{-\infty}^{\infty} x(\tau)h(t - \tau)\, d\tau \tag{3.34}$$

The symbol $*$ is usually employed to simplify the expression of Equation 3.34 as

$$y(t) = x(t) * h(t)^{+}$$

In this book, we also use the symbol $*$ to indicate convolution of functions. For example, $z(t) * [x(t) * y(t)]$ means

$$z(t) * [x(t) * h(t)] = \int_{-\infty}^{\infty} z(t - u) \int_{-\infty}^{\infty} x(\tau)h(u - \tau)\, d\tau du$$

As an example, suppose an analog function $y(t)$ is equal to a convolution of analog functions $x(t)$ and $h(t)$ as shown in Fig. 3.8. From Equation 3.34, $y(t)$ at $t = 0$ is given by $\int_{-\infty}^{\infty} x(\tau)h(-\tau)\, d\tau$. Turning $h(\tau)$ 180° with respect to the vertical axis at $t = 0$, we will have $h(-\tau)$ as shown in Fig. 3.9(a). Multiplying this with $x(\tau)$ gives a result $x(\tau)h(-\tau)$ as shown in Fig. 3.9(b). Clearly, $\int_{-\infty}^{\infty} x(\tau)h(-\tau)\, d\tau$ is 0. Hence, $y(0) = 0$.

For $0 < t < 1$, $h(t - \tau)$ is obtained by shifting $h(-\tau)$ t unit to the right as shown in Fig. 3.10(a). Hence, $x(\tau)h(t - \tau)$ is shown in Fig. 3.10(b). Thus

+ Because $\int_{-\infty}^{\infty} x(\tau)h(t - \tau)\, d\tau = \int_{-\infty}^{\infty} x(t - \tau)h(\tau)\, d\tau$, $x(t) * h(t)$ and $h(t) * x(t)$ give the same result.

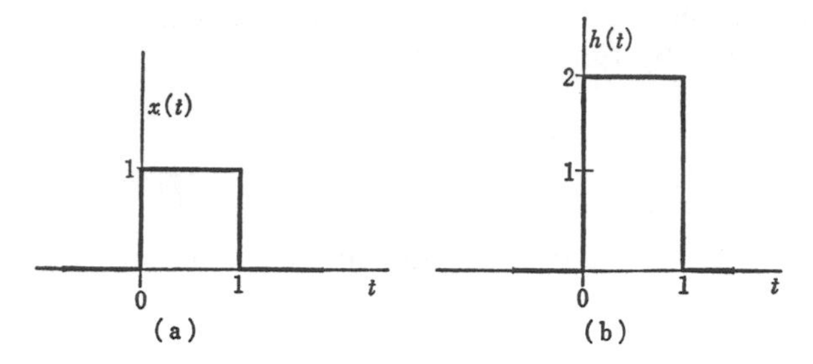

Figure 3.8 Analog Functions $x(t)$ and $h(t)$. (a) $x(t)$; (b) $h(t)$.

$\int_{-\infty}^{\infty} x(\tau)h(t - \tau)\, d\tau$ is the meshed area shown in Fig. 3.10(b). For example, when $t = 0.5$, the area is 1. Hence, $y(0.5) = 1$. When $t = 1$, the area produced by $x(\tau)h(1 - \tau)$ becomes 2. Hence, $y(1) = \int_{-\infty}^{\infty} x(\tau)h(1 - \tau)\, d\tau = 2$. When $t = 1.5$, $h(t - \tau)$ becomes one shown in Fig. 3.11. Hence, $\int_{-\infty}^{\infty} x(\tau)h(1.5 - \tau)\, d\tau$ becomes 1 or $y(1.5) = 1$. It can be seen easily that as t increases further, the area formed by $x(\tau)h(t - \tau)$ decreases and when $t = 2$, as shown in Fig. 3.12, $\int_{-\infty}^{\infty} x(\tau)h(2 - \tau)\, d\tau$ becomes 0. When $t > 2$, the area formed by $x(\tau)h(t - \tau)$ is 0. This means $y(t) = \int_{-\infty}^{\infty} x(\tau)h(t - \tau)\, d\tau = 0$ for $t > 2$. The result of the convolution of $x(t)$ and $h(t)$ is, thus, shown in Fig. 3.13.

What will be the result of convolution of $x(t)$ and $h(t)$ when $h(t)$ is an impulse function $\delta(t - T)$? From Equation 3.27,

$$\int_{-\infty}^{\infty} x(\tau)\delta(t - T - \tau)\, d\tau = x(t - T) \tag{3.35}$$

This means that the convolution of $x(t)$ and $\delta(t - T)$ makes $x(t)$ delay as shown in Fig. 3.14.

When $h(t)$ is an impulse series $\sum_{n=-\infty}^{\infty} \delta(t - nT)$, the convolution of $x(t)$ and $h(t)$ becomes

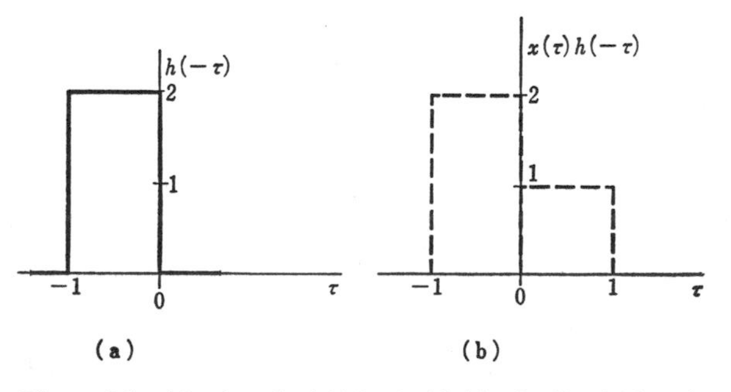

Figure 3.9 $h(-\tau)$ and $x(\tau)h(-\tau)$. (a) $h(-\tau)$; (b) $x(\tau)h(-\tau)$.

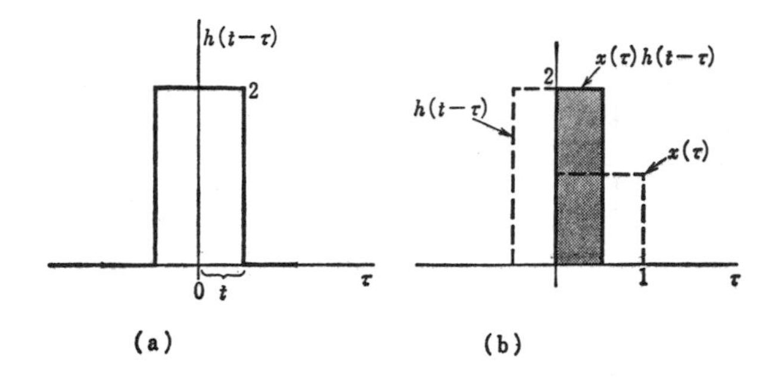

Figure 3.10 $x(\tau)h(t - \tau)$ for $0 < t < 1$. (a) $h(t - \tau)$; (b) $x(\tau)h(t - \tau)$.

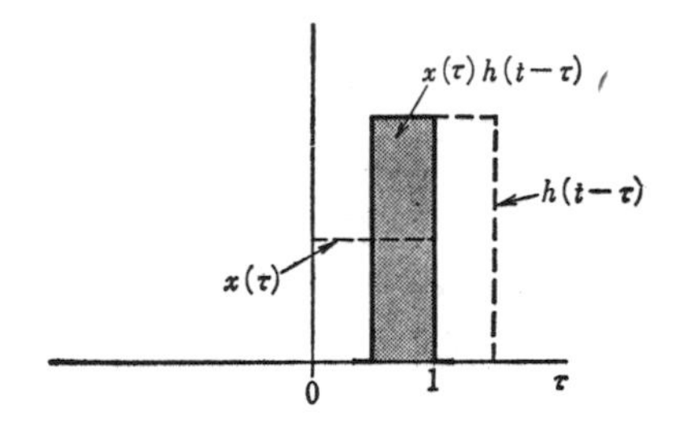

Figure 3.11 $x(\tau)h(t - \tau)$ for $1 < t < 2$

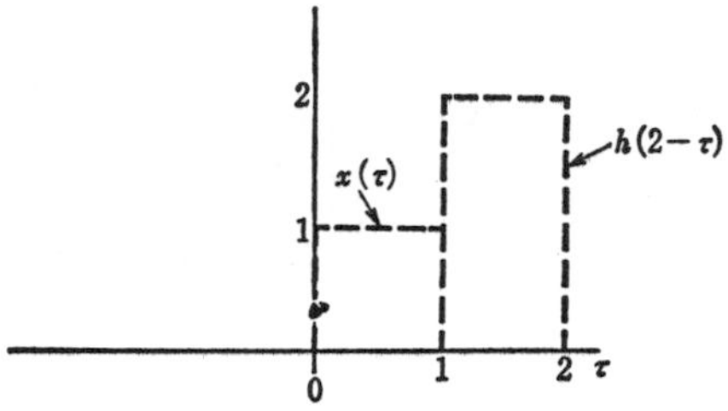

Figure 3.12 $x(\tau)h(t - \tau)$ for $t = 2$

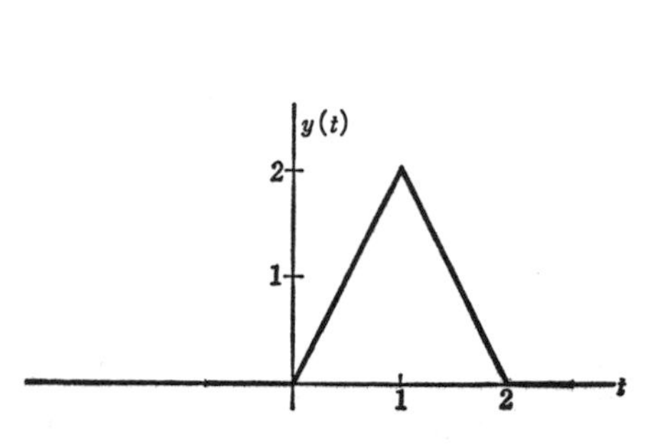

Figure 3.13 $y(t)$

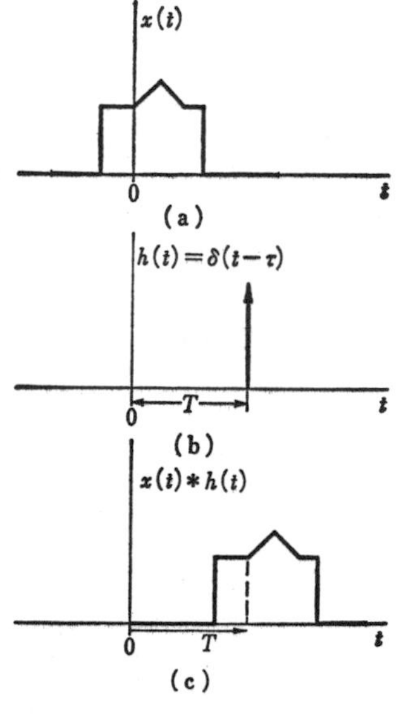

Figure 3.14 $x(t) * \delta(t - T)$

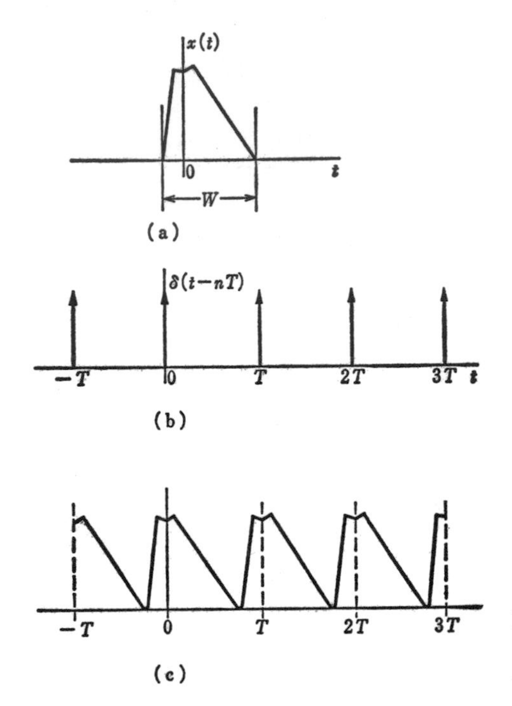

Figure 3.15 $x(t) * (\Sigma\ \delta(t - nT)$ when $T \geq W$. (a) Width W of $x(t)$; (b) $\Sigma\ \delta(t - nT)$; (c) $x(t) * (\Sigma\ \delta(t - nT)$.

$$\int_{-\infty}^{\infty} x(\tau) \sum_{n=-\infty}^{\infty} \delta(t - nT - \tau)\, d\tau$$
$$= \sum_{n=-\infty}^{\infty} \int_{-\infty}^{\infty} x(\tau)\delta(t - nT - \tau)\, d\tau = \sum_{n=-\infty}^{\infty} x(t - nT) \tag{3.36}$$

Hence, when the width W of $x(t)$ is smaller than T as shown in Fig. 3.15(a) and (b), a shape of $x(t)$ will be repeated as shown in Fig. 3.15(c). Conversely, if the width W of $x(t)$ is larger than T, repeated $x(t)$ will overlap as shown in Fig. 3.16, which is known as aliasing.

Let us investigate the Fourier transformation of convolution of $x(t)$ and $h(t)$. From Equation 3.21, the Fourier transformation of $x(t) * h(t)$ is

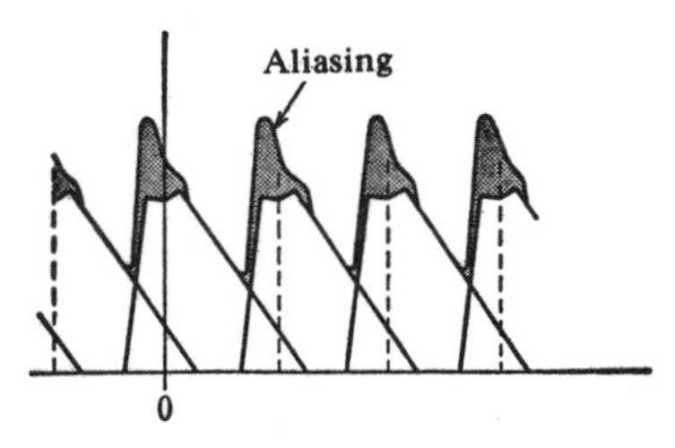

Figure 3.16 $x(t) * (\Sigma\ \delta(t - nT))$ when $T < W$

$$\mathscr{F}[x(t) * h(t)] = \int_{-\infty}^{\infty} \left[\int_{-\infty}^{\infty} x(\tau)h(t-\tau)\, d\tau\right] e^{-j\omega t}\, dt$$

$$= \int_{-\infty}^{\infty} x(\tau) \int_{-\infty}^{\infty} h(t-\tau)e^{-j\omega t}\, dt d\tau$$

which can be written as

$$\mathscr{F}[x(t) * h(t)] = \int_{-\infty}^{\infty} x(\tau) \int_{-\infty}^{\infty} h(t-\tau)e^{-j\omega(t-\tau)}\, dt e^{-j\omega\tau}\, d\tau$$

$$= \int_{-\infty}^{\infty} x(\tau)e^{-j\omega\tau}\, d\tau \int_{-\infty}^{\infty} h(u)e^{-j\omega u}\, du$$

Hence, we have

$$\mathscr{F}[x(t) * h(t)] = \mathscr{F}[x(t)]\mathscr{F}[h(t)] = X(j\omega)H(j\omega) \tag{3.37}$$

where $X(j\omega)$ is the Fourier transformation of $x(t)$, and $H(j\omega)$ is the Fourier transformation of $h(t)$. This shows that the Fourier transformation of the convolution $x(t) * h(t)$ is equal to the product of the Fourier transformations of each $x(t)$ and $h(t)$. This is a very important property.

Similarly, the inverse Fourier transformation of convolution $X(j\omega) * H(j\omega)$ is given as

$$\mathscr{F}^{-1}[X(j\omega) * H(j\omega)] = \mathscr{F}^{-1}[X(j\omega)]\mathscr{F}^{-1}[H(j\omega)] = 2\pi x(t)h(t) \tag{3.38}$$

where $X(j\omega)$ and $H(j\omega)$ are the Fourier transformation of $x(t)$ and $h(t)$, respectively. This shows that the inverse Fourier transformation of convolution of $X(j\omega)$ and $H(j\omega)$ is equal to 2π times the product of $x(t)$ and $h(t)$. As an example, consider $x(t)$ and $h(t)$ in Fig. 3.17(a). Because $x(t)$ is

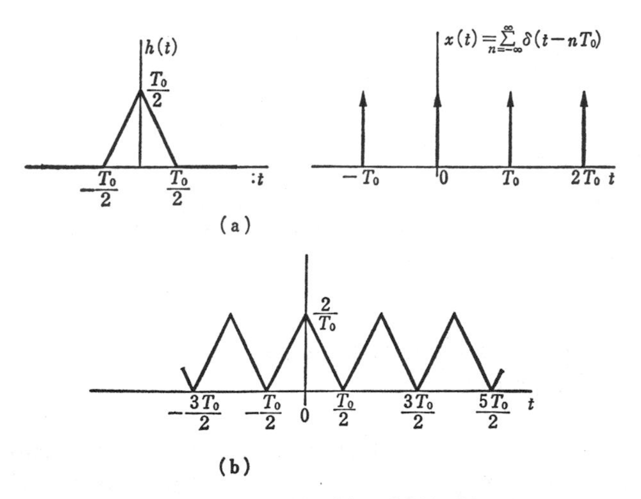

Figure 3.17 $x(t) * h(t)$. (a) $x(t)$ and $h(t)$; (b) $x(t) * h(t)$.

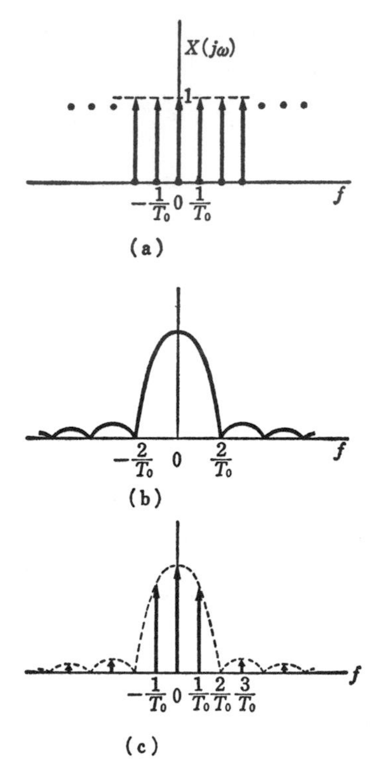

Figure 3.18 $\mathscr{F}[x(t) * h(t)] = X(j\omega)H(j\omega)$. (a) $X(j\omega)$; (b) $H(j\omega)$; (c) $X(j\omega)H(j\omega)$.

$$x(t) = \sum_{n=-\infty}^{\infty} \delta(t - nT_0) \tag{3.39}$$

the convolution $x(t) * h(t)$ is a repetition of a triangular configuration as shown in Fig. 3.17(b). By Equation 3.31, the Fourier transformation of $x(t)$ is shown in Fig. 3.18(a). Fig. 3.18(b) shows the Fourier transformation of $h(t)$. Hence, the product $X(j\omega)H(j\omega)$ becomes the one shown in Fig. 3.18(c). Equation 3.37 indicates that the inverse Fourier transformation of $X(j\omega)H(j\omega)$ is $x(t) * h(t)$. It must be noticed that Fig. 3.18(a) shows coefficients in the impulse series $X(j\omega)$. Similarly, Fig. 3.18(c) shows coefficients in the impulse series $X(j\omega)H(j\omega)$.

3.4 RELATIONSHIP BETWEEN ANALOG AND DISCRETE SIGNALS

Suppose $x(nT)$ is a discrete signal obtained by sampling an analog signal $x(t)$ with interval T. By using a unit sample u_I in Equation 2.4, $x(nT)$ can be expressed as

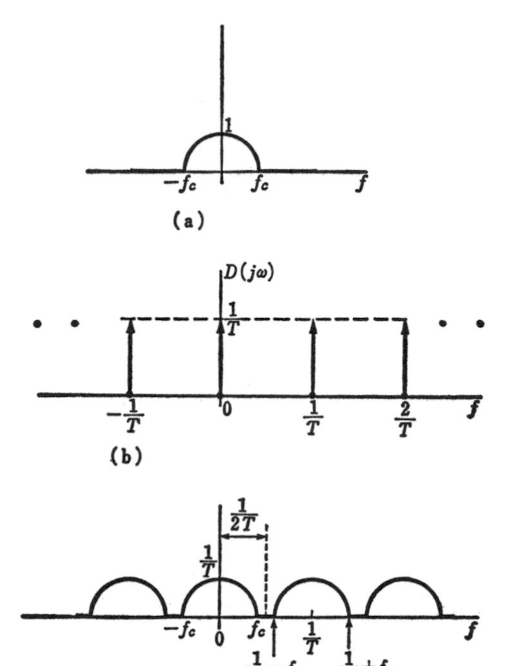

Figure 3.19 $X(j\omega) * D(j\omega)$. (a) $X(j\omega)$; (b) Coefficient of Impulse Series $D(j\omega)$; (c) $X(j\omega) * D(j\omega)$.

$$x(nT) = x(t) \sum_{n=-\infty}^{\infty} u_I(t - nT) \tag{3.40}$$

Replacing a unit sample u_I with an impulse δ, we will have a kind of discrete signal as

$$\hat{x}(nT) = x(t) \sum_{n=-\infty}^{\infty} \delta(t - nT) \tag{3.41}$$

This $\hat{x}(nT)$ will be infinite when $t = nT$ because of impulse $\delta(t - nT)$. Hence, $\hat{x}(nT)$ will not be equal to $x(nT)$. Coefficients of $\delta(t - nT)$ indicate $x(nT)$, however. Using this $\hat{x}(nT)$, we would like to show whether we can reproduce $x(t)$ from $x(nT)$.

Suppose the Fourier transformation $X(j\omega)$ of $x(t)$ is the one as shown in Fig. 3.19(a). By Equation 3.31, the Fourier transformation $D(j\omega)$ of $\sum_{n=-\infty}^{\infty} \delta(t - nT)$ is

$$D(j\omega) = \frac{1}{T} \sum_{n=-\infty}^{\infty} \delta\left(f - \frac{n}{T}\right) \tag{3.42}$$

The coefficients in the impulse series $D(j\omega)$ is the one shown in Fig. 3.19(b). Suppose a graph in Fig. 3.19(c) is the result of the convolution

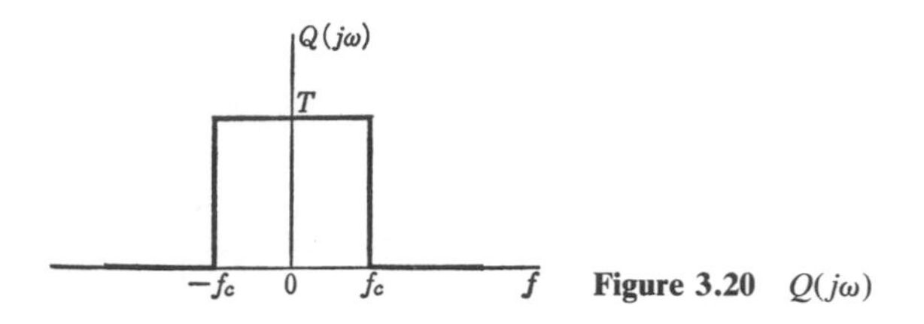

Figure 3.20 $Q(j\omega)$

$X(j\omega) * D(j\omega)$. Then by Equation 3.38, the inverse Fourier transformation of $X(j\omega) * D(j\omega)$ is $\hat{x}(nT)$. In other words, the Fourier transformation of $\hat{x}(nT)$ is $X(j\omega) * D(j\omega)$.

Conversely, the inverse Fourier transformation of $X(j\omega)$ is $x(t)$. Hence, if we can obtain $X(j\omega)$ from $X(j\omega) * D(j\omega)$, we can reproduce $x(t)$ from $X(j\omega)$. In Fig. 3.19(c), there is $(1/T)X(j\omega)$ between $-f_c$ and f_c. How can we take this $(1/T)X(j\omega)$ out? One easy way is to multiply $Q(j\omega)$ as shown in Fig. 3.20 to $X(j\omega) * D(j\omega)$. An important question is whether $X(j\omega) * D(j\omega)$ always has a shape as shown in Fig. 3.19(c) so that multiplying $Q(j\omega)$ in Fig. 3.20 to $X(j\omega) * D(j\omega)$ produces $X(j\omega)$.

Fig. 3.19(c) shows the result of convolution of $X(j\omega)$ and $D(j\omega)$. Because $D(j\omega)$ is $1/T \sum_{n=-\infty}^{\infty} \delta(f - n/T)$, $(1/T)X(j\omega)$ between f_c and $-f_c$ in the figure is produced by convolution of one term $1/T\ \delta(f)$ of $D(j\omega)$ and $X(j\omega)$. Also an adjacent shape $1/T\ X(j2\pi(f - 1/T))$ is produced by convolution of one term $1/T\ \delta(f - 1/T)$ of $D(j\omega)$ and $X(j\omega)$. If these adjacent shapes are too close to each other as shown in Fig. 3.21, these will overlap to produce aliasing, which has been discussed previously. Under this circumstance, $X(j\omega)$ cannot be extracted by multiplying $Q(j\omega)$ in Fig. 3.20. Hence, it is necessary that f_c shown in Fig. 3.19(a) and (c) should not be larger than $1/2T$ to extract $X(j\omega)$ by the use of $Q(j\omega)$. In other words, satisfying

$$2f_c \leq \frac{1}{T} \tag{3.43}$$

is necessary and sufficient for obtaining $X(j\omega)$ by multiplying $Q(j\omega)$ to

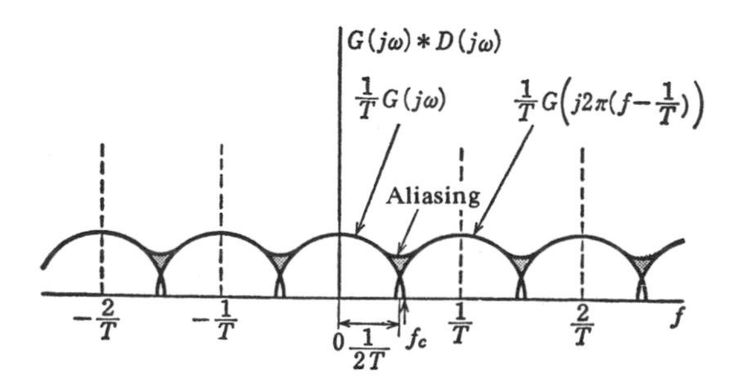

Figure 3.21 Aliasing

$X(j\omega) * D(j\omega)$. This f_c is known as the "cutoff frequency," which is the smallest frequency such that $X(j\omega)$ contains no frequencies equal to or higher than this frequency. In other words, all frequencies (whose magnitude are nonzero) in the Fourier transformation $X(j\omega)$ of $x(t)$ must be lower than f_c.

Rewriting Equation 3.43, we have

$$T \leq \frac{1}{2f_c} \tag{3.44}$$

$1/2f_c$ in this equation is called the "Nyquist sampling interval." Also the inverse of the Nyquist sampling interval $2f_c$ is called the "Nyquist sampling rate." Now we know that $x(t)$ can be reproduced from $x(nT)$ if the sampling interval T is not larger than the Nyquist sampling interval. In other words, if the highest frequency component of $x(t)$ is not finite, it is impossible to reproduce $x(t)$ from a discrete signal $x(nT)$ obtained by sampling $x(t)$. Next, we will study how we can reproduce $x(t)$ from $x(nT)$.

We assume that $x(t)$ has a finite cutoff frequency, and $x(nT)$ is obtained from $x(t)$ using a sampling interval T satisfying Equation 3.44. $X(j\omega)$ is the Fourier transformation of $x(t)$, and $D(j\omega)$ is the one given by Equation 3.43, which is the Fourier transformation of $\sum_{n=-\infty}^{\infty} \delta(t - nT)$. Then $X(j\omega) * D(j\omega)$ is the Fourier transformation of $x(nT)$ given by Equation 3.13. It has been discussed that $X(j\omega)$ can be obtained by multiplying $Q(j\omega)$ in Fig. 3.20 to $X(j\omega) * D(j\omega)$, and the inverse Fourier transformation of $X(j\omega)$ is $x(t)$. That is,

$$x(t) = \mathcal{F}^{-1}\{[X(j\omega) * D(j\omega)]Q(j\omega)\} \tag{3.45}$$

By Equation 3.37, this can be changed as

$$x(t) = \mathcal{F}^{-1}[X(j\omega) * D(j\omega)] * \mathcal{F}^{-1}[Q(j\omega)]$$

Also $\mathcal{F}^{-1}[X(j\omega) * D(j\omega)]$ is

$$\mathcal{F}^{-1}[X(j\omega) * D(j\omega)] = x(t) \sum_{n=-\infty}^{\infty} \delta(t - nT) = \hat{x}(nT)$$

but

$$x(t) \sum_{n=-\infty}^{\infty} \delta(t - nT) = \sum_{n=-\infty}^{\infty} x(nT)\delta(t - nT)$$

Hence, Equation 3.45 becomes

$$\begin{aligned} x(t) &= \mathcal{F}^{-1}\{[X(j\omega) * D(j\omega)]Q(j\omega)\} = \hat{x}(nT) * q(t) \\ &= \sum_{n=-\infty}^{\infty} x(nT)\delta(t - nT) * q(t) = \sum_{n=-\infty}^{\infty} x(nT)q(t - nT) \end{aligned} \tag{3.46}$$

Here $q(t)$ is

$$q(t) = \mathscr{F}^{-1}[Q(j\omega)] = T\,\frac{\sin 2\pi f_c t}{\pi t}$$

Hence, Equation 3.46 is

$$x(t) = T \sum_{n=-\infty}^{\infty} x(nT)\,\frac{\sin 2\pi f_c(t - nT)}{\pi(t - nT)} \tag{3.47}$$

which shows how to reproduce $x(t)$ from $x(nT)$.

3.5 FREQUENCY RESPONSE

A drawing in Fig. 3.22 shows the relationship between an input signal and an output signal of a linear shift invariant system where $X(e^{j\omega})$ is an input signal, $Y(e^{j\omega})$ is an output signal, and $H(e^{j\omega})$ is known as a frequency response. The symbol ω is the angular velocity equal to $\omega = 2\pi f$, which has been discussed in Section 3.2.

To lead the relationship between input and output signals given by

$$Y(e^{j\omega}) = H(e^{j\omega})X(e^{j\omega}), \tag{3.48}$$

we will explain a frequency response $H(e^{j\omega})$ first. For a linear shift invariant system, we have already known that the output signal $y(n)$ when the input signal is $x(n)$ is given by

$$y(n) = \sum_{k=-\infty}^{\infty} h(k)x(n - k)$$

Taking $e^{j\omega n}$ as $x(n)$, $y(n)$ becomes

$$y(n) = \left(\sum_{k=-\infty}^{\infty} h(k)e^{-j\omega k}\right) e^{j\omega n}$$

This suggests that we can define a frequency response $H(e^{j\omega})$ as

$$H(e^{j\omega}) = \sum_{k=-\infty}^{\infty} h(k)\,e^{-j\omega k} \tag{3.49}$$

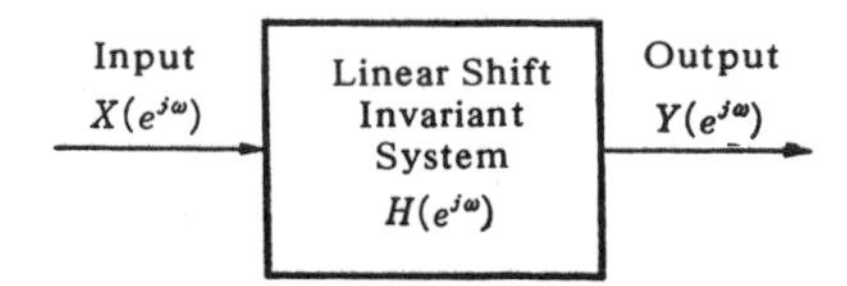

Figure 3.22 Frequency-Domain Representation

For example, if a unit sample response $h(n)$ is

$$h(n) = 0.5^n$$

then the frequency response $H(e^{j\omega})$ is

$$H(e^{j\omega}) = \sum_{k=-\infty}^{\infty} 0.5^k e^{-j\omega k} = \sum_{k=-\infty}^{\infty} (0.5e^{-j\omega})^k$$

Let us take another example. Consider a difference equation

$$y(n) = x(n) + ay(n-1) + by(n-2) \qquad \text{where } a, b > 0 \tag{3.50}$$

with initial conditions $y(-1) = 0$ and $y(-2) = 0$. To obtain a general solution, we assume $y(n) = c\alpha^n$ and substitute it in $y(n) - ay(n-1) - by(n-2) = 0$ and obtain

$$ca^n - ac\alpha^{n-1} - bc\alpha^{n-2} = 0$$

From this, we have

$$\alpha_1 = \frac{1}{2}(a + \sqrt{a^2 + 4b})$$

and

$$\alpha_2 = \frac{1}{2}(a - \sqrt{a^2 + 4b})$$

Hence, the general solution $y_g(n)$ will be

$$y_g(n) = c_1\alpha_1^n + c_2\alpha_2^n$$

To obtain a unit sample response $h(n)$, we take $x(n) = u_I(n)$. From Equation 3.50 with $n = 0$, $y(0)$ is

$$y(0) = 1$$

By using this $y(0)$ as the initial condition, we do not need to obtain a particular solution. The solution of Equation 3.50 will be

$$y(n) = y_g(n)$$

By setting $n = 1$, Equation 3.50 gives $y(1) = a$. Hence, using initial conditions $y(0) = 1$ and $y(1) = a$, the constants c_1 and c_2 can be obtained from $y(n) = c_1\alpha_1^n + c_2\alpha_2^n$ as follows:

Because

$$y(0) = c_1 + c_2 = 1$$

and

$$y(1) = c_1\alpha_1 + c_2\alpha_2 = a$$

we have

$$c_1 = \frac{\alpha_2 - a}{\alpha_2 - \alpha_1}, \qquad c_2 = \frac{a - \alpha_1}{\alpha_2 - \alpha_1}$$

With these constants, the unit sample response $h(n)$ becomes

$$h(n) = \frac{\alpha_2 - a}{\alpha^2 - \alpha_1} \alpha_1^n + \frac{a - \alpha_1}{\alpha_2 - \alpha_1} \alpha_2^n \qquad n \geq 0$$

Hence, by Equation 3.49, the frequency response $H(e^{j\omega})$ will be

$$H(e^{j\omega}) = \sum_{k=0}^{\infty} h(k) e^{-j\omega k} = \sum_{k=0}^{\infty} \left[\frac{\alpha_2 - a}{\alpha_2 - \alpha_1} (\alpha_1 e^{-j\omega})^k + \frac{a - \alpha^1}{\alpha_2 - \alpha_1} (\alpha_2 e^{-j\omega})^k \right]$$

$e^{-j\omega k}$ appeared in Equation 3.49 is known as a phasor. Let us investigate properties of a phasor. From Equation a-6, $e^{-j\theta}$ can be expressed as

$$e^{-j\theta} = \cos\theta - j\sin\theta \tag{3.51}$$

Hence, it is clear that $e^{-j\theta}$ can be represented by a drawing in Fig. 3.23. Changing θ by ωk as shown in Fig. 3.24, we will have a phasor $e^{-j\omega k}$.

Let us investigate the case when $\omega = \pi/4$. That is, $e^{-j(\pi/4)k}$ when k changes from 0 to 1, then to 2, and so on, the phaseor $e^{-j(\pi/4)k}$ will change from e^{-j0} to $e^{-j(\pi/4)}$, then to $e^{-j(\pi/2)}$ and so on. From Fig. 3.24, we can see that the vector representing $e^{-j(\pi/4)k}$ will rotate in clockwise as k increases.

To understand how $e^{-j\omega k}$ will vary when ω is slightly larger than 2π, we will take an example when $\omega = 2\pi + (\pi/4)$. By changing k from 0 to 1 continuously, $e^{-j(2\pi+\pi/4)k}$ will move from e^{-j0} in the clockwise direction to complete one turn then to $e^{-j(2\pi+\pi/4)}$ as shown in Fig. 3.25(a). The continuously changing k from 1 to 2 makes it complete one turn from $e^{-j(2\pi+\pi/4)}$ in the clockwise direction and moves to $e^{-j(4\pi+\pi/2)}$ as shown in Fig. 3.25(b). Hence, by considering the position of $e^{-j(2\pi+\pi/4)k}$ at $k = 0, 1, 2, \ldots$, we can see that the position of $e^{-j(2\pi+\pi/4)k}$ will move in clockwise as $k = 0, 1, 2, \ldots$ which is the same as that of $e^{-j(\pi/4)k}$.

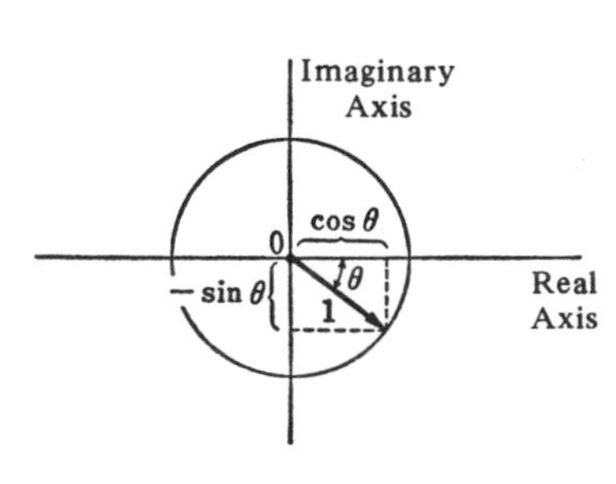

Figure 3.23 $e^{-j\theta}$

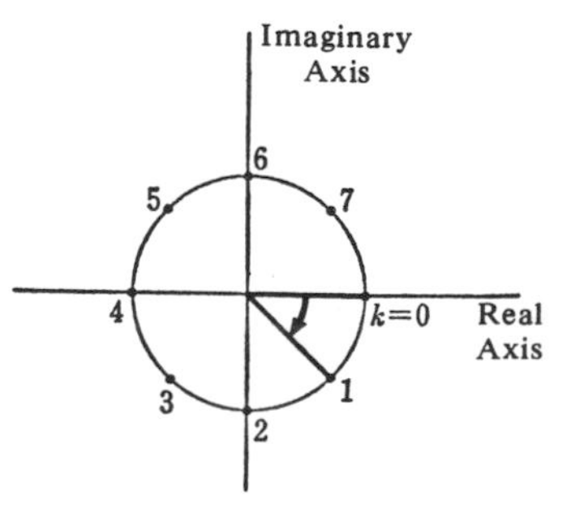

Figure 3.24 Movement of $e^{-j(\pi/4)k}$

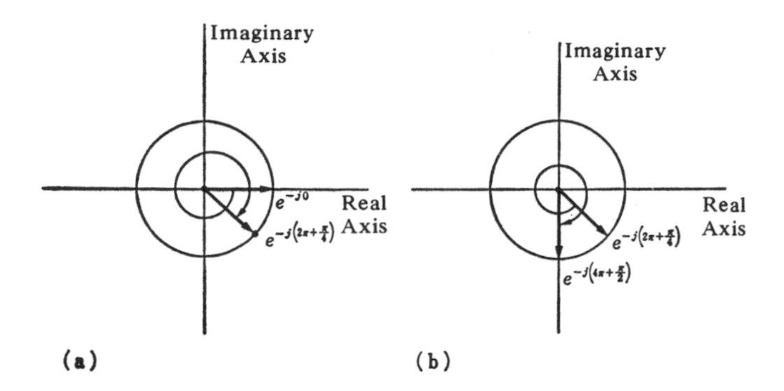

Figure 3.25 $e^{-j(2\pi+\pi/4)k}$. (a) From e^{-j0} to $e^{-j(2\pi+\pi/4)}$; (b) From $e^{-j(2\pi+\pi/4)}$ to $e^{-j(4\pi+\pi/4)}$.

When ω is slightly smaller than 2π, $e^{-j\omega k}$ moves differently. To see this we will take a case when $\omega = 2\pi - (\pi/4)$. When k is changing from 0 to 1 continuously, $e^{-j\omega k}$ moves from e^{-j0} in the clockwise direction to reach the point equal to $e^{-j(2\pi-\pi/4)}$ as shown in Fig. 3.26(a). Changing k from 1 to 2 continuously moves it from $e^{-j(2\pi-\pi/4)}$ to $e^{-j(4\pi-\pi/2)}$ in the clockwise direction as shown in Fig. 3.26(b). When by the position of $e^{-j(2\pi-\pi/4)k}$ only at $k = 0, 1, 2, \ldots,$ however, $e^{-j\omega k}$ is changing its position as if it is moving in the counterclockwise direction as shown in Fig. 3.27. Because $e^{-j\omega k}$ can be expressed as

$$e^{-j\omega k} = \cos \omega k - j \sin \omega k$$

by Equation 3.51, when ω is changed to $2\pi \pm \omega$, we have

$$e^{-j(2\pi\pm\omega)k} = \cos (2\pi k \pm \omega k) - j \sin (2\pi k \pm \omega k) \tag{3.52}$$

When k is an integer, Equation 3.52 can be changed as

$$e^{-j(2\pi\pm\omega)k} = \cos (\pm\omega k) - j \sin (\pm\omega k) = e^{-j(\pm\omega)k}$$

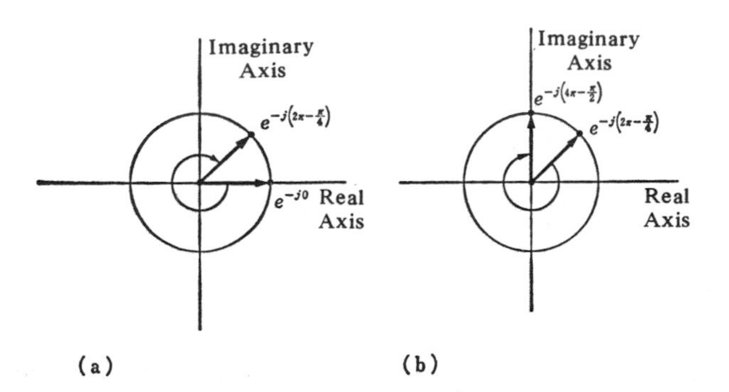

Figure 3.26 $e^{-j(2\pi-\pi/4)}$. (a) From e^{-j0} to $e^{-j(2\pi-\pi/4)}$; (b) From $e^{-j(2\pi-\pi/4)}$ to $e^{-j(4\pi-\pi/4)}$.

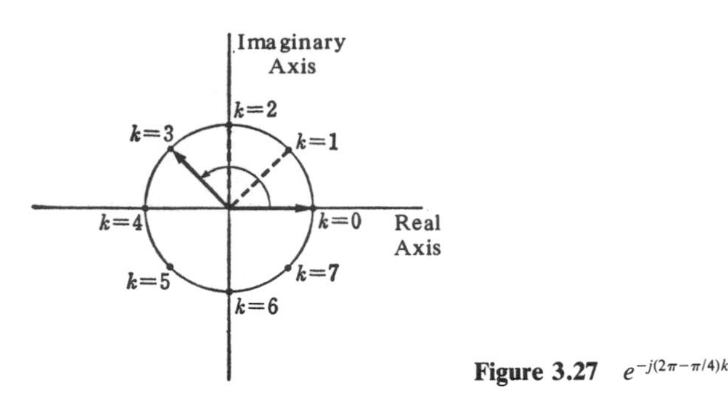

Figure 3.27 $e^{-j(2\pi-\pi/4)k}$

From this expression, we can easily see that $e^{-j(\pm\omega)k}$ is a periodic function with 2π as its period. Multiplying a constant $h(k)$, function $e^{-j\omega k}$ is a periodic function of period 2π. Because summing two periodic functions having the same period results in a periodic function of the same period, a frequency response $H(e^{j\omega})$ in Equation 3.49 is a periodic function of period 2π. This is one of the very important properties of a frequency response $H(e^{j\omega})$.

As an example, consider a unit sample response

$$h(n) = \frac{\alpha_2 - a}{\alpha_2 - \alpha_1}\,\alpha_1^n + \frac{a - \alpha_1}{\alpha_2 - \alpha_1}\,\alpha_2^n$$

which we have obtained previously. The frequency response $H(e^{j\omega})$ calculated from this unit sample response is

$$H(e^{j\omega}) = \sum_{k=0}^{\infty}\left[\frac{\alpha_2 - a}{\alpha_2 - \alpha_1}\,(\alpha_1 e^{-j\omega})^k + \frac{a - \alpha_1}{\alpha_2 - \alpha_1}\,(\alpha_1 e^{-j\omega})^k\right]$$

Suppose we change ω by $\omega + 2\pi m$ in this frequency response $H(e^{j\omega})$, then we have

$$H(e^{j(\omega+2\pi m)}) = \sum_{k=0}^{\infty}\left[\frac{\alpha_2 - a}{\alpha_2 - \alpha_1}\,(\alpha_1 e^{-j\omega})^k + \frac{a - \alpha_1}{\alpha_2 - \alpha_1}\,(\alpha_2 e^{-j\omega})^k\right] e^{-j2\pi mk}$$

Because m and k in $e^{-j2\pi mk}$ are both integers, $e^{-j2\pi mk}$ can be expressed as

$$e^{-j2\pi mk} = \cos(2\pi mk) - j\sin(2\pi mk) = 1$$

Hence,

$$H(e^{j(\omega+2\pi m)}) = H(e^{j\omega})$$

This shows that a frequency response is a periodic function of period 2π.

3.6 DISCRETE FUNCTION FOURIER TRANSFORMATION

The Fourier exponential series of a function $x(t)$

$$x(t) = \sum_{k=-\infty}^{\infty} C_k e^{j2\pi k f_0 t} \tag{3.53}$$

is a periodic function of a period $1/f_0$. Changing $2\pi k f_0 t$ by $-\omega k$, the right-hand side of $x(t)$ in Equation 3.53 becomes $\sum_{k=-\infty}^{\infty} C_k e^{-j\omega k}$. This C_k is the $h(k)$ in the right-hand side of Equation 3.53. That is

$$H(e^{j\omega}) = \sum_{k=-\infty}^{\infty} h(k) e^{-j\omega k} \tag{3.54}$$

can be considered as a transformation of $h(k)$. Hence, a coefficient C_k of Equation 3.53 is (by Equation 3.13)

$$C_k = \frac{1}{T} \int_{T_1}^{T_2} x(t) e^{-j2\pi k f_0 t}\, dt \qquad \text{where } T_2 - T_1 = T \tag{3.55}$$

Because changing $2\pi k f_0 t$ by $-\omega k$ and $x(t)$ by $H(e^{j\omega})$, C_k is clearly $h(k)$, Equation 3.55 becomes

$$h(k) = \frac{1}{T} \int_{\omega_1}^{\omega_2} H(e^{j\omega}) e^{j\omega k} \frac{T}{2\pi}\, d\omega$$

Because $H(e^{j\omega})$ is a periodic function of a period 2π, by changing ω_1 by $-\pi$ and ω_2 by π, $h(k)$ becomes

$$h(k) = \frac{1}{2\pi} \int_{-\pi}^{\pi} H(e^{j\omega}) e^{j\omega\kappa}\, d\omega \tag{3.56}$$

Equation 3.54 is called the discrete function Fourier transformation of a unit sample response $h(k)$, and Equation 3.56 is the inverse discrete function Fourier transformation of $H(e^{j\omega})$.

For example, suppose one period of $H(e^{j\omega})$ from $-\pi$ to π is

$$H(e^{j\omega}) = \begin{cases} 1, & |\omega| \leq \dfrac{\pi}{2} \\ 0, & \dfrac{\pi}{2} < |\omega| \leq \pi \end{cases}$$

(See Fig. 3.28[a]). Using the inverse discrete function Fourier transformation, we can obtain $h(k)$ as

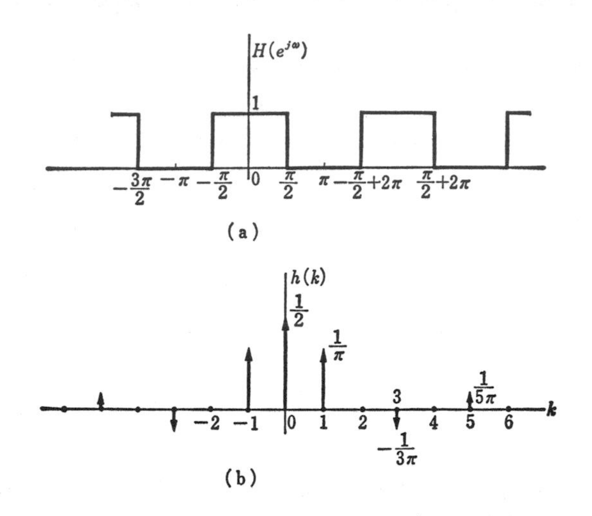

Figure 3.28 $H(e^{j\omega})$ and $h(t)$. (a) $H(e^{j\omega})$; (b) $h(k)$.

$$h(k) = \frac{1}{2\pi}\int_{-\pi/2}^{\pi/2} e^{j\omega k}\, d\omega = \frac{\sin\left(\frac{\pi}{2}k\right)}{\pi k}$$

which is shown in Fig. 3.28(b).

By Equations 3.54 and 3.56, the discrete function Fourier transformation and the inverse discrete function Fourier transformation of an input signal $x(t)$ in a system shown in Fig. 3.22 are

$$X(e^{j\omega}) = \sum_{n=-\infty}^{\infty} x(n)e^{-j\omega n} \tag{3.57}$$

and

$$x(n) = \frac{1}{2\pi}\int_{-\pi}^{\pi} X(e^{j\omega})e^{j\omega n}\, d\omega \tag{3.58}$$

Because an output signal of a linear shift invariant system is by Equation 2.23

$$y(n) = \sum_{k=-\infty}^{\infty} h(n-k)x(k)$$

the discrete function Fourier transformation of an output signal $Y(e^{j\omega})$ is given by

$$Y(e^{j\omega}) = \sum_{n=-\infty}^{\infty} \left(\sum_{k=-\infty}^{\infty} h(n-k)x(k) \right) e^{-j\omega n}$$

Changing $n - k$ by m, this expression can be rewritten as

$$\begin{aligned} Y(e^{j\omega}) &= \sum_{m=-\infty}^{\infty} \sum_{k=-\infty}^{\infty} h(m)x(k)e^{-j\omega(m+k)} \\ &= \sum_{m=-\infty}^{\infty} h(m)e^{-j\omega m} \sum_{k=-\infty}^{\infty} x(k)e^{-j\omega k} \end{aligned} \tag{3.59}$$

The right-hand side of this equation is equal to $H(e^{j\omega})X(e^{j\omega})$. Hence, we have obtained Equation 3.48. Also using the inverse discrete function Fourier transformation given by Equation 3.56, an output signal $y(n)$ can be expressed as

$$y(n) = \frac{1}{2\pi} \int_{-\pi}^{\pi} Y(e^{j\omega})\, e^{j\omega n}\, d\omega = \frac{1}{2\pi} \int_{-\pi}^{\pi} H(e^{j\omega})X(e^{j\omega})e^{j\omega n}\, d\omega \tag{3.60}$$

As an example, consider a unit sample response $h(n)$

$$h(n) = u_I(n) + u_I(n-1) + u_I(n-2)$$

For an input signal $x(n)$

$$x(n) = u_I(n-1) + u_I(n-2),$$

the output signal $y(n)$ can be obtained by the use of Eq. 3.60 as follows: First, we calculate the discrete function Fourier transformation of $h(n)$ as

$$H(e^{j\omega}) = \sum_{k=-\infty}^{\infty} (u_I(k) + u_I(k-1) + u_I(k-2))e^{-j\omega k} = 1 + e^{-j\omega} + e^{-j2\omega}$$

and the discrete function Fourier transformation of $x(n)$ as

$$X(e^{j\omega}) = \sum_{k=-\infty}^{\infty} (u_I(k-1) + u_I(k-2))e^{-j\omega k} = e^{-j\omega} + e^{-j2\omega}$$

Then by Equation 3.49 the discrete function Fourier transformation of the output signal $y(n)$ can be calculated as

$$\begin{aligned} Y(e^{j\omega}) &= H(e^{j\omega})X(e^{j\omega}) = (1 + e^{-j\omega} + e^{-j2\omega})(e^{-j\omega} + e^{-j2\omega}) \\ &= e^{-j\omega} + 2e^{-j2\omega} + 2e^{-j3\omega} + e^{-j4\omega} \end{aligned}$$

Finally, the use of the inverse discrete function Fourier transformation in Equation 3.60 gives

$$y(n) = \frac{1}{2\pi}\int_{-\pi}^{\pi} (e^{-j\omega} + 2e^{-j2\omega} + 2e^{-j3\omega} + e^{-j4\omega})e^{j\omega n}\,d\omega$$

$$= \frac{1}{2\pi}\int_{-\pi}^{\pi} (e^{j\omega(n-1)} + 2e^{j\omega(n-2)} + 2e^{j\omega(n-3)} + e^{-j\omega(n-4)})\,d\omega$$

$$= u_I(n-1) + 2u_I(n-2) + 2u_I(n-3) + u_I(n-4)$$

3.7 SUMMARY

1. The Fourier series is

$$x(t) = \frac{a_0}{2} + \sum_{k=1}^{\infty} [a_k \cos(2\pi k f_0 t) + b_k \sin(2\pi k f_0 t)]$$

where

$$a_k = \frac{2}{T}\int_{T_1}^{T_1} x(t)\cos(2\pi k f_0 t)\,dt, \qquad b_k = \frac{2}{T}\int_{T_1}^{T_2} x(t)\sin(2\pi k f_0 t)\,dt$$

$f_0 = 1/T, \qquad T = T_2 - T_1$, which is a period of $x(t)$

2. The Fourier exponential series is

$$x(t) = \sum_{m=-\infty}^{\infty} C_m e^{j2\pi m f_0 t}$$

where

$$C_m = \frac{1}{T}\int_{T_1}^{T_2} x(t)e^{-j2\pi m f_0 t}\,dt, \qquad f_0 = 1/T, \qquad T = T_2 - T_1,$$

which is a period of $x(t)$.

3. When a coefficient C_m of the Fourier exponential series of a signal $x(t)$ is expressed as $C_m = |C_m|e^{j\phi m}$, then C_m is called an amplitude spectrum and Φ_m is called a phase spectrum of $x(t)$.

4. The Fourier transformation $\mathcal{F}[x(t)]$ of a signal $x(t)$ is

$$\mathcal{F}[x(t)] = X(j\omega) = \int_{-\infty}^{\infty} x(t)e^{-j\omega t}\,dt$$

5. The inverse Fourier transformation $\mathcal{F}^{-1}[X(j\omega)]$ of $X(e^{j\omega})$ is

$$\mathcal{F}^{-1}[X(j\omega)] = x(t) = \int_{-\infty}^{\infty} X(j\omega)e^{j\omega t}\,df \qquad \text{where } 2\pi f = \omega$$

6. An impulse function $\delta(t - t_0)$ is

$$\int_{-\infty}^{\infty} \delta(t - t_0)\, dt = 1$$

which satisfies

$$\delta(t - t_0) = 0$$

for all t except $t = t_0$.

7. Some of the well known functions representing an impulse function are

$$\text{a. } \delta(t) = \lim_{a\to\infty} \int_{-a}^{a} e^{\pm j\omega t}\, df$$

$$\text{b. } \delta(t) = \lim_{a\to\infty} \int_{-a}^{a} \cos \omega t df, \qquad \text{where } \omega = 2\pi f$$

$$\text{c. } \delta(t) = \lim_{a\to\infty} \frac{\sin(2\pi a t)}{\pi t}$$

8. An important property of an impulse function is

$$\int_{-\infty}^{\infty} \Phi(t)\delta(t - t_0)\, dt = \Phi(t_0)$$

9. The Fourier transformation of an impulse series $\sum_{n=-\infty}^{\infty} \delta(t - nT)$ is

$$\mathcal{F}\left[\sum_{n=-\infty}^{\infty} \delta(t - nT)\right] = \frac{1}{T}\sum_{n=-\infty}^{\infty} \delta\left(f - \frac{n}{T}\right)$$

10. Convolution of analog signals $x(t)$ and $h(t)$ is

$$x(t) * h(t) = \int_{-\infty}^{\infty} x(\tau)h(t - \tau)\, d\tau = \int_{-\infty}^{\infty} x(t - \tau)h(\tau)\, d\tau$$

11. The Fourier transformation of the convolution of $x(t)$ and $h(t)$ is

$$\mathcal{F}[x(t) * h(t)] = \mathcal{F}[x(t)]\mathcal{F}[h(t)] = X(j\omega)H(j\omega)$$

12. The Fourier transformation of product $x(t)h(t)$ is

$$\mathcal{F}[x(t)h(t)] = \frac{1}{2\pi}\mathcal{F}[x(t)] * \mathcal{F}[h(t)]$$

13. A unit sample $u_I(t - t_0)$ is

$$u_I(t - t_0) = \begin{cases} 1, & t = t_0 \\ 0, & t \neq t_0 \end{cases}$$

14. The cutoff frequency symbolized by f_c of an analog signal is the smallest frequency such that the frequency of any nonzero frequency component of the analog signal is smaller than f_c.

15. The Nyquist sampling interval T is a sampling interval satisfying $T \leq 1/2f_c$, which must be satisfied so that a discrete signal obtained by sampling with interval T of an analog signal can reproduce the analog signal.

16. The discrete function Fourier transformation of a discrete function $h(n)$ is

$$H(e^{j\omega}) = \mathcal{F}[h(n)] = \sum_{n=-\infty}^{\infty} h(n)e^{-j\omega n}$$

17. The inverse discrete function Fourier transformation of $H(e^{j\omega})$ is

$$h(n) = \mathcal{F}^{-1}[H(e^{j\omega})] = \frac{1}{2\pi}\int_{-\pi}^{\pi} H(e^{j\omega})e^{j\omega n}\, d\omega$$

18. The discrete function Fourier transformation of the convolution of discrete signals $x(n)$ and $h(n)$ is

$$\mathcal{F}[x(n)*h(n)] = \mathcal{F}[x(n)]\mathcal{F}[h(n)] = X(e^{j\omega})H(e^{j\omega})$$

where $x(n)*h(n) = \sum_{k=-\infty}^{\infty} x(n-k)h(k) = \sum_{k=-\infty}^{\infty} x(k)h(n-k)$

3.8 PROBLEMS

1. Calculate the Fourier exponential series of periodic functions shown in Fig. 3.29.

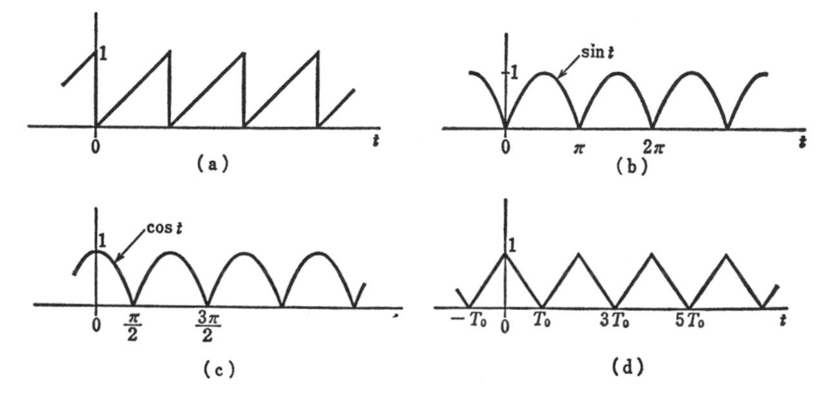

Figure 3.29 Periodic Functions

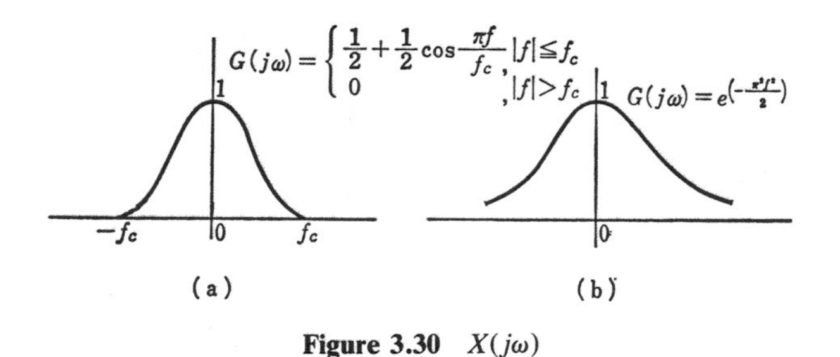

Figure 3.30 $X(j\omega)$

2. Obtain the magnitude spectrum and the phase spectrum of functions in Fig. 3.29.

3. Calculate the Fourier transformation of the following functions.

a. $x(t) = \dfrac{\sin 2\pi t}{\pi t}$ **b.** $x(t) = \begin{cases} \cos 2\pi f_0 t, & |t| < \dfrac{T_0}{2} \\ 0, & |t| > \dfrac{T_0}{2} \end{cases}$

c. $x(t) = \begin{cases} \sin 2\pi f_0 t, & |t| < T_0 \\ 0, & |t| > T_0 \end{cases}$ **d.** $x(t) = \begin{cases} e^{-\alpha t}, & t > 0 \\ 0, & t < 0 \end{cases}$

where $T_0 = 1/f_0$, $\omega_0 = 2\pi f_0$

4. Obtain the inverse Fourier transformation of functions given in Fig. 3.30.

5. Calculate the inverse Fourier transformation of the following functions:

a. $X(j\omega) = K$ **b.** $X(j\omega) = \delta(f - f_0) + \delta(f + f_0)$

c. $X(j\omega) = \begin{cases} 1, & |f| < f_0 \\ 0, & |f| > f_0 \end{cases}$ where $\omega = 2\pi f$

6. Obtain the convolution of a pair of functions shown in Fig. 3.31.

7. Calculate $x(t) * h(t)$ of the following $x(t)$ and $h(t)$:

a. $x(t) = \begin{cases} 1, & 0 < t < 1 \\ 0, & \text{Otherwise} \end{cases}$ **b.** $x(t) = \begin{cases} t, & 0 < t < 1 \\ 0, & \text{Otherwise} \end{cases}$

$h(t) = \begin{cases} e^t, & 0 < t < \dfrac{1}{2} \\ 0, & \text{Otherwise} \end{cases}$ $h(t) = \begin{cases} 2, & 0 < t < 1 \\ 0, & \text{Otherwise} \end{cases}$

8. When a unit sample response $h(n)$ of a linear shift invariant system and an input signal $x(n)$ are

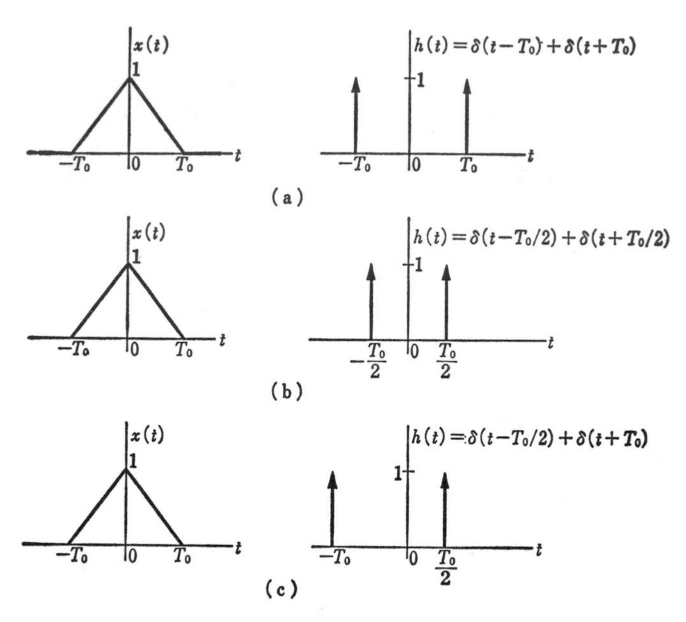

Figure 3.31 $x(t)$ and $h(t)$

a. $h(n) = \begin{cases} 0.5^n, & n \geq 0 \\ 0, & n < 0 \end{cases}$ **b.** $h(n) = \begin{cases} 0.2^n, & n \geq 0 \\ 0, & n < 0 \end{cases}$

$x(n) = \begin{cases} 0.4^n, & n = 0, 1 \\ 0, & \text{Otherwise} \end{cases}$ $x(n) = \begin{cases} 1, & n = 1, 2 \\ 0, & \text{Otherwise} \end{cases}$

Calculate an output signal $y(n)$ by the use of the

a. Relationship $y(n) = \sum_{k=-\infty}^{\infty} h(n - k)x(k)$

b. Discrete function Fourier transformation and the inverse discrete function Fourier transformation

4 Laplace Transformation and Z-Transformation

4.1 LAPLACE TRANSFORMATION

In Chapter 3, it is shown that an inverse Fourier transformation $\mathscr{F}^{-1}[H(j\omega)X(j\omega)]$ of product of Fourier transformations $H(j\omega)$ and $X(j\omega)$ of functions $h(t)$ and $x(t)$ is equal to the convolution of $h(t)$ and $x(t)$. When it is necessary to obtain an output signal $y(t)$ from an input signal $x(t)$ and a transfer function $h(t)$* by the use of computers, applying the Fourier transformations and the inverse Fourier transformation is often simpler than calculating the convolution of $h(t)$ and $x(t)$. There are many functions, however, that cannot be handled by transformations given by Equations 3.21 and 3.22.

Suppose a function $x(t)$ cannot be Fourier transformable. By multiplying $e^{-\alpha t}$ to $x(t)$ and taking α large enough so that $\int_{-\infty}^{\infty} x(t)e^{-\alpha t}e^{-j\omega t}\,dt$ will neither be 0 nor be ∞, the Fourier transformation $X(j\omega)$ of $x(t)e^{-\alpha t}$ can be obtained. By knowing α, we can apply the inverse Fourier transformation on $X(j\omega)$ to obtain the original function $x(t)$. This technique is also applicable for a function that can be Fourier transformable. Hence, we multiply $e^{-\alpha t}$ to a function before applying the Fourier transformation as

* It is also known as an "impulse response."

$$\int_{-\infty}^{\infty} x(t)e^{-\alpha t}e^{-j\omega t}\,dt = \int_{-\infty}^{\infty} x(t)e^{-(\alpha+j\omega)t}\,dt$$

Changing $\alpha + j\omega$ by a complex variable s, we have

$$X(s) = \int_{-\infty}^{\infty} x(t)e^{-st}\,dt \tag{4.1}$$

which is called the "two-sided Laplace transformation." When a system satisfies the initial rest conditions such as electronic circuits, the range of integration can be from 0^- to ∞ as

$$X(s) = \int_{0^-}^{\infty} x(t)e^{-st}\,dt \tag{4.2}$$

which is generally known as the "one-sided Laplace transformation" or simply the "Laplace transformation." The symbol 0^- indicates the time just before 0. In other words, 0^- is not 0 but is a negative number that is the nearest to 0. For convenience, we use the symbol $X(s)$ or $\mathcal{L}[x(t)]$ to represent the Laplace transformation of a function $x(t)$. As an example, consider a function $x(t)$ as

$$x(t) = \begin{cases} e^{-at}, & t \geq 0 \\ 0, & t < 0 \end{cases} \tag{4.3}$$

The Laplace transformation of this $x(t)$ is

$$\mathcal{L}[x(t)] = \int_{0^-}^{\infty} x(t)e^{-st}\,dt = \int_{0^-}^{\infty} e^{-at}e^{-st}\,dt = \frac{1}{s+a} \tag{4.4}$$

Fig. 4.1 shows $x(t)$ in Equation 4.3 when the value of the coefficient a is 1.0, 0.5, 0.25, and 0.01. From Fig. 4.1, we can see that $x(t)$ is 1 for $t \geq 0$ and 0 for $t < 0$ when $a = 0$. We call such a function a "unit step" and indicate it by the symbol u_s. That is, a unit step is defined as

$$u_s(t) = \begin{cases} 1, & t \geq 0 \\ 0, & t < 0 \end{cases} \tag{4.5}$$

and the Laplace transformation of a unit step is

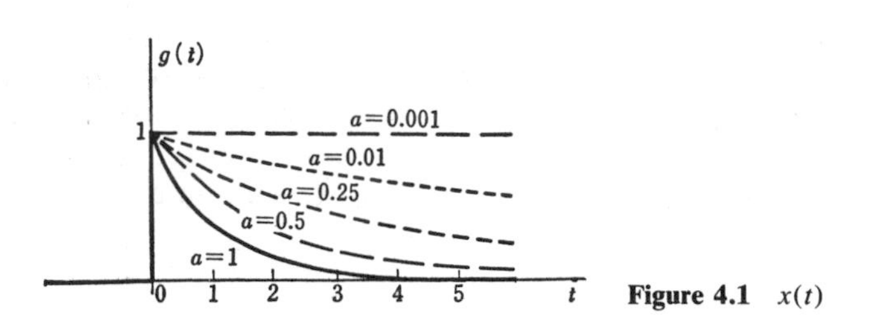

Figure 4.1 $x(t)$

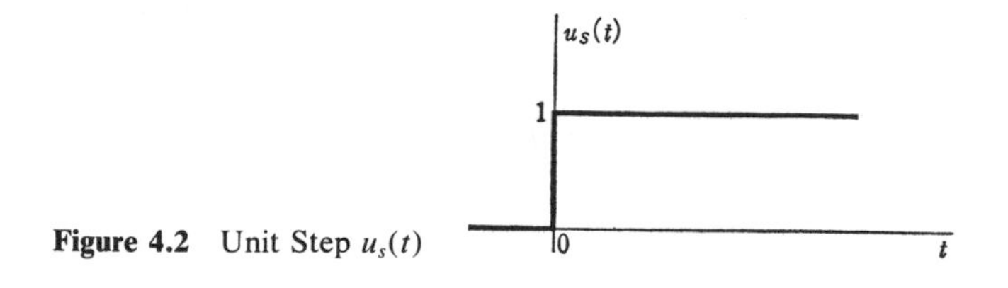

Figure 4.2 Unit Step $u_s(t)$

$$\mathcal{L}[u_s(t)] = \frac{1}{s} \tag{4.6}$$

(by Equation 4.4).

A unit step (Fig. 4.2) is very useful when we express a function that is 0 for $t < 0$. For example, a function $x(t)$ in Equation 4.3 can be expressed as $x(t) = e^{-at}u_s(t)$ and the Laplace transformation of this function becomes as

$$\mathcal{L}[e^{-at}u_s(t)] = \frac{1}{s + a} \tag{4.7}$$

A derivative of the Laplace transformation of a function $x(t)$ with respect to s is

$$\frac{d}{ds}\int_{0^-}^{\infty} x(t)e^{-st}\,dt = \int_{0^-}^{\infty} x(t)\frac{d}{ds}\,e^{-st}\,dt = -\int_{0^-}^{\infty} tx(t)e^{-st}\,dt$$

The right-hand side of this equation is equal to -1 times the Laplace transformation of $tx(t)$. Hence,

$$\mathcal{L}[tx(t)] = -\frac{d}{ds}\,\mathcal{L}[x(t)] \tag{4.8}$$

As an example, the Laplace transformation of a function $x(t) = te^{-at}u_s(t)$ can be expressed as

$$\mathcal{L}[te^{-at}u_s(t)] = -\frac{d}{ds}\,\mathcal{L}[e^{-at}u_s(t)]$$

This $\mathcal{L}[e^{-at}u_s(t)]$ is $1/(s + a)$ by Equation 4.7. Hence,

$$\mathcal{L}[te^{-at}u_s(t)] = -\frac{d}{ds}\left(\frac{1}{s + a}\right) = \frac{1}{(s + a)^2}$$

The Laplace transformation of $x(t) = \dfrac{t^k}{k!}\,e^{-at}u_s(t)$ can be obtained by the use of the above technique successively as

$$\mathcal{L}\left[\frac{t^k}{k!}\,e^{-at}u_s(t)\right] = \frac{1}{(s + a)^{k+1}} \tag{4.9}$$

The Laplace transformation of a derivative of a function $x(t)$ with respect to t is

$$\mathcal{L}\left[\frac{d}{dt}x(t)\right] = \int_{0^-}^{\infty} \frac{dx(t)}{dt} e^{-st}\, dt = \int_{0^-}^{\infty} e^{-st}\, dx(t) \tag{4.10}$$

By choosing

$$\left.\begin{aligned} u &= e^{-st} \\ dv &= dx(t) \end{aligned}\right\}$$

and using the relationship

$$\int_a^b u dv = uv \Big|_a^b - \int_a^b v du$$

Equation 4.10 becomes

$$\mathcal{L}\left[\frac{d}{at}x(t)\right] = e^{-st}x(t)\Big|_{0^-}^{\infty} + s\int_{0^-}^{\infty} x(t)e^{-st}\, dt$$

The second term in the right-hand side of this equation is product of s and $\mathcal{L}[x(t)]$. Hence, we have

$$\mathcal{L}\left[\frac{d}{dt}x(t)\right] = s\mathcal{L}[x(t)] - x(0^-) \tag{4.11}$$

This $x(0^-)$ is the value of $x(t)$ at $t = 0^-$ and is called an initial value of $x(t)$. Using this equation, let us obtain the Laplace transformation of an impulse function $\delta(t)$, which is very useful in analysis of analog systems. By Equation 3.23,

$$\int_{-\infty}^{\infty} \delta(t)\, dt = \int_{-\infty}^{0^-} \delta(t)\, dt + \int_{0^-}^{\infty} \delta(t)\, dt = 1 \tag{4.12}$$

Because $\delta(t) = 0$ for $t < 0$,

$$\int_{-\infty}^{0^-} \delta(t)\, dt = 0$$

Hence, by Equation 4.12, we have

$$\int_{0^-}^{\infty} \delta(t)\, dt = 1 \tag{4.13}$$

This means that integrating $\delta(t)$ produces a unit step $u_s(t)$. Conversely, derivative $du_s(t)/dt$ of a unit step becomes an impulse. Hence, the Laplace transformation of an impulse $\delta(t)$ is

$$\mathcal{L}[\delta(t)] = \mathcal{L}\left[\frac{d}{dt}u_s(t)\right]$$

By Equation 4.15, the right-hand side of this equation becomes

$$\mathcal{L}\left[\frac{d}{dt}\,u_s(t)\right] = s\mathcal{L}[u_s(t)] - u_s(0^-) = 1$$

which shows that the Laplace transformation of an impulse function $\delta(t)$ is 1.

The Laplace transformation of a function $x_1(t) + ax_2(t)$ is

$$\begin{aligned}\mathcal{L}[x_1(t) + ax_2(t)] &= \int_{0^-}^{\infty} [x_1(t) + ax_2(t)]e^{-st}\,dt \\ &= \mathcal{L}[x_1(t)] + a\mathcal{L}[x_2(t)]\end{aligned} \tag{4.14}$$

which shows that the Laplace transformation is linear. Suppose we would like to solve the following differential equation

$$\frac{d^2y(x)}{dt^2} + b\,\frac{dy(t)}{dt} + cy(t) = x(t)$$

One way is to obtain the Laplace transformation of this equation as

$$\mathcal{L}\left[\frac{d^2y(t)}{dt^2} + b\,\frac{dy(t)}{dt} + cy(t)\right] = \mathcal{L}[x(t)]$$

By Equation 4.14, the left-hand side of this becomes

$$\mathcal{L}\left[\frac{d^2y(t)}{dt^2}\right] + b\mathcal{L}\left[\frac{dy(t)}{dt}\right] + c\mathcal{L}[y(t)] = \mathcal{L}[x(t)] \tag{4.15}$$

Next, we will take the Laplace transformation of each of the terms in the left-hand side of the equation. Hence, we need the Laplace transformation of the second derivative of a function that we will develop first.

The Laplace transformation of the second derivative of a function $y(t)$ is

$$\mathcal{L}\left[\frac{d^2}{dt^2}\,y(t)\right] = \mathcal{L}\left[\frac{d}{dt}\left(\frac{d}{dt}\,y(t)\right)\right]$$

By letting $z(t) = (d/dt)y(t)$, this equation can be written as

$$\mathcal{L}\left[\frac{d^2}{dt^2}\,y(t)\right] = \mathcal{L}\left[\frac{d}{dt}\,z(t)\right]$$

By Equation 4.10, the right-hand side of this equation is equal to

$$\mathcal{L}\left[\frac{d^2}{dt^2}\,y(t)\right] = s\mathcal{L}[z(t)] - z(0^-) = s\mathcal{L}\left[\frac{d}{dt}\,y(t)\right] - y^{(1)}(0^-)$$

Again using Equation 4.10, we have

$$\mathcal{L}\left[\frac{d^2}{dt^2}y(t)\right] = s[s\mathcal{L}[y(t)] - y(0^-)] - y^{(1)}(0^-)$$

$$= s^2\mathcal{L}[y(t)] - sy(0^-) - y^{(1)}(0^-)$$

In general, the nth derivative of a function $y(t)$ can be expressed as

$$\mathcal{L}\left[\frac{d^n}{dt^n}y(t)\right] = s^n\mathcal{L}[y(t)] - s^{n-1}y(0^-) - s^{n-2}y^{(1)}(0^-) - \cdots - y^{(n-1)}(0^-) \tag{4.16}$$

where $y^{(k)}(0^-)$ is $\left.\frac{d^k}{dt^k}y(t)\right|_{t=0^-}$ which is the value of $\frac{d^k}{dt^k}y(t)$ at $t = 0^-$.

By the use of Equation 4.16, Equation 4.15 can be rewritten as

$$\{s^2\mathcal{L}[y(t)] - sy(0^-) - y^{(1)}(0^-)\} + b\{s\mathcal{L}[y(t)] - y(0^-)\} + c\mathcal{L}[y(t)] = \mathcal{L}[x(t)]$$

which can be rewritten as

$$(s^2 + bs + c)\mathcal{L}[y(t)] - (s + b)y(0^-) - y^{(1)}(0^-) = \mathcal{L}[x(t)]$$

Hence, $\mathcal{L}[y(t)]$ is equal to

$$\mathcal{L}[y(t)] = \frac{\mathcal{L}[x(t)] + (s + b)y(0^-) + y^{(1)}(0^-)}{s^2 + bs + c} \tag{4.17}$$

The inverse Laplace transformation of $\mathcal{L}[y(t)]$ gives the solution to the given differential equation.

Suppose we use the symbol $Y(s)$ to indicate the Laplace transformation $\mathcal{L}[y(t)]$ of a function $y(t)$. Then the inverse Laplace transformation of $Y(s)$ is

$$\mathcal{L}^{-1}[Y(s)] = y(t) = \frac{1}{2\pi j}\int_c Y(s)e^{st}\,ds \tag{4.18}$$

where the symbol c indicates the Jordan curve, which must enclose all singularities. Instead of using this equation, there is a simple way of obtaining the inverse Laplace transformation of $Y(s)$, which will be shown in the next section. Here, we will use Equation 4.18 to show a linearity of this transformation.

Suppose $Y(s)$ is $Y_1(s) + aY_2(s)$. Then the inverse Laplace transformation of $Y(s)$ is

$$\begin{aligned}\mathcal{L}^{-1}[Y_1(s) + aY_2(s)] &= \frac{1}{2\pi j}\int_c [Y_1(s) + aY_2(s)]e^{st}\,ds \\ &= \frac{1}{2\pi j}\int_c Y_1(s)e^{st}\,dt + a\frac{1}{2\pi j}\int_c Y_2(s)e^{st}\,ds \\ &= \mathcal{L}^{-1}[Y_1(s)] + a\mathcal{L}^{-1}[Y_2(s)]\end{aligned} \tag{4.19}$$

which shows that the inverse Laplace transformation is linear. For example, the inverse Laplace transformation in Equation 4.17 can be expressed by the use of Equation 4.19 as

$$\mathcal{L}^{-1}\left[\frac{\mathcal{L}[x(t)] + (s + b)y(0^-) + y^{(1)}(0^-)}{s^2 + bs + c}\right]$$

$$= \mathcal{L}^{-1}\left[\frac{\mathcal{L}[x(t)]}{s^2 + bs + c}\right] + \mathcal{L}^{-1}\left[\frac{(s + b)y(0^-)}{s^2 + bs + c}\right] + \mathcal{L}^{-1}\left[\frac{y^{(1)}(0^-)}{s^2 + bs + c}\right] \quad (4.20)$$

The right-hand side of this equation is a quotient of polynomial of a variable s. One way of changing a quotient of polynomial in a simpler form is known as a partial fraction expansion, which is very important for obtaining the inverse Laplace transformation.

4.2 PARTIAL FRACTION EXPANSION

A pole of a quotient of polynomial $Y(s)$

$$Y(s) = \frac{a_m s^m + a_{m-1}s^{m-1} + \cdots + a_0}{b_n s^n + b_{n-1}s^{n-1} + \cdots + b_0} \quad (4.21)$$

is a value of s at which $Y(s)$ becomes ∞. For example, the following $Y(s)$

$$Y(s) = \frac{s + a}{s(s - b)}$$

has two poles $s = 0$ and $s = b$. If $Y(s)$ is

$$Y(s) = \frac{s(s + a)}{(s - b)}$$

then its poles are $s = b$ and $s = \infty$.

Suppose $r_1, r_2, \ldots$ and r_n are the roots of an equation obtained by setting the denominator $b_n s^n + b_{n-1}s^{n-1} + \cdots + b_0$ of $Y(s)$ in Equation 4.21 to 0. Then the denominator can be written with these roots as

$$b_n s^n + b_{n-1}s^{n-1} + \cdots + b_0 = b_n(s - r_1)(s - r_2)\cdots(s - r_n)$$

Similarly, the numerator of $Y(s)$ can be expressed as

$$a_m s^m + a_{m-1}s^{m-1} + \cdots + a_0 = a_m(s - z_1)(s - z_2)\cdots(s - z_m)$$

When any $r_i (1 \le i \le n)$ is not equal to $z_j (1 \le j \le m)$, then $Y(s)$ is said to be a "minimum form."

Hereafter, we will assume that a given $Y(s)$ is a minimum form. By expressing $Y(s)$ in Equation 4.21 as

$$Y(s) = \frac{a_m s^m + a_{m-1}s^{m-1} + \cdots + a_0}{b_n(s - r_1)(s - r_1)\cdots(s - r_n)} \quad (4.22)$$

it is clear that for $m < n$, these roots r_1, r_2, . . . and r_m are entire poles of $Y(s)$. If a pole r_i is different from all other poles, then the pole r_i is called a "simple pole." Conversely, if there are k poles ($k > 1$) having the same value, then these poles are called "k-multiple poles" or simply "multiple poles." For example, a quotient of polynomial $Y(s)$

$$Y(s) = \frac{a_5 s^5 + a_4 s^4 + a_3 s^3 + a_2 s^2 + a_1 s + a_0}{(s + 1)^2(s + 2)(s + 3)^3}$$

has two-multiple poles at $s = -1$, a simple pole at $s = -2$, and three-multiple poles at $s = -3$.

When r_1 is a simple pole of $Y(s)$, then $Y(s)$ can be expressed as the sum of $k_1/(s - r_1)$ and a quotient of polynomial $Y_1(s)$, which has no pole at $s = r_1$, as

$$Y(s) = \frac{K_1}{s - r_1} + Y_1(s) \tag{4.23}$$

where

$$K_1 = Y(s)(s - r_1)|_{s=r_1} \tag{4.24}$$

This can be seen by multiplying $(s - r_1)$ to both sides of Equation 4.23 as

$$Y(s)(s - r_1) = K_1 + Y_1(s)(s - r_1)$$

and letting $s = r_1$. Because $Y_1(s)$ does not have a pole at $s = r_1$, $Y_1(s)(s - r_1)|_{s=r_1}$ becomes 0, and we will have the value of K_1 as shown in Equation 4.24. For example, consider a quotient of polynomial

$$Y(s) = \frac{s}{(s - 1)(s - 2)^2(s - 3)^3}$$

Because $r_1 = 1$ is a simple pole, $Y(s)$ can be expressed by using r_1 as

$$Y(s) = \frac{K_1}{s - 1} + Y_1(s)$$

By Equation 4.24, K_1 is

$$K_1 = Y(s)(s - 1)\bigg|_{s=1} = \frac{s}{(s - 2)^2(s - 3)^3}\bigg|_{s=1} = \frac{1}{-8}$$

$Y_1(s)$ is a quotient of polynomial obtained from $Y(s)$ by removing $K_1/(s - 1)$ or

$$Y_1(s) = \frac{s}{(s - 1)(s - 2)^2(s - 3)^3} - \frac{-\frac{1}{8}}{s - 1}$$

Hence, $Y(s)$ can be expressed as

$$\frac{s}{(s-1)(s-2)^2(s-3)^3} = \frac{\frac{-1}{8}}{s-1} + Y_1(s)$$

When r_2 is two-multiple poles of $Y(s)$, then $Y(s)$ can be expressed as sum of

$$\left\{\frac{K_2}{(s-r_2)^2} + \frac{K_1}{(s-r_2)}\right\}$$

and a quotient of polynomial $Y_2(s)$, which contains no poles at $s = r_2$, as

$$Y(s) = \frac{K_2}{(s-r_2)^2} + \frac{K_1}{s-r_2} + Y_2(s) \tag{4.25}$$

These constants K_1 and K_2 are

$$K_2 = Y(s)(s-r_2)^2|_{s=r_2} \tag{4.26}$$

$$K_1 = \frac{d}{ds} Y(s)(s-r_2)^2|_{s=r_2} \tag{4.27}$$

We can obtain these equations for k_1 and k_2 by first multiplying $(s - r_2)^2$ to both sides of Equation 4.2 as

$$Y(s)(s-r_2)^2 = K_2 + K_1(s-r_2) + Y_2(s)(s-r_2)^2 \tag{4.28}$$

then set $s = r_2$. Because $Y_2(s)$ has no poles at $s = r_2$, $Y_2(s)(s - r_2)^2$ will be 0 when $s = r_2$. Also, $K_1(s - r_2)$ will be 0 at $s = r_2$. Hence, we can obtain Equation 4.26. Next, we differentiate Equation 4.28 as

$$\frac{d}{ds} Y(s)(s-r_2)^2 = K_1 + 2\,Y_2(s)(s-r_2) + (s-r_2)^2 \frac{d}{ds} Y_2(s)$$

then set $s = r_2$. Because $Y_2(s)(s - r_2)$ will be 0 at $s = r_2$ and $(s - r_2)^2 \dfrac{dY_2(s)}{dt}$ will also be 0 at $s = r_2$, Equation 4.27 can be obtained. As an example, consider a quotient of polynomial $Y(s)$ used in the previous sample which is

$$Y(s) = \frac{s}{(s-1)(s-2)^2(s-3)^3}$$

Because $r_2 = 2$ is two-multiple poles, we can expand $Y(s)$ using r_2 as

$$Y(s) = \frac{K_{22}}{(s-2)^2} + \frac{K_{21}}{s-2} + Y_2(s)$$

The coefficient K_{22} can be calculated by Equation 4.26 as

$$K_{22} = \frac{s}{(s-1)(s-2)^2(s-3)^3}(s-2)^2\bigg|_{s=2} = -2$$

and K_{21} can be calculated by Equation 4.27 as

$$K_{21} = \frac{d}{ds} \frac{s}{(s-1)(s-3)^3} \bigg|_{s=2} = -5$$

Hence, $Y(s)$ becomes

$$\frac{s}{(s-1)(s-2)^2(s-3)^3} = \frac{-2}{(s-2)^2} + \frac{-5}{s-2} + Y_2(s)$$

where $Y_2(s)$ is

$$Y_2(s) = \frac{s}{(s-1)(s-2)^2(s-3)^3} - \frac{-2}{(s-2)^2} - \frac{-5}{s-2}$$

In general, $Y(s)$ can be expressed with q-multiple poles at $s = r_3$ as

$$Y(s) = \frac{K_q}{(s-r_3)^q} + \frac{K_{q-1}}{(s-r_3)^{q-1}} + \cdots + \frac{K_{q-u}}{(s-r_3)^{q-u}} + \cdots + \frac{K_1}{s-r_3} + Y_3(s) \quad (4.29)$$

where $Y_3(s)$ has no poles at $s = r_3$ and K_{q-u} is equal to

$$K_{q-u} = \frac{1}{u!} \frac{d^u}{ds^u} Y(s)(s-r_3)^q \bigg|_{s=r_3} \quad (4.30)$$

As an example, consider a quotient of polynomial $Y(s)$ used in the previous example, which is

$$Y(s) = \frac{s}{(s-1)(s-2)^2(s-3)^3}$$

Because $r_3 = 3$ is three-multiple poles, $Y(s)$ can be expanded as

$$Y(s) = \frac{K_{33}}{(s-3)^3} + \frac{K_{32}}{(s-3)^2} + \frac{K_{31}}{s-3} + Y_3(s)$$

where K_{33} can be obtained from Equation 4.30 with $t = 0$ as

$$K_{33} = \frac{s}{(s-1)(s-2)^2(s-3)^3}(s-3)^3 \bigg|_{s=3} = \frac{3}{2}$$

K_{32} is from Equation 4.30 with $u = 1$ as

$$K_{32} = \frac{d}{ds}\left(\frac{s}{(s-1)(s-2)^2}\right)\bigg|_{s=3} = \frac{-13}{4}$$

and K_{31} can be calculated from Equation 4.30 with $u = 2$ as

$$K_{31} = \frac{1}{2!} \frac{d^2}{ds^2}\left(\frac{s}{(s-1)(s-2)^2}\right)\bigg|_{s=3} = \frac{41}{8}$$

Hence, $Y(s)$ can be expanded as

$$\frac{s}{(s-1)(s-2)^2(s-3)^3} = \frac{\frac{3}{2}}{(s-3)^3} + \frac{-\frac{13}{4}}{(s-3)^2} + \frac{\frac{41}{8}}{(s-3)} + Y_3(s)$$

When $m < n$ of $Y(s)$ in Equation 4.22, we can expand $Y(s)$ using all poles as

$$Y(s) = \frac{K_{p_1q_1}}{(s-r_{p_1})^{q_1}} + \frac{K_{p_1q_1-1}}{(s-r_{p_1})^{q_1-1}} + \cdots + \frac{K_{p_11}}{s-r_{p_1}} + \frac{K_{p_2q_2}}{(s-r_{p_2})^{q_2}} + \cdots \tag{4.31}$$

which is called the "partial fraction expansion." For example, a quotient of polynomial $Y(s)$ used in the previous example, which is

$$Y(s) = \frac{s}{(s-1)(s-2)^2(s-3)^3}$$

can be expanded by using $r_1 = 1$, $r_2 = 2$, and $r_3 = 3$ as

$$\frac{s}{(s-1)(s-2)^2(s-3)^3} = \frac{K_{11}}{2-1} + \frac{K_{22}}{(s-2)^2} + \frac{K_{21}}{s-2} + \frac{K_{33}}{(s-3)^3} + \frac{K_{32}}{(s-3)^2} + \frac{K_{31}}{s-3} = \frac{-\frac{1}{8}}{s-1} + \frac{-2}{(s-2)^2} + \frac{-5}{s-2} + \frac{\frac{3}{2}}{(s-3)^3} + \frac{-\frac{13}{4}}{(s-3)^2} + \frac{\frac{41}{8}}{s-3}$$

which is the partial fraction expansion of $Y(s)$.

When $Y(s)$ in Equation 4.22 satisfies $m \geq n$, then $Y(s)$ can be expanded as

$$Y(s) = c_k s^k + c_{k-1}s^{k-1} + \cdots + c_1 s + c_0 + Y_m(s)$$

where

$$k = m - n$$

and the degree of the numerator of $Y_m(s)$ is smaller than that of the denominator of $Y_m(s)$. Hence, any quotient of polynomial $Y(s)$ can be expanded as

$$Y(s) = c_k s^k + c_{k-1}s^{k-1} + \cdots + c_1 s + c_0 + \frac{K_{p_1q_1}}{(s-r_{p_1})^{q_1}} + \frac{K_{p_1q_1-1}}{(s-r_{p_1})^{q_1-1}} + \cdots + \frac{K_{p_2q_2}}{(s-p_2)^{q_2}} + \cdots \tag{4.32}$$

which is known as the "generalized partial fraction expansion." As an example, consider

$$Y(s) = \frac{2s^4 + 7s^3 + 9s^2 + 26s + 20}{s^2 + 4s + 4}$$

Because the degree of the numerator in this $Y(s)$ is larger than that of the denominator, we can divide the numerator by the denominator as

$$Y(s) = 2s^2 - s + 5 + \frac{10s}{s^2 + 4s + 4}$$

$Y_m(s)$ of the preceding $Y(s)$ is

$$Y_m(s) = \frac{10s}{s^2 + 4s + 4} = \frac{10s}{(s + 2)^2}$$

Because there are two-multiple poles at $s = -2$, the partial fraction expansion of $Y_m(s)$ is

$$Y_m(s) = \frac{-20}{(s + 2)^2} + \frac{10}{(s + 2)}$$

Hence, the generalized partial fraction expansion of $Y(s)$ is

$$Y(s) = 2s^2 - s + 5 + \frac{-20}{(s + 2)^2} + \frac{10}{s + 2}$$

4.3 SOLVING DIFFERENTIAL EQUATIONS

Consider a differential equation

$$\frac{d^2}{dt^2} y(t) + b \frac{d}{dt} y(t) + cy(t) = x(t)$$

To obtain a solution, first we can take the Laplace transformation of a differential equation and obtain $\mathscr{L}[y(t)]$ as

$$\mathscr{L}[y(t)] = \frac{\mathscr{L}[x(t)] + (s + b)y(0^-) + y^{(1)}(0^-)}{s^2 + bs + c}$$

Then taking the inverse Laplace transformation of this $\mathscr{L}[y(t)]$ using Equation 4.20 we have

$$\mathscr{L}^{-1}\{\mathscr{L}[y(t)]\} = y(t) = \mathscr{L}^{-1}\left[\frac{\mathscr{L}[x(t)]}{s^2 + bs + c}\right] + \mathscr{L}^{-1}\left[\frac{(s + b)y(0^-)}{s^2 + bs + c}\right] + \mathscr{L}^{-1}\left[\frac{y^{(1)}(0^-)}{s^2 + bs + c}\right] \tag{4.33}$$

Since $\mathcal{L}^{-1}\{\mathcal{L}[y(t)]\}$ is $y(t)$, the inverse Laplace transformation of every term in the right-hand side of Equation 4.33 gives a solution of a given differential equation. To show a simple method of taking the inverse Laplace transformation by the use of the partial fraction expansion, consider the last term in the right-hand side of Equation 4.33. Notice that $y^{(1)}(0^-)$ is an initial value that is a known constant.

Assume that $\sqrt{b^2 - 4c} \neq 0$, poles r_1 and r_2 of $y^{(1)}(0^-)/(s^2 + bs + c)$ are $r_1 = \frac{1}{2}(-b + \sqrt{b^2 - 4c}$ and $r_1 = \frac{1}{2}(-b - \sqrt{b^2 - 4c})$. Hence,

$$\frac{y^{(1)}(0^-)}{s^2 + bs + c} = \frac{y^{(1)}(0^-)/\sqrt{b^2 - 4c}}{s + \frac{1}{2}(b - \sqrt{b^2 - 4c})} + \frac{-y^{(1)}(0^-)/\sqrt{b^2 - 4c}}{s + \frac{1}{2}(b + \sqrt{b^2 - 4c})}$$

Thus by Equation 4.23, the last term in the right-hand side of Equation 4.33 can be expressed as

$$\mathcal{L}^{-1}\left[\frac{y^{(1)}(0^-)}{s^2 + bs + c}\right] = [(y^{(1)}(0^-)/\sqrt{b^2 - 4c}]\mathcal{L}^{-1}\left[\frac{1}{s + \frac{1}{2}(b - \sqrt{b^2 - 4c})}\right]$$
$$-(y^{(1)}(0^-)/\sqrt{b^2 - 4c})\mathcal{L}^{-1}\left[\frac{1}{s + \frac{1}{2}(b + \sqrt{b^2 - 4c})}\right] \qquad (4.34)$$

Each term in the right-hand side is of the form $\mathcal{L}^{-1}\left[\frac{1}{s + a}\right]$. Because $\frac{1}{s + a}$ is the Laplace transformation of a function $e^{-at}u_s(t)$ (Equation 4.7), Equation 4.34 becomes

$$\mathcal{L}^{-1}\left[\frac{u^{(1)}(0^-)}{s^2 + bs + c}\right] = (y^{(1)}(0^-)/\sqrt{b^2 - 4c})e^{-\frac{1}{2}(b-\sqrt{b^2-4c})t}u_s(t)$$
$$- (y^{(1)}(0^-)/\sqrt{b^2 - 4c})e^{-\frac{1}{2}(b+\sqrt{b^2-4c})t}u_s(t)$$

Let us take the second term in the right-hand side of equation 4.33. First, the partial fraction expansion of a quotient of polynomial $(s + b)y(0^-)/(s^2 + bs + c)$ is

$$\mathcal{L}^{-1}\left[\frac{(s + b)y(0^-)}{s^2 + bs + c}\right] = \mathcal{L}^{-1}\left[\frac{\left[-\frac{1}{2}(b - \sqrt{b^2 - 4c}) + b\right]\Big/\sqrt{b^2 - 4c}}{s + \frac{1}{2}(b - \sqrt{b^2 - 4c})}y(0^-)\right.$$
$$\left. - \frac{\left[-\frac{1}{2}(b + \sqrt{b^2 - 4c}) + b\right]\Big/\sqrt{b^2 - 4c}}{s + \frac{1}{2}(b + \sqrt{b^2 - 4c})}y(0^-)\right]$$

$$= \frac{-\frac{1}{2}(b - \sqrt{b^2 - 4c}) + b}{\sqrt{b^2 - 4c}} y(0^-)\mathcal{L}^{-1}\left[\frac{1}{s + \frac{1}{2}(b - \sqrt{b^2 - 4c})}\right]$$

$$- \frac{-\frac{1}{2}(b + \sqrt{b^2 - 4c}) + b}{\sqrt{b^2 - 4c}} y(0^-)\mathcal{L}^{-1}\left[\frac{1}{s + \frac{1}{2}(b + \sqrt{b^2 - 4c})}\right]$$

Then using Equation 4.7, we have

$$\mathcal{L}^{-1}\left[\frac{(s + b)y(0^-)}{s^2 + bs + c}\right] = \frac{-\frac{1}{2}(b - \sqrt{b^2 - 4c}) + b}{\sqrt{b^2 - 4c}} y(0^-)e^{-\frac{1}{2}(b - \sqrt{b^2 - 4c})t}\, u_s(t)$$

$$- \frac{-\frac{1}{2}(b + \sqrt{b^2 - 4c}) + b}{\sqrt{b^2 - 4c}} y(0^-)e^{-\frac{1}{2}(b + \sqrt{b^2 - 4c})t}\, u_s(t)$$

For the first term $\mathcal{L}^{-1}\left[\frac{\mathcal{L}[x(t)]}{s^2 + bs + c}\right]$ in the right-hand side of Equation 4.33, suppose $\mathcal{L}[x(t)]$ is given. Then the inverse Laplace transformation of this term can be obtained as the same way as other terms in the equation. For example, if $x(t) = e^{-t}u_s(t)$ then

$$\mathcal{L}[x(t)] = \frac{1}{s + 1}$$

Hence, this term becomes

$$\mathcal{L}^{-1}\left[\frac{\mathcal{L}[x(t)]}{s^2 + bs + c}\right] = \mathcal{L}^{-1}\left[\frac{1}{(s + 1)(s^2 + bs + c)}\right]$$

Thus,

$$\mathcal{L}^{-1}\left[\frac{\mathcal{L}[e^{-t}u_s(t)]}{s^2 + bs + c}\right] = \frac{e^{-t}u_s(t)}{\left[-1 + \frac{1}{2}(b - \sqrt{b^2 - 4c})\right]\left[-1 + \frac{1}{2}(b + \sqrt{b^2 - 4c})\right]}$$

$$+ \frac{1}{\left[-\frac{1}{2}(b - \sqrt{b^2 - 4c}) + 1\right]\sqrt{b^2 - 4c}} e^{-\frac{1}{2}(b - \sqrt{b^2 - 4c})t}u_s(t)$$

$$+ \frac{1}{\left[-\frac{1}{2}(b + \sqrt{b^2 - 4c}) + 1\right]\sqrt{b^2 - 4c}} e^{-\frac{1}{2}(b + \sqrt{b^2 - 4c})t}u_s(t)$$

4.4 Z-TRANSFORMATION

We have learned that for those analog signals that are not Fourier transformable, the Laplace transformation was developed. We also learned that the discrete function Fourier transformation can be employed if Equation 3.54 converges. Because there are discrete signals for which the discrete functions Fourier transformation cannot be applied, there is a transformation called the z-transformation for discrete signals similar to the Laplace transformation for analog signals.

Suppose

$$X(e^{j\omega}) = \sum_{n=-\infty}^{\infty} x(n)e^{-j\omega n} \tag{4.35}$$

does not converge. By changing $e^{-j\omega n}$ to $e^{-(\alpha+j\alpha)n}$, the right-hand side of Equation 4.35 becomes $\sum_{n=-\infty}^{\infty} (x(n)e^{-\alpha n})e^{-j\omega n}$, which can be used as the transformation of $x(n)$ if there is a value α that makes $\sum_{n=-\infty}^{\infty} (x(n)e^{-\alpha n})e^{-j\omega n}$ converge. Using z, which is equal to $e^{(\alpha+j\omega)}$, this equation can be written as

$$X(z) = \sum_{n=-\infty}^{\infty} X(n)z^{-n} \tag{4.36}$$

This is called the "two-sided z-transformation" of $x(n)$, which is very useful in analysis and design of discrete systems. Especially for a system satisfying initially rest conditions, it is easier to use the transformation by changing the summation in Equation 4.36 from 0 to ∞ as

$$X(z) = \sum_{n=0}^{\infty} x(n)z^{-n} \tag{4.37}$$

This transformation is known as the "one-sided z-transformation" or simply the "z-transformation." The symbol $\mathfrak{Z}[x(n)]$ or $X(z)$ is commonly used for indicating the z-transformation of a discrete signal $x(n)$.

Suppose a discrete signal $x(n)$ is

$$x(n) = \begin{cases} e^{-an}, & n \geq 0 \\ 0 \quad , & n < 0 \end{cases}$$

Then the z-transformation of $x(n)$ can be obtained by Equation 4.37 as

$$\mathfrak{Z}[x(n)] = \sum_{n=0}^{\infty} e^{-an}z^{-n} = \sum_{n=0}^{\infty} (e^{a}z)^{-n} = \frac{z}{z - e^{-a}} \tag{4.38}$$

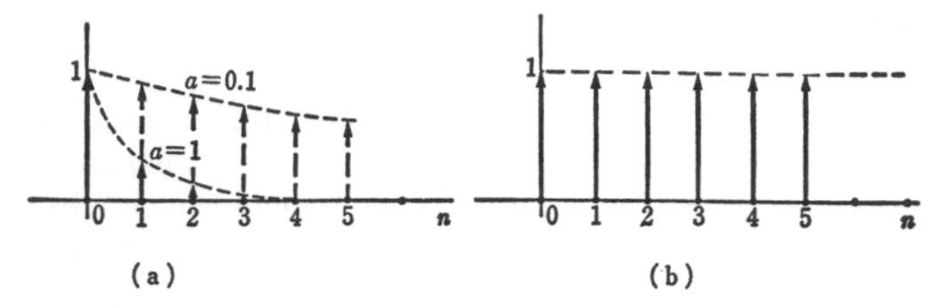

Figure 4.3 Discrete Signal. (a) e^{-an}, $n \geq 0$; (b) $u_s(n)$.

Fig. 4.3(a) shows this signal $x(n)$ for $a = 1$ and $a = 0.1$. when $a = 0$, $x(n)$ becomes a unit step $u_s(n)$ as shown in Fig. 4.3(b). Hence, by letting $a = 0$ in Equation 4.38, we will have the z-transformation of a unit step $u_s(n)$ as

$$\mathfrak{Z}[u_s(n)] = \frac{z}{z-1} \tag{4.39}$$

Because a unit step $u_s(n)$ is

$$u_s(n) = \begin{cases} 1, & n \geq 0 \\ 0, & n < 0 \end{cases}$$

a discrete function $x(n)$

$$x(n) = \begin{cases} x_1(n), & n \geq 0 \\ 0 \quad , & n < 0 \end{cases}$$

can be expressed as $x(n) = x_1(n)u_s(n)$.

By Equation 4.36, the z-transformation of a derivative of a discrete function $x(n)$ can be expressed as

$$\frac{d}{dz}\mathfrak{Z}[x(n)] = \frac{d}{dz}\sum_{n=-\infty}^{\infty} x(n)z^{-n} = -\sum_{n=-\infty}^{\infty} nx(n)z^{-n-1} = -z^{-1}\mathfrak{Z}[nx(n)]$$

which gives

$$\mathfrak{Z}[nx(n)] = -z\frac{d}{dz}\mathfrak{Z}[x(n)]$$

Using this equation, the z-transformation of $x(n) = ne^{-an}u_s(n)$ is

$$\mathfrak{Z}[ne^{-an}u_s(n)] = -z\frac{d}{dz}\mathfrak{Z}[e^{-an}u_s(n)]$$

Because $\mathfrak{Z}[e^{-an}u_s(n)]$ in the right-hand side is $z/(z - e^{-a})$ by Equation 4.38, we have

$$\mathfrak{Z}[ne^{-an}u_s(n)] = -z\frac{d}{dz}\left(\frac{z}{z-e^{-a}}\right) = \frac{e^{-a}z}{(z-e^{-a})^2} \tag{4.40}$$

By letting $a = 0$, we have

$$\mathscr{Z}[nu_s(n)] = \frac{z}{(z-1)^2}$$

The z-transformation of a discrete function $x(n) = a^n u_s(n)$ $(a > 0)$ can be obtained from Equation 4.38 by changing e^{-a} to a as

$$\mathscr{Z}[a^n u_s(n)] = \frac{z}{z-a} \tag{4.41}$$

Also from Equation 4.40, we have

$$\mathscr{Z}[na^n u_s(n)] = \frac{az}{(z-a)^2} \tag{4.42}$$

To obtain a solution of a difference equation by the use of the z-transformation, we need the z-transformation of a function $x(n-k)$. By Equation 4.37, the z-transformation of $x(n-k)$ is

$$\mathscr{Z}[x(n-k)] = \sum_{n=0}^{\infty} x(n-k)z^{-n}$$

Changing $n - k$ by m, we have

$$\mathscr{Z}[x(n-k)] = \sum_{m=-k}^{\infty} x(m)z^{-m-k}$$

Expanding the right-hand side for $m = -k, -k+1, -k+2, \ldots, -1$, it becomes

$$\mathscr{Z}[x(n-k)] = x(-k) + z^{-1}x(-k+1) + \cdots + z^{-k+1}x(-1) + z^{-k}\sum_{m=0}^{\infty} x(m)z^{-m}$$

Because $\sum_{m=0}^{\infty} x(m)z^{-m}$ is $\mathscr{Z}[x(m)]$, this equation can be written as

$$\mathscr{Z}[x(n-k)] = z^{-k}\mathscr{Z}[x(n)] + z^{-k+1}x(-1) + z^{-k+1}x(-2) + \cdots + z^{-1}x(-k+1) + x(-k) \tag{4.43}$$

A linearity of the z-transformation can be seen by the use of a discrete signal $x_1(n) + x_2(n)$ where α is a constant as

$$\mathscr{Z}[x_1(n) + \alpha x_2(n)] = \sum_{n=0}^{\infty} (x_1(n) + \alpha x_2(n))z^{-n} = \mathscr{Z}[x_1(n)] + \alpha\mathscr{Z}[x_2(n)] \tag{4.44}$$

For an example of applying the z-transformation, consider a difference equation

$$y(n-2) + 6y(n-1) + 8y(n) = x(n)$$

with $x(n) = 0.5u_s(n)$, $y(-1) = 0$, and $y(-2) = 0$. The z-transformation of this equation is

$$\mathfrak{Z}[y(n-2) + 6y(n-1) + 8y(n)] = \mathfrak{Z}[x(n)]$$

By Equation 4.44, this equation can be expressed as

$$\mathfrak{Z}[y(n-2)] + 6\mathfrak{Z}[y(n-1)] + 8\mathfrak{Z}[y(n)] = \mathfrak{Z}[x(n)]$$

Using Equation 4.43, each term in the left side of this equation can be changed as

$$z^{-2}\mathfrak{Z}[y(n)] + z^{-1}y(-1) + y(-2) + 6\{z^{-1}\mathfrak{Z}[y(n)] + y(-1)\} + 8\mathfrak{Z}[y(n)] = \mathfrak{Z}[0.5^n u_s(n)]$$

Using initial conditions $y(-1) = 0$ and $y(-2) = 0$, we can obtain $\mathfrak{Z}[y(n)]$ as

$$\mathfrak{Z}[y(n)] = \frac{\mathfrak{Z}[0.5^n u_s(n)] - 1}{z^{-2} + 6z^{-1} + 8}$$

Because $\mathfrak{Z}[0.5^n u_s(n)]$ is $z/(z - 0.5)$ by Equation 4.41, $\mathfrak{Z}[y(n)]$ becomes

$$\mathfrak{Z}[y(n)] = \frac{\dfrac{z}{z-0.5} - 1}{z^{-2} + 6z^{-1} + 8} = \frac{0.5z^2}{(z-0.5)(8z^2 + 6z + 1)}$$

$$= \frac{0.5z^2}{8(z-0.5)(z+0.25)(z+0.5)}$$

To obtain a solution $y(n)$, we should know how to obtain the inverse z-transformation of $\mathfrak{Z}[y(n)]$.

4.5 SOLVING DIFFERENCE EQUATIONS

Let $X(z)$ be the z-transformation of $x(n)$. Then the inverse z-transformation $\mathfrak{Z}^{-1}[X(z)]$ is

$$\mathfrak{Z}^{-1}[X(z)] = x(n) = \frac{1}{2\pi j}\int_c X(z)z^{n-1}\,dz \tag{4.45}$$

where a Jordan curve c must be chosen so that it enclose all singularities. There is a simpler way of obtaining the inverse z-transformation than using Equation 4.45, which we will study later. Here, we will use Equation 4.45 only to learn the linearity property of the inverse z-transformation. Suppose $X(z) = X_1(z) + \alpha X_2(z)$. Then the inverse z-transformation will be

$$\mathcal{Z}^{-1}[X_1(z) + \alpha X_2(z)] = \frac{1}{2\pi j}\int_c [X_1(z) + \alpha X_2(z)]z^{n-1}\,dz$$

$$= \frac{1}{2\pi j}\int_c X_1(z)z^{n-1}\,dz + \frac{\alpha}{2\pi j}\int_c X_2(z)z^{n-1}\,dz \tag{4.46}$$

$$= \mathcal{Z}^{-1}[X_1(z)] + \alpha\mathcal{Z}^{-1}[X_2(z)]$$

This shows the linearity property of the inverse z-transformation. Using this property, a solution $y(n)$ can be obtained from Equation 4.46 as follows: First we take the inverse z-transformation of Equation 4.46 as

$$y(n) = \mathcal{Z}^{-1}\left[\frac{0.5z^2}{8(z - 0.5)(z + 0.25)(z + 0.5)}\right]$$

In the case of the inverse Laplace transformation, the inside of the bracket in the right-hand side of the equation is changed to the form of the partial fraction expansion. By investigating Equations 4.38, 4.39, and 4.41 carefully, however, we can see that there is z in the numerator of these transformations. Hence, it would be desirable to expand the inside of the bracket in the form of $K_p z/(z + a)^p$. To accomplish this, we remove one z and change the remaining in the form of a partial fraction expansion. That is, it is necessary to change $0.5z/8(z - 0.5)(z + 0.25)(z + 0.5)$ as

$$\mathcal{Z}^{-1}\left[\frac{0.5z^2}{8(z - 0.5)(z + 0.25)(z + 0.5)}\right]$$

$$= \mathcal{Z}^{-1}\left[z\left(\frac{K_1}{z - 0.5} + \frac{K_2}{z + 0.25} + \frac{K_3}{z + 0.5}\right)\right]$$

where K_1, K_2, and K_3 are

$$K_1 = \frac{0.5z}{8(z - 0.5)(z + 0.25)(z + 0.5)}(z - 0.5)\bigg|_{z=0.5} = \frac{1}{24}$$

$$K_2 = \frac{0.5z}{8(z - 0.5)(z + 0.25)(z + 0.5)}(z + 0.25)\bigg|_{z=-0.25} = \frac{1}{12}$$

$$K_3 = \frac{0.5z}{8(z - 0.5)(z + 0.25)(z + 0.5)}(z + 0.5)\bigg|_{z=-0.5} = \frac{-1}{8}$$

Hence, by Equation 4.45, a solution will be

$$y(n) = \mathcal{Z}^{-1}\left[\frac{\frac{1}{24}z}{z - 0.5} + \frac{\frac{1}{12}z}{z + 0.25} + \frac{-\frac{1}{8}z}{z + 0.5}\right]$$

$$= \frac{1}{24}\mathcal{Z}^{-1}\left[\frac{z}{z - 0.5}\right] + \frac{1}{12}\mathcal{Z}^{-1}\left[\frac{z}{z + 0.25}\right] - \frac{1}{8}\mathcal{Z}^{-1}\left[\frac{z}{z + 0.5}\right]$$

By Equation 4.41, $z/(z - a)$ is the z-transformation of $a^n u_s(n)$. Hence, the preceding equation becomes

$$y(n) = \left[\frac{1}{24}\,0.5^n + \frac{1}{12}\,(-0.25)^n - \frac{1}{8}\,(-0.5)^n\right] u_s(n)$$

which is a solution of a difference equation

$$y(n - 2) + 6y(n - 1) + 8y(n) = x(n)$$

In the preceding example, an input signal $x(n)$ is prespecified as $x(n) = 0.5^n u_s(n)$. Even if an input signal $x(n)$ is unspecified, we can obtain a solution as follows: Taking the z-transformation of a difference equation becomes

$$z^{-1}\mathcal{Z}[y(n)] + z^{-1}y(-1) + y(-2) + 6\{z^{-1}\mathcal{Z}[y(n)] + y(-1)\} + 8\mathcal{Z}[y(n)] = \mathcal{Z}[x(n)]$$

from which $\mathcal{Z}[y(n)]$ is

$$\mathcal{Z}[y(n)] = \frac{\mathcal{Z}[x(n)] - 1}{z^{-2} + 6z^{-1} + 8}$$

Hence, $y(n)$ is

$$\begin{aligned} y(n) &= \mathcal{Z}^{-1}\left[\frac{\mathcal{Z}[x(n)] - 1}{z^{-2} + 6z^{-1} + 8}\right] \\ &= \mathcal{Z}^{-1}\left[\frac{\mathcal{Z}[x(n)]}{z^{-2} + 6z^{-1} + 8}\right] - \mathcal{Z}^{-1}\left[\frac{1}{z^{-2} + 6z^{-1} + 8}\right] \end{aligned}$$

The second term in the right-hand side of this equation is

$$\begin{aligned} \mathcal{Z}^{-1}\left[\frac{1}{z^{-2} + 6z^{-1} + 8}\right] &= \mathcal{Z}^{-1}\left[\frac{-z}{8(z + 0.25)}\right] + \mathcal{Z}^{-1}\left[\frac{z}{4(z + 0.5)}\right] \\ &= -\frac{1}{8}\,(-0.25)^n + \frac{1}{4}\,(-0.5)^n \end{aligned}$$

The first term in the right-hand side can be expressed as

$$\mathcal{Z}^{-1}\left[\frac{\mathcal{Z}[x(n)]}{z^{-2} + 6z^{-1} + 8}\right] = \mathcal{Z}^{-1}\left[\left(\frac{1}{z^{-2} + 6z^{-1} + 8}\right)\mathcal{Z}[x(n)]\right] \tag{4.47}$$

The right-hand side indicates the inverse z-transformation of product of $1/(z^{-2} + 6z^{-1} + 8)$ and $\mathcal{Z}[x(n)]$, which are the z-transformations of discrete functions. In the case of Laplace transformations, the inverse Laplace transformation of $H(s)X(s)$ is equal to convolution of $h(t)$ and $x(t)$, where $H(s)$ and $X(s)$ are the Laplace transformations of $h(t)$ and $x(t)$, respectively. Do we have the same property under z-transformation? In other words, is the inverse z-transformation of $H(z)X(z)$ equal to the convolution of $h(n)$ and $x(n)$, where $H(z)$ and $X(z)$ are z-transformations of $h(n)$ and $x(n)$? To investigate this, consider a discrete signal $y(n)$, which is the convolution of $h(n)$ and $x(n)$, that is,

$$y(n) = \sum_{m=-\infty}^{\infty} h(m)x(n-m)$$

The z-transformation of $y(n)$ is by Equation 4.37

$$\mathscr{Z}[y(n)] = \mathscr{Z}\left[\sum_{m=-\infty}^{\infty} h(m)x(n-m)\right] = \sum_{n=0}^{\infty}\sum_{m=-\infty}^{\infty} h(m)x(n-m)z^{-n}$$

$$= \sum_{m=-\infty}^{\infty} h(m)\sum_{n=0}^{\infty} x(n-m)z^{-n} = \mathscr{Z}[h(n)]\mathscr{Z}[x(n)]$$

where $h(n)$ and $x(n)$ are assumed to be 0 for $t < 0$. This shows that the inverse z-transformation of $\mathscr{Z}[h(n)]\mathscr{Z}[x(n)]$ is equal to $h(n) * x(n)$ (convolution of $h(n)$ and $x(n)$).

Because of the preceding property, the inverse z-transformation of product $\mathscr{Z}[h(n)]\mathscr{Z}[x(n)]$ is equal to convolution of $h(n)$ and $x(n)$. Since $\mathscr{Z}^{-1}\left[\frac{1}{z^{-1} + 6x^{-1} + 8}\right]$ in Equation 4.47 is $-\frac{1}{8}(-0.25)^n + \frac{1}{4}(-0.5)^n$, we can see that $\mathscr{Z}^{-1}\left[\frac{\mathscr{Z}[x(n)]}{z^{-2} + 6z^{-1} + 8}\right]$ is equal to the convolution of $-\frac{1}{8}(-0.25)^n + \frac{1}{4}(-0.5)^n$ and $x(n)$. Hence,

$$y(n) = \sum_{m=-\infty}^{\infty}\left[-\frac{1}{8}(-0.25)^m + \frac{1}{4}(-0.5)^m\right]x(n-m)$$
$$+ \frac{1}{8}(-0.25)^n - \frac{1}{4}(-0.5)^n$$

4.6 SUMMARY

1. The Laplace transformation and the inverse Laplace transformation of $x(t)$ are

$$X(s) = \mathscr{L}[x(t)] = \int_{0^-}^{\infty} x(t)e^{-st}\,dt$$

$$x(t) = \mathscr{L}^{-1}[X(s)] = \frac{1}{2\pi j}\int_c X(s)e^{st}\,ds$$

2. The z-transformation and the inverse z-transformation are

TABLE 4.1 Transformations $\mathcal{L}[x(t)]$ and $\mathcal{Z}[x(n)]$

$x(t)$	$\mathcal{L}[x(t)]$	$x(n)$	$\mathcal{Z}[x(n)]$
$u_s(t)$	$\frac{1}{s}$	$u_s(n)$	$\frac{z}{z-1}$
$tu_s(t)$	$\frac{1}{s^1}$	$nu_s(n)$	$\frac{z}{(z-1)^2}$
$e^{-at}u_s(t)$	$\frac{1}{s+a}$	$e^{-an}u_s(n)$	$\frac{z}{z-e^{-a}}$
$te^{-at}u_s(t)$	$\frac{1}{(s+a)^2}$	$ne^{-an}u_s(n)$	$\frac{e^{-a}z}{(z-e^{-a})^2}$
$a^t u_s(t)$	$\frac{1}{s-l_n a}$	$a^n u_s(n)$	$\frac{x}{x-a}$
$\delta(t)$	1	$u_I(n)$	1
$\mathcal{L}[tx(t)] = -\frac{d}{ds}\mathcal{L}[x(t)]$		$\mathcal{Z}[nx(n)] = -z\frac{d}{dz}\mathcal{Z}[x(n)]$	
$\mathcal{L}[x_1(t) + ax_2(t)] = \mathcal{L}[x_1(t)] + a\mathcal{L}[x_2(t)]$		$\mathcal{Z}[x_1(n) + \alpha x_2(n)] = \mathcal{Z}[x_1(n)] + \alpha\mathcal{Z}[x_2(n)]$	
$\mathcal{L}[x_1(t) * x_2(t)] = \mathcal{L}[x_1(t)]\mathcal{L}[x_2(t)]$		$\mathcal{Z}[x_1(n) * x_2(n)] = \mathcal{Z}[x_1(n)]\mathcal{Z}[x_2(n)]$	
$\mathcal{L}[x_1(t)x_2(t)] = \mathcal{L}[x_1(t)] * \mathcal{L}[x_2(t)]$		$\mathcal{Z}[x_1(n)x_2(n)] = \mathcal{Z}[x_1(n)] * \mathcal{Z}[x_2(n)]$	
$\mathcal{L}\left[\frac{dx(t)}{dt}\right] = s\mathcal{L}[x(t)] - x(0)$		$\mathcal{Z}[x(n-1)] = z^{-1}\mathcal{Z}[x(n)] + x(-1)$	
$\mathcal{L}\left[\frac{d^k x(t)}{dt^k}\right] = s^k\mathcal{L}[x(t)] - s^{k-1}x(0^-) - s^{k-2}x^{(1)}(0^-) - \cdots - x^{(k-1)}(0^-)$		$\mathcal{Z}[x(n-k)] = z^{-k}\mathcal{Z}[x(n)] + z^{-k+1}x(-1) + z^{-k+2}x(-2) + \cdots + x(-k)$	

$$X(z) = \mathcal{Z}[x(n)] = \sum_{n=0}^{\infty} x(n)z^{-n}, \qquad \text{where } z = e^{(\alpha+j\omega)}$$

$$x(n) = \mathcal{Z}^{-1}[X(z)] = \frac{1}{2\pi j}\int_c X(z)z^{n-1}\,dz$$

3. Table 4.1 shows the Laplace transformation and the z-transformation.
4. A pole of a rational function $Y(s)$ is the value of s, which makes the absolute value of $Y(s) = \infty$.
5. When r_p is q-multiple pole, a rational function $Y(s)$ can be expressed as

$$Y(s) = \frac{K_q}{(s-r_p)^q} + \frac{K_{q-1}}{(s-r_p)^{q-1}} + \cdots + \frac{K_{q-m}}{(s-r_p)^{q-m}} + \cdots + \frac{K_1}{s-r_p} + Y_1(s) \tag{4.48}$$

where $Y_1(s)$ does not have r_p as a pole, and K_{q-m} is

$$K_{q-m} = \frac{1}{m!}\frac{d^m}{ds^m}\left[Y(s)(s - r_p)^q\right]\Big|_{s=r_p}$$

except when $m = 0$

$$K_q = Y(s)(s - r_p)^q\big|_{s=r_p}$$

When the degree m of the numerator is smaller than the degree n of the denominator, expressing $Y(s)$ by using Equation 4.48 for all poles of $Y(s)$ is called a "partial fraction expansion" of $Y(s)$.

When $m > n$, dividing the numerator by the denominator to obtain

$$Y(s) = K_u s^u + K_{u-1}s^{u-1} + \cdots + K_0 + Y_1(s)$$

where $u = m - n$. Then $Y_1(s)$ can be expressed in the form of a partial fraction expansion which is called a "generalized partial fraction expansion."

4.7 PROBLEMS

1. Obtain the Laplace transformation of the following functions:

a. $t^2 u_s(t)$ **b.** $t^3 e^{-st} u_s(t)$ **c.** $5 \sin(2t) u_s(t)$
d. $\sin(5 + 2t) u_s(t)$ **e.** $\cos^2(at) u_s(t)$ **f.** $t \sin(2t) u_s(t)$
g. $(t - 2e^t) u_s(t)$ **h.** $[\sin 2t - \cos 2t] u_s(t)$

2. Obtain the inverse Laplace transformation of the following:

a. $\dfrac{s^2}{(s + a)^3}$ **b.** $\dfrac{s}{s^2 + 2s + 2}$ **c.** $\dfrac{s + 1}{s(s + 2)}$

d. $\dfrac{s + 2}{s^2 + s}$ **e.** $\dfrac{s + 2}{s^2 + 2s + 2}$ **f.** $\dfrac{s^3 + 2s^2 + s}{s^3 + 3s^2 + 2s}$

g. $\dfrac{s^3 + s^2 + s + 1}{s^4 + 2s^3 + s^2 + 2s}$ **h.** $\dfrac{s}{s^2 + 1}$ **i.** $\dfrac{s^2}{s^3 + s^2 + s + 1}$

3. Solve the following differential equations by using the Laplace transformation:

a. $\dfrac{d^2y(t)}{dt^2} + 3\dfrac{dy(t)}{dt} + 2y(t) = e^{3t}u_s(t), \quad y(0^-) = 1,\ y^{(1)}(0^-) = 0$

b. $\dfrac{d^2y(t)}{dt^2} + 3\dfrac{dy(t)}{dt} + 2y(t) = e^{3t}u_s(t), \quad y(0^-) = 0,\ y^{(1)}(0^-) = 1$

c. $\frac{d^2y(t)}{dt^2} + 3\frac{dy(t)}{dt} + 2y(t) = e^{3t}u_s(t), \quad y(0^-) = 2,\ y^{(1)}(0^-) = 1$

d. $\frac{d^2y(t)}{dt^2} + 5\frac{dy(t)}{dt} + 4y(t) = 5u_s(t), \quad y(0^-) = 0,\ y^{(1)}(0^-) = 1$

e. $\frac{d^2y(t)}{dt^2} + 5\frac{dy(t)}{dt} + 4y(t) = 5u_s(t), \quad y(0^-) = 1,\ y^{(1)}(0^-) = 0$

f. $\frac{d^2y(t)}{dt^2} + 2y(t) = 0, \quad y(0^-) = 0,\ y^{(1)}(0^-) = 1$

g. $\frac{d^2y(t)}{dt^2} + y(t) = 1, \quad y(0^-) = 0,\ y^{(1)}(0^-) = 0$

4. Obtain the z-transformation of the following discrete functions:

a. $n^2u_s(n)$ b. $(e^{-an} + n)u_s(n)$
c. $\sin(an)u_s(n)$ d. $2\cos(3n)u_s(n)$
e. $5n\sin(2n)u_s(n)$ f. $n^3e^{2n}u_s(n)$
g. $(\sin n + \cos(2n))u_s(n)$ h. $e^{2n}\sin(3n)u_s(n)$
i. $ne^{2n}\sin(2n)u_s(n)$

5. Obtain the inverse z-transformation of the following:

a. $\frac{z}{z-5}$ b. $\frac{z}{z-e^2}$ c. $\frac{2z}{(z-2)^2}$

d. $\frac{z^2}{(z-3)^2}$ e. $\frac{z^3 - z}{(z-3)(z^2-4z+3)}$ f. $\frac{z(z-2)}{(z-1)(z-3)}$

g. $\frac{1}{(z-3)(z+2)}$ h. $\frac{(5z-3)z}{z^2-4}$ i. $\frac{\sin 5}{z^{-1}-5}$

6. Solve the following difference equations by using z-transformations:

a. $y(n-2) + 3y(n-1) + 2y(n) = 3^nu_s(n), \quad y(-1) = -1,\ y(-2) = 0$
b. $y(n-2) + 3y(n-1) + 2y(n) = 3^nu_s(n), \quad y(-1) = 0,\ y(-2) = 1$
c. $y(n-2) + 3y(n-1) + 2y(n) = 3^nu_s(n), \quad y(-1) = 0,\ y(-2) = 0$
d. $y(n-2) + 5y(n-1) + 4y(n) = 5^nu_s(n), \quad y(-1) = 0,\ y(-2) = 0$
e. $y(n-2) + 5y(n-1) + 4y(n) = 5^nu_s(n), \quad y(-1) = 1,\ y(-2) = 0$
f. $y(n-2) + 2y(n) = 0, \quad y(-1) = 1,\ y(-2) = 0$
g. $y(n-2) - y(n) = u_s(n), \quad y(-1) = 0,\ y(-2) = 0$
h. $y(n-2) - 4y(n) = 2^nu_s(n), \quad y(-1) = 0,\ y(-2) = 0$

5
Transfer Function and Signal Flow Graph

5.1 SIGNAL FLOW GRAPH

To show that signal flow graphs are useful in analysis and design of digital systems, we will review a signal flow graph that readers may employ in analyses of analog systems. A signal flow graph consists of nodes and branches where directions are indicated by arrows. A node x_r indicates a signal x_r, and there is a constant or a variable assigned to each branch called a weight of the branch. A signal is transmitted by a branch in the direction given by the arrow of the branch and the amount of the signal transmitted by the branch is equal to a signal entering to the branch times the weight of the branch. For convenience, we may use a symbol a_i assigned to each branch so that we can use it as the name of a branch as well as the weight of the branch. Let us take a signal flow graph in Fig. 5.1. The amount of a signal transmitted by branch a_1 to node x_4 is a_1x_1 and that of a signal transmitted by branch a_2 to node x_4 is a_2x_2. This signal flow graph consists of nodes x_1, x_2, x_3, and x_4, and branches whose weights are a_1, a_2, a_3, b_1, b_2, b_3, and b_4.

The total signal received at a node x_p is the sum of signals transferred by the all branches entering at the node x_p. A signal that exits from a node x_p is x_p, which must be equal to the total signal entering at the node x_p. In a signal flow graph in Fig. 5.1, the signals entering at node x_4 are a_1x_1 from

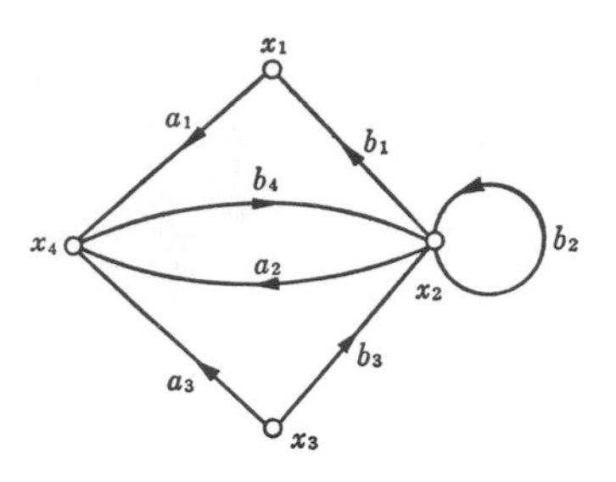

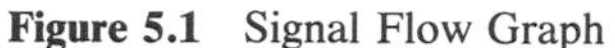

Figure 5.1 Signal Flow Graph

node x_1, a_2x_2 from the node x_2, and a_3x_3 from the node x_3. Hence, the total signal entering the node x_4 is $a_1x_1 + a_2x_2 + a_3x_3$, which must be equal to the signal exit from the node x_4. Thus

$$x_4 = a_1x_1 + a_2x_2 + a_3x_3$$

which is the equation associated with the node x_4. Similarly, b_1x_2 is the only signal entering at node x_1. Hence, the equation associated with node x_1 is

$$x_1 = b_1x_2$$

Likewise, the equation associated with node x_2 is

$$x_2 = b_2x_2 + b_3x_3 + b_4x_4$$

Notice that there is a branch b_2 that is leaving from node x_2 and entering at the same node x_2. Such a branch is called a "self-loop," and the right-hand side of the equation associated with node x_2 contains a term b_2x_2 because b_2x_2 is one of the signals entering the node x_2.

This time let us construct a signal flow graph from a set of equations. Suppose a set of simultaneous equations

$$\left.\begin{aligned} x_0 &= x_1 + 2x_2 + 3x_3 \\ x_1 &= 4x_1 + 5x_2 + 8x_3 \\ x_2 &= 7x_1 + 6x_2 + 9x_3 \end{aligned}\right\} \tag{5.1}$$

is given. There are four variables x_0, x_1, x_2, and x_3 in these equations; hence, we use four nodes x_0, x_1, x_2, and x_3 as shown in Fig. 5.2(a). Taking the first equation $x_0 = x_1 + 2x_2 + 3x_3$, $1x_1$ in the right-hand side of the equation indicates that there is a branch from node x_1 to node x_0 with its weight equal to 1 as shown in Fig. 5.2(b).

The second term $2x_2$ in the right-hand side of the equation $x_0 = x_1 + 2x_2 + 3x_3$ means that there is a branch from node x_2 to node x_0 with the weight 2. The last term $3x_3$ in $x_0 = x_1 + 2x_2 + 3x_3$ indicates that a branch with the weight 3 must be located from node x_3 to node x_0 as shown in Fig. 5.2(c), and there are no branches other than those three branches entering at the node x_0.

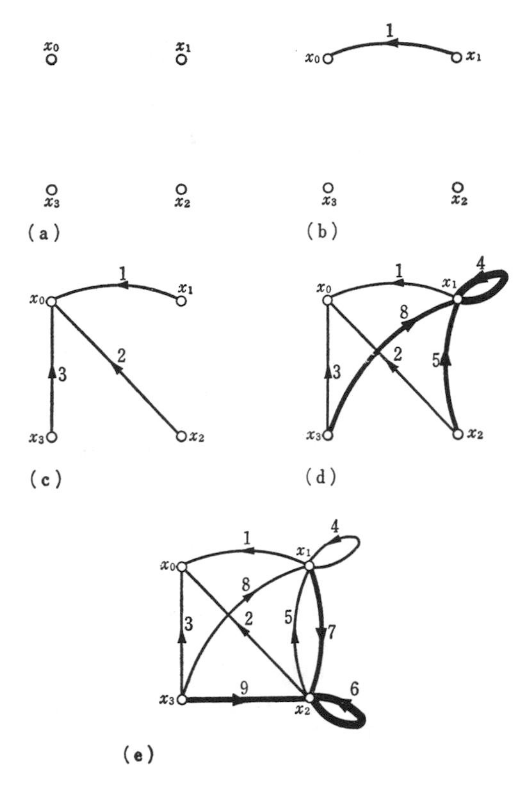

Figure 5.2 Signal Flow Graph (a) Vertices x_0, x_1, x_2, and x_3; (b) Edge from x_1 to x_0; (c) $x_0 = x_1 + 2x_2 + 3x_3$; (d) Signal Flow Graph Representing Two Equations; (e) Signal Flow Graph Representing Equation 5.1.

For the second equation, the first term $4x_1$ in the right hand side of $x_1 = 4x_1 + 5x_2 + 8x_3$ indicates that there is a branch whose weight is 4 and that is connected from node x_1 to the same node x_1. For the second term $5x_2$, a branch whose weight is 5 must be connected from node x_2 to node x_1. For the last term $8x_3$, a branch should be from node x_3 to node x_1 with the weight 8. Notice that no other branches must be entered at node x_1 as shown in Fig. 5.2(d) so that corresponding equation at node x_1 is $x_1 = 4x_1 + 5x_2 + 8x_3$.

By the last equation $x_2 = 7x_1 + 6x_2 + 9x_3$, there is a branch whose weight is 7 which is connected from node x_1 to node x_2, there is a self-loop connected at the node x_2 whose weight is 6, and there is a branch of the weight 9 connected from node x_3 to node x_2 as shown in Fig. 5.2(e). Because equations associated with nodes x_0, x_1, and x_2 in this signal flow graph are the given set of equations, a signal flow graph in Fig. 5.1(e) is the signal flow graph of the given simultaneous equations. Notice that there are no branches entering at the vertex x_3 in this signal flow graph. Hence, there is no equation associated with the node x_3. Such a node is called a "source." In other words, a source is a node in a signal flow graph at which all branches connected are those leaving from the node. A property of signal flow graphs is that there is at least one source in a signal flow graph.

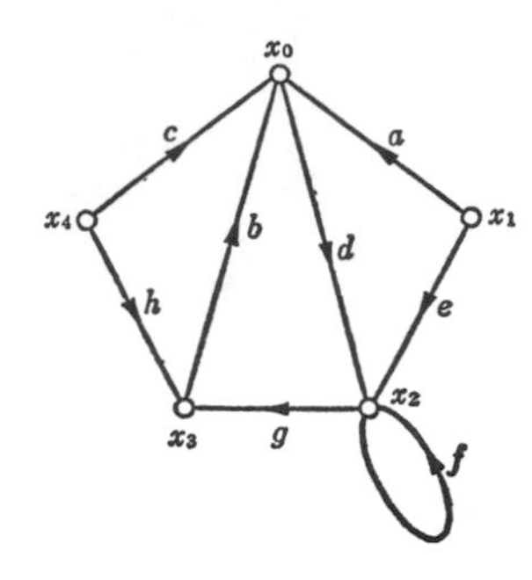

Figure 5.3 Signal Flow Graph

Suppose there are n_s sources in a signal flow graph having n_v nodes. Because there is one equation associated with each vertex that is not a source, the signal flow graph represents $n_v - n_s$ simultaneous equations. For example, a signal flow graph in Fig. 5.2(e) consists of four nodes. Because all branches connected at the node x_3 are those leaving from the node x_3, the node x_3 is a source. There are no other nodes that are sources in this signal flow graph. Hence, this signal flow graph represents 4 − 1 simultaneous equations.

Consider a signal flow graph in Fig. 5.3 where there are five nodes x_0, x_1, x_2, x_3, and x_4. Because x_1 and x_4 are sources, the signal flow graph represents 5 − 2 simultaneous equations as

$$x_0 = ax_1 + bx_3 + cx_4$$

$$x_2 = dx_0 + ex_1 + fx_2$$

$$x_3 = gx_2 + hx_4$$

Again, consider a signal flow graph in Fig. 5.2(e). All branches connected at a node x_0 are entering at the node x_0. Such a node is called a "sink." In other words, a sink is a node at which all branches are entering. A signal flow graph in Fig. 5.3 has no sink.

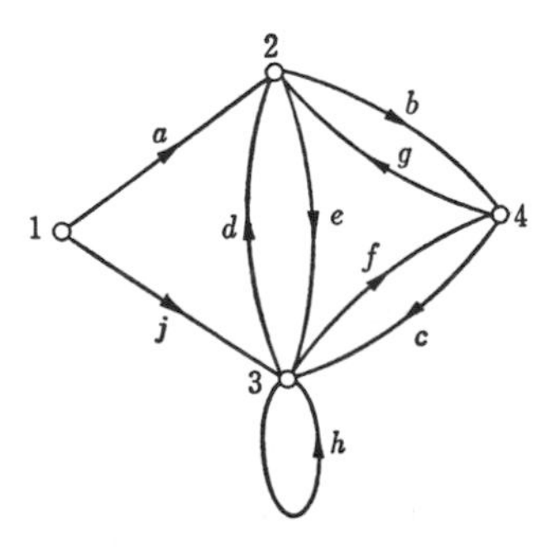

Figure 5.4 Graph G

5.2 MASON'S FORMULA

Suppose x_m is a source, and x_0 is not a source in a signal flow graph G. Then a ratio

$$x_0 = Hx_m \tag{5.2}$$

is called a ''transfer function''.* A formula by which a transfer function H can be obtained directly from a signal flow graph G is known as the ''Mason's formula.'' Before explaining the Mason's formula, we will study necessary terminologies.

From a chosen node p in a graph G, if we trace branches one by one in the direction coinciding with the arrow of the branch and reach at another chosen node q such that each node can not be passed more than once, the collection of traced branches is called a ''forward path'' from the node p to the node q.

For example, let us obtain a forward path from node 1 to node 4 in a signal flow graph G in Fig. 5.4. From node 1, suppose we trace the branch a and reach at the node 2. Then tracing branch b, we can reach at node 4. Hence, a forward path from 1 to 4 is a set of branches a and b as shown in Fig. 5.5(a). After tracing branch a to reach at the vertex 2, if we trace branch e to reach at the vertex 3, then trace branch d to reach at the vertex 2 and trace branch b, however, we will reach the vertex 4 as shown in Fig. 5.5(b). The set of branches a, e, d, and b is not a forward path from 1 to 4 because the node 2 is passed twice by this trace. After tracing branch a and branch e, if we trace branch f to reach at the vertex 4 as shown in Fig. 5.5(c), the set of traced branches a, e, and f is a forward path from 1 to 4. Another forward path from 1 to 4 consists of branches j, d, and b, the branches j and f form the other forward path from 1 to 4.

For a forward path from a node p to a node q, *the node p* is called the ''initial node,'' and the node q is called the ''final node.'' When the initial

* It is also known as a ''transfer response.''

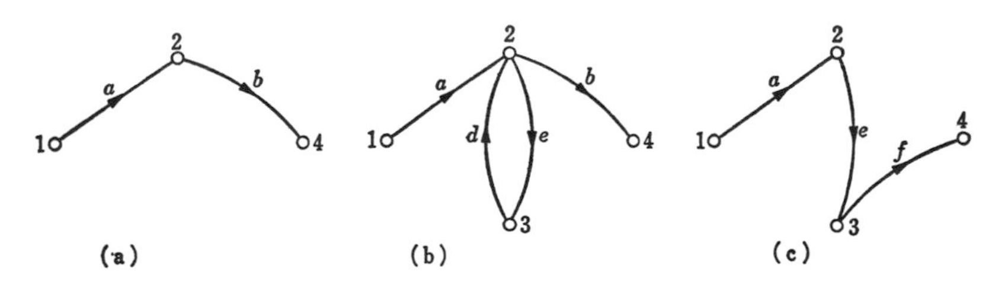

Figure 5.5 Forward Path and Non-Forward Path. (a) Forward Path; (b) Not a Forward Path; (c) Forward Path.

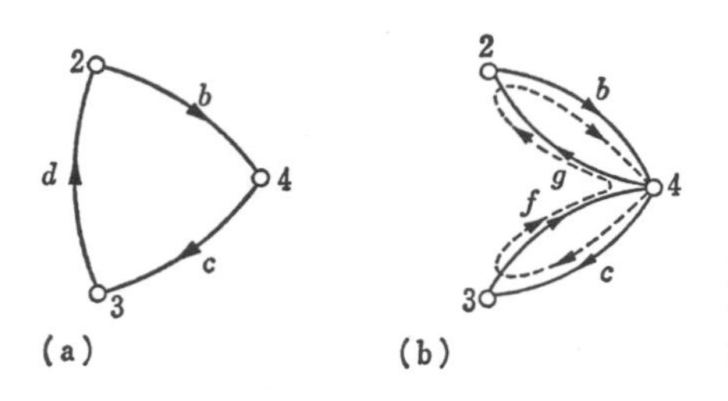

Figure 5.6 Closed Loop and Non-Closed Loop. (a) Closed Loop; (b) Not a Closed Loop.

node and the final node of a forward path are the same, the set of branches in this forward path is called a "closed loop." In other words, starting from an initial node, if we trace branches one by one in the direction coinciding with the arrow of each branch and back to the initial node such that each node cannot be passed more than once, then a set of traced branches is a closed loop.

Let us obtain closed loops in a signal flow graph in Fig. 5.4. From a node 2, suppose we trace the branch b and reach at the node 4. Then we trace the branch c to the node 3 and trace branch d by which we will reach at the initial node 2 as shown in Fig. 5.6(a). Clearly these edges b, c, and d form a closed loop. After tracing branch b from the node 2, if we trace branches c, f, and g, we can reach at the initial node 2 as shown in Fig. 5.6(b). This trace passes the vertex 4 more than once, however. Hence, the edges b, c, f, and g will not form a closed loop. Fig. 5.7 shows all closed loops in a signal flow graph in Fig. 5.4. Notice that a self-loop is a closed loop.

The Mason's formula for a transfer function H in Equation 5.3 is

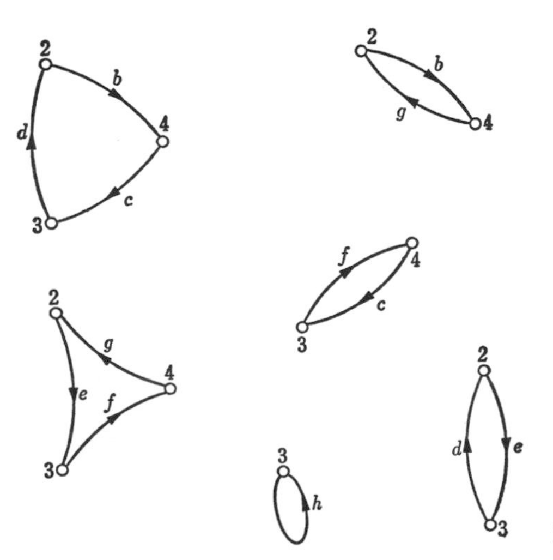

Figure 5.7 Closed Loops

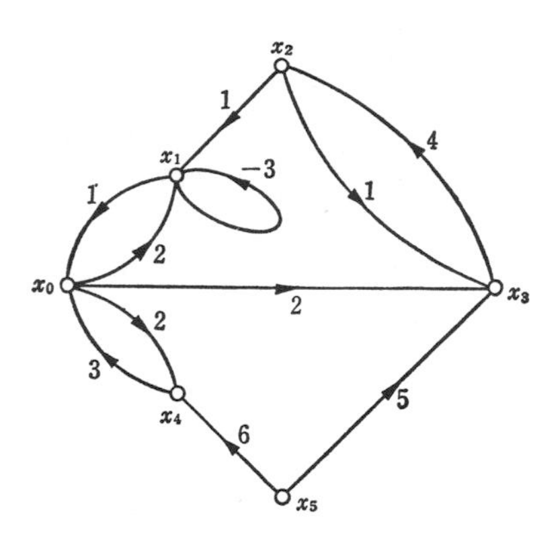

Figure 5.8 Signal Flow Graph

$$\frac{x_0}{x_m} = H = \frac{\sum\limits_{(k)} \Pi P_{m0k}[1 - C_1 + C_2 - C_3 + \cdots \text{ of } G(\overline{P}_{m0k})]}{1 - C_1 + C_2 - C_3 + \cdots + (-1)^r C_r + \cdots} \tag{5.4}$$

where the symbols used in this formula are as follows: The symbol C_r in the denominator, $1 - C_1 + C_2 - \cdots + (-1)^r C_r + \cdots$ is

$$C_r = \sum_{(k)} (\Pi L_{rk}) \tag{5.5}$$

where the symbol L_{rk} is a set of r closed loops, any two of which cannot have a node in common. The symbol ΠL_{rk} is the product of the weights of all branches in L_{rk} and $\sum\limits_{(k)}$ indicates the summation of all ΠL_{rk}. When $r = 1$, Equation 5.5 becomes

$$C_1 = \sum_{(k)} (\Pi L_{1k})$$

where ΠL_{1k} is the product of the weights of all branches in a closed loop L_{1k}.

As an example, consider a signal flow graph G in Fig. 5.8. All closed loops in G are L_{11}, L_{12}, L_{13}, L_{14}, and L_{15} shown in Fig. 5.9. Hence, C_1 is

$$C_1 = \sum_{k=1}^{5} (\Pi_{1k}) = (1)(2) + (-3) + (1)(4) + (2)(3) + (1)(1)(2)(4) = 17$$

C_2 in the denominator of the Mason's formula is

$$C_1 = \sum_{(k)} (\Pi L_{rk})$$

L_{2k} is a set of two closed loops that cannot have a node in common. Hence, there are four sets L_{21}, L_{22}, L_{23}, and L_{24} in the signal flow graph in Fig. 5.8 as shown in Fig. 5.10. These are obtained by taking two of L_{11}, L_{12}, L_{13}, L_{14}, and L_{15} in Fig. 5.9. For example, L_{21} is L_{11} and L_{13}. Because L_{11} and L_{12}

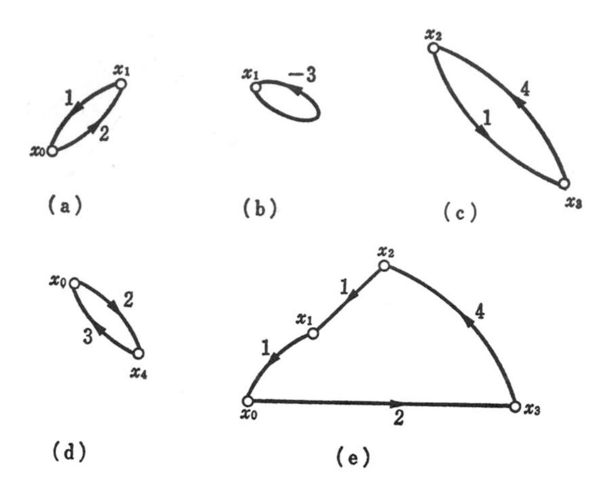

Figure 5.9 Closed Loops L_{1k}. (a) L_{11}; (b) L_{12}; (c) L_{13}; (d) L_{14}; (e) L_{15}.

have the node x_1 in common, they cannot be a L_{2k}. From these four L_{2k}, C_2 is

$$C_2 = \sum_{k=1}^{4} (\Pi L_{2k}) = (1)(2)(1)(4) + (-3)(1)(4) + (-3)(2)(3) + (2)(3)(1)(4) = 2$$

L_{3k} is a set of three closed loops, any two of which cannot have a node

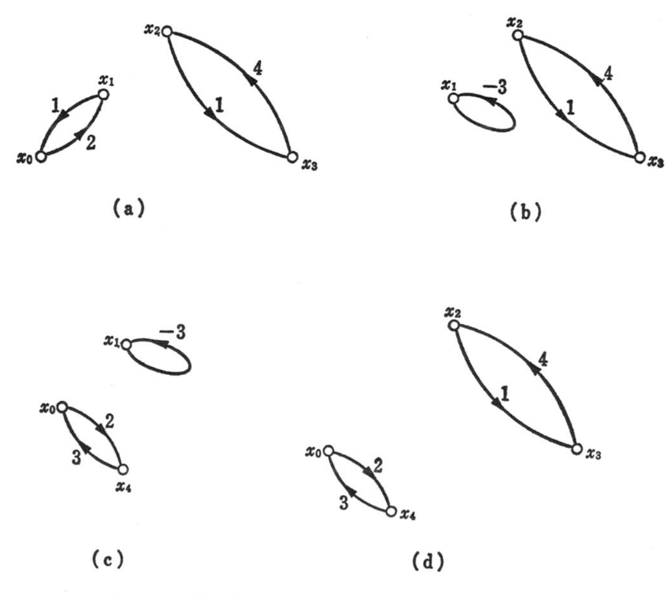

Figure 5.10 L_{2k}. (a) L_{21}; (b) L_{22}; (c) L_{23}; (d) L_{24}.

in common. In the signal flow graph G in Fig. 5.8, there is one L_{31} formed by L_{12}, L_{13}, and L_{14} as shown in Fig. 5.11. Hence, C_3 is

$$C_3 = (-3)(1)(4)(2)(3) = -72$$

There are no L_{4k} in G in Fig. 5.8, which is a set of four closed loops, any two of which do not have a node in common. It is clear that if there are no L_{mk}, then there are no L_{rk} for $r > m$. Hence, $C_m = 0$ means that $C_r = 0$ for $r > m$. Because there are no L_{4k} in G in Fig. 5.8, $C_4 = 0$ and $C_r = 0$ for $r > 4$. Thus, the denominator of Equation 5.4 is

$$1 - C_1 + C_2 - C_3 + \cdots = 1 - 17 + 2 - (-72) = 58$$

The numerator of the Mason's formula in Equation 5.4 is

$$\sum_{(k)} \Pi P_{m0_k}[1 - C_1 + C_2 - \cdots \text{ of } G(\overline{P}_{m0_k})]$$

where P_{m0_k} is a forward path from x_m to x_0, and ΠP_{m0_k} is the product of the weight of all branches in P_{m0_k}. The symbol $G(\overline{P}_{m0_k})$ is a signal flow graph obtained from G by deleting all nodes in P_{m0_k}. Notice that deleting nodes means to delete nodes and all branches connected at these nodes. The symbol $\sum_{(k)}$ is the summation of all $\Pi P_{m0_k}[1 - C_1 + C_2 - \cdots \text{ of } G(\overline{P}_{m0_k})]$.

As an example, let us obtain the numerator of the Mason's formula for the signal flow graph in Fig. 5.8. First, there are two forward paths P_{50_1} and P_{50_2} from the node x_5 to the node x_0 as shown in Fig. 5.12. A signal flow graph $G(\overline{P_{50_1}})$ is obtained from G in Fig. 5.8 by deleting nodes x_0, x_4, and x_5 and all branches connected at these nodes. Hence, $G(\overline{P_{50_1}})$ is one shown in Fig. 5.13(a). To obtain $1 - C_1 + C_2 - \cdots$ of $G(\overline{P_{50_1}})$, we must find all closed loops in $G(\overline{P_{50_1}})$. There are L_{11} and L_{12} as shown in Fig. 5.13(b). Hence, C_1 is

$$C_1 = (-3) + (1)(4) = 1$$

Because these closed loops L_{11} and L_{12} have no nodes in common, they can form L_{21}. Hence C_2 is

$$C_2 = (-3)(1)(4) = -12$$

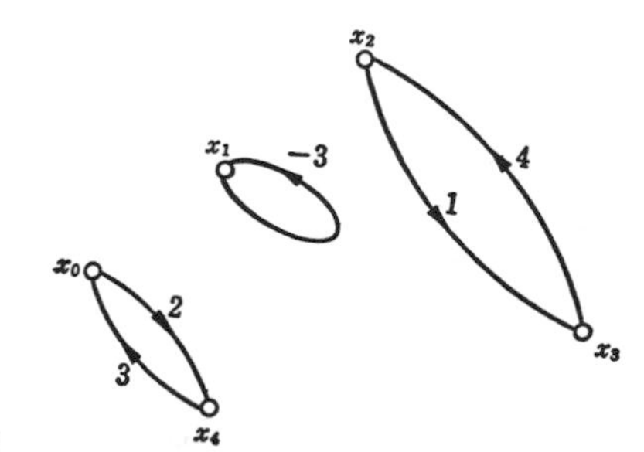

Figure 5.11 L_{31}

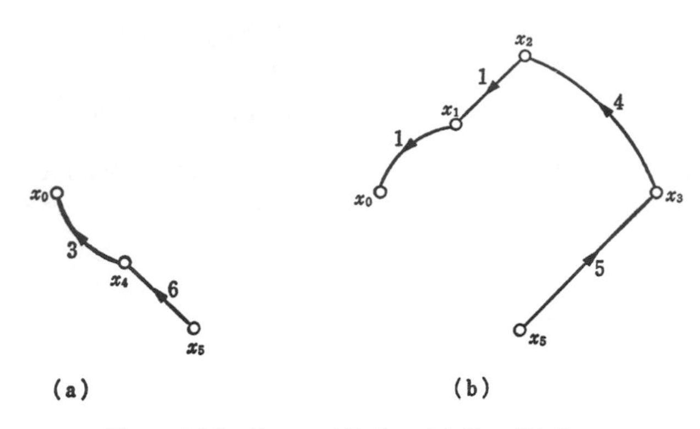

Figure 5.12 Forward Paths. (a) P_{50_1}; (b) P_{50_2}.

Thus, $1 - C_1 + C_2 - \cdots$ of $G(\overline{P_{50_1}})$ is

$$1 - C_1 + C_2 \text{ of } G(\overline{P_{50_1}}) = 1 - (1) + (-12) = -12$$

and $\Pi P_{50_1}[1 - C_1 + C_2 - \cdots$ of $G(\overline{P_{50_1}})]$ is

$$\Pi P_{50_1}[1 - C_1 + C_2 - \cdots \text{ of } G(\overline{P_{50_1}})] = (3)(6)(-12) = -216$$

Because the forward path P_{50_2} contains all nodes of G in Fig. 5.8, $G(\overline{P_{50_2}})$ is empty. Hence, $1 - C_1 + C_2 - \cdots$ of $G(\overline{P_{50_2}})$ is 1. Thus, $\Pi P_{50_2}[1 - C_1 + C_2 - \cdots$ of $G(\overline{P_{50_2}})]$ is

$$\Pi P_{50_2}[1 - C_1 + C_2 - \cdots \text{ of } G(\overline{P_{50_2}})] = (5)(4)(1)(1) = 20$$

From these results, the numerator of the Mason's formula for the signal flow graph in Fig. 5.8 is

$$\sum_{k=1}^{2} \Pi P_{50_k}[1 - C_1 + C_2 - \cdots \text{ of } G(\overline{P_{50_h}})] = -216 + 20 = -196$$

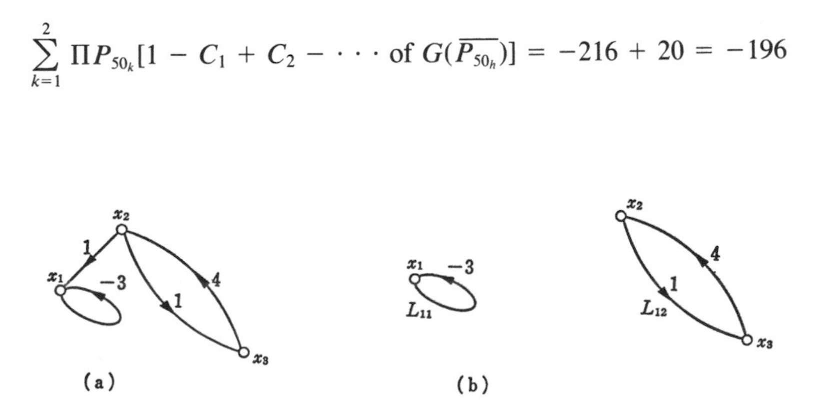

Figure 5.13 $G(\overline{P}_{50_1})$ and Closed Loops. (a) $G(\overline{P}_{50_1})$; (b) Closed Loops in $G(\overline{P}_{50_1})$.

Thus, a transfer function $H = x_0/x_5$ of the signal flow graph in Fig. 5.8 is (by Equation 5.4)

$$\frac{x_0}{x_5} = \frac{\sum_{k=1}^{2} P_{50_k}[1 - C_1 + C_2 - \cdots \text{ of } G(\overline{P_{50_k}})]}{1 - C_1 + C_2 - \cdots} = \frac{-196}{58}$$

5.3 SIGNAL FLOW GRAPH FOR DISCRETE TRANSFER FUNCTION

It is very important to know the process necessary to calculate a solution $y(n)$ of a difference equation for $n = 0, 1, 2, \ldots$ to design a suitable system of obtaining $y(n)$. It is known that by the use of graphic symbols for addition, a constant multiplier, and a unit delay as shown in Fig. 5.14, we can draw a diagram showing the process for obtaining a solution $y(n)$ of a difference equation. For example, a difference equation

$$y(n) = ay(n-1) + by(n-2) + cx(n) \tag{5.6}$$

can be indicated by a diagram in Fig. 5.15.

In Table 4.1, showing z-transformation, z^{-1} corresponds to a unit delay. By the use of z^{-1} as a constant to indicate a unit delay, the diagram in Fig. 5.15 can be simplified as a graph in Fig. 5.16. The symbol "+" showing an addition in a diagram is just a node in a signal flow graph. Hence, employing a signal flow graph to express a difference equation, a graph in Fig. 5.16 can be simplified as shown in Fig. 5.17. Furthermore, we will make a rule that when a weight of a branch is 1, the weight need not be indicated. In

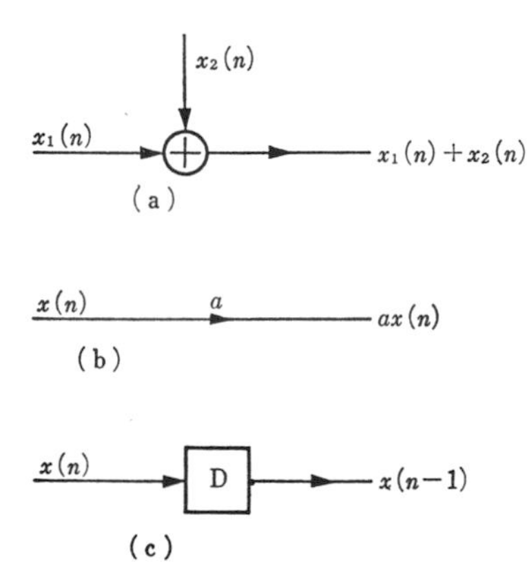

Figure 5.14 Basic Operations. (a) Addition; (b) Constant Multiplication; (c) Unit Delay.

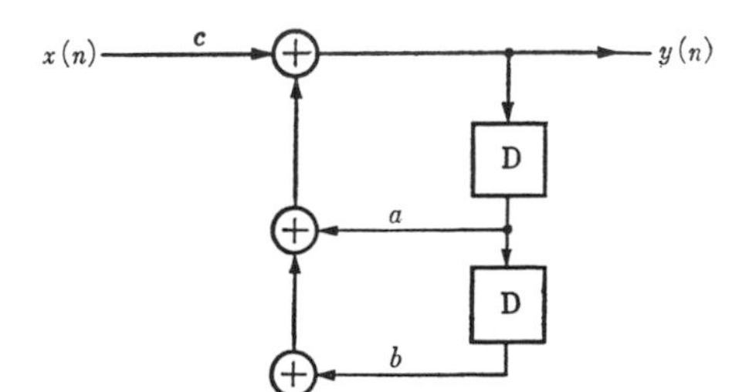

Figure 5.15 Diagram of Difference Equation

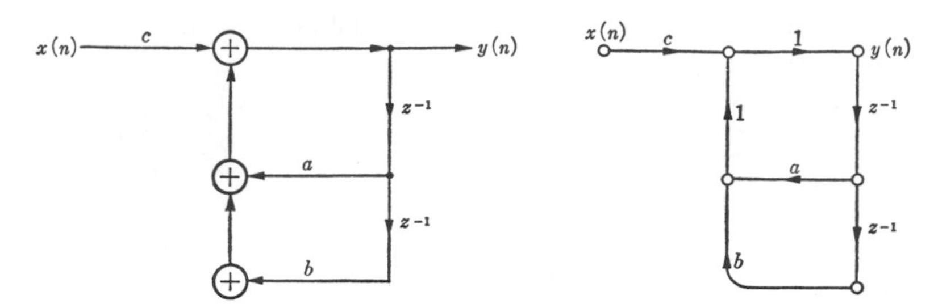

Figure 5.16 Replacing a Unit Delay by z^{-1}

Figure 5.17 Signal Flow Graph

other words, if a branch in a signal flow graph has no weight, it means that the weight of the branch is 1. By this rule, a signal flow graph in Fig. 5.17 can be simplified further as in Fig. 5.18, which corresponds to the difference equation in Equation 5.6.

Changing structure of a signal flow graph by maintaining the same relationship between an input signal x and an output signal y is known as an "equivalent transformation." Notice that a signal flow graph G and a signal flow graph G' obtained from G by applying an equivalent transformation will have different processes of obtaining an output signal y. Hence, it would be possible to obtain a signal flow graph possessing a desirable process of calculating an output signal y by the use of equivalent transformations.

Fig. 5.19 shows typical equivalent transformations, which I will explain one by one.

Type I changes two branches in series into one branch whose weight is equal to the product of the weights of two branches. To show that this change is valid, let x be the common node of the two branches in series. The equation at the node x is $x = ax_1$, and the equation at the node x_j is

$$x_j = bx + \cdots\cdots$$

Substituting $x = ax_i$ into this equation, we have

$$x_j = abx_i + \cdots\cdots$$

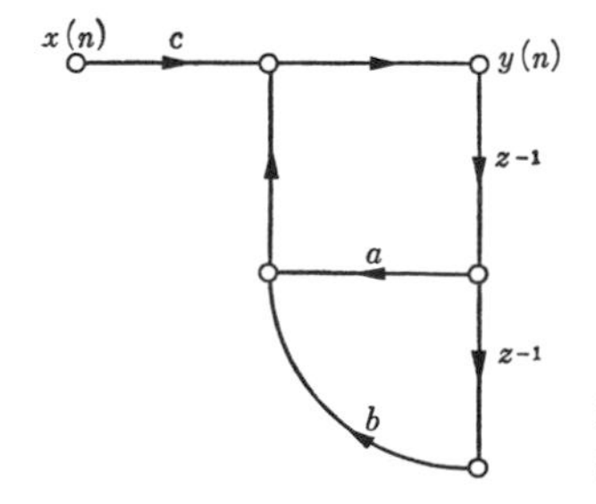

Figure 5.18 Signal Flow Graph of Difference Equation

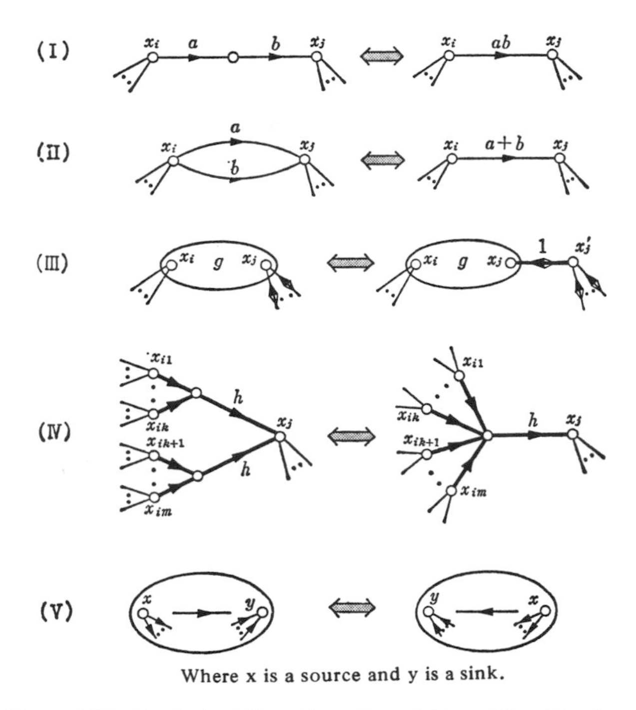

Figure 5.19 Equivalent Transformation of Signal Flow Graph

It is clear that when we replace these two branches by one with its weight equal to ab, the equation at node x_j is equal to the preceding equation; equations at all other remaining nodes are unaltered. Hence, this change is an equivalent transformation.

Type II replaces two branches in parallel to one branch whose weight is equal to the sum of the weights of these two branches. The equation at node x_j before changing two branches to one is

$$x_j = ax_i + bx_i + \cdots$$

and the equation at node x_j after changing two branches to one branch is

$$x_j = (a + b)x_i + \cdots$$

Because these two equations are the same, this change is an equivalent transformation.

Type III is when branches from the outside of a subgraph g connected at a node x are either all entering at the node x_j or all leaving from the node x_j. A signal flow graph G in the left side of Fig. 5.20(a) shows a case in which all branches $b_1, \ldots,$ and b_k from the nodes $x_{01}, \ldots,$ and x_{0k} outside of a subgraph g are entering at the node x_j. The equation at the node x_j in G is

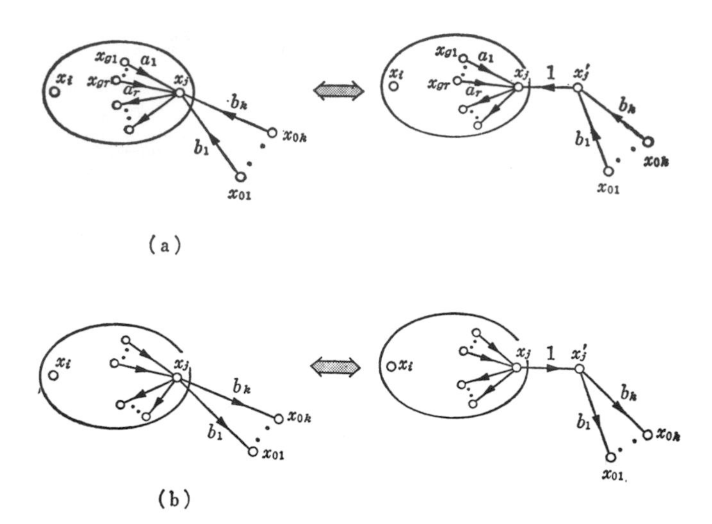

Figure 5.20 Equivalent Transformation of Type III. (a) All Edges from x_{01}, . . . , and x_{0k} Are Entering at x_i; (b) All Edges from x_{10}, . . . , and x_{1k} are Leaving from x_i.

$$x_j = \sum_{n=1}^{r} a_n x_{gn} + \sum_{m=1}^{k} b_m x_{0m} \tag{5.7}$$

The equation at the node x_j of a signal flow graph G' in the right-hand side of Fig. 5.20(a) is

$$x_j = x_j' + \sum_{n=1}^{r} a_n x_{gn} \tag{5.8}$$

and the equation at the node x_j' is

$$x_j' = \sum_{m=1}^{k} b_m x_{0m} \tag{5.9}$$

Substituting the equation for x_j into Equation 5.8 produces an equation identical with the one in Equation 5.7. Hence, inserting a node x_j' and reconnecting all branches b_1, . . . , and b_k from x_j to x_j', and connecting a new branch from x_j' to x_j is an equivalent transformation.

Similarly, when all branches b_1, . . . , and b_k are leaving from node x_j as shown in the left-hand side of Fig. 5.20(b), obtaining a signal flow graph by inserting one node x_j' and relocating all branches b_1, . . . , and b_k leaving from the node x_j', and inserting a new branch connecting from the node x_j to the node x_j' as shown in the right-hand side of Fig. 5.20(b) is an equivalent transformation.

Type IV is a case that combines two branches having the same

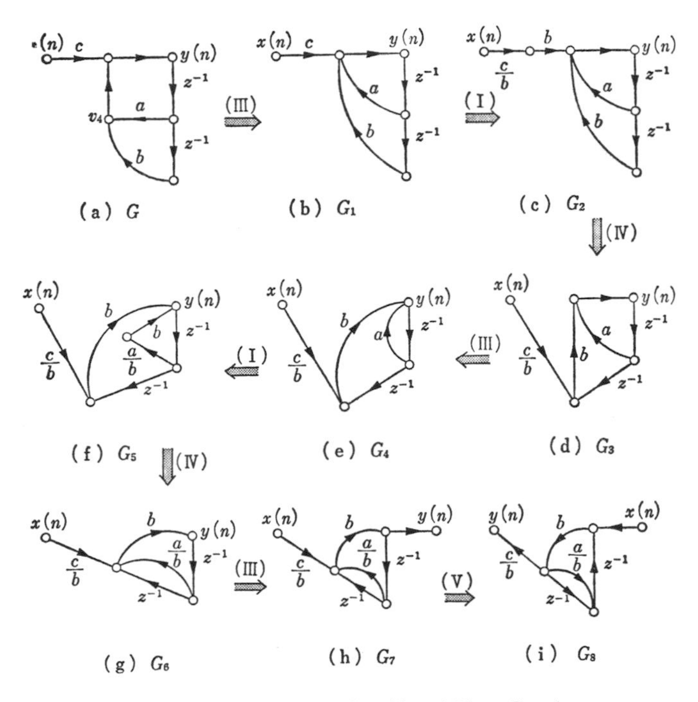

Figure 5.21 Changing Signal Flow Graph

weight. It can be seen that product of the weights of all branches in each forward path from a node x_p to a node x_q of a signal flow graph in the left-hand side and that of a signal flow graph in the right-hand side will be the same. Hence, by Mason's formula, the transformation of type IV is an equivalent transformation.

Type V changes the orientation of all branches, and interchanges a sink and a source. Because the orientation of every branch is reversed, it can be seen easily that each forward path from a source to a sink of a signal flow graph in the left-hand side becomes a forward path from a source to a sink of a signal flow graph in the right-hand side. Also a closed loop will be a closed loop when the orientation of all branches are reversed. Hence, by Mason's formula, we can see that this type is an equivalent transformation.

We have seen that for each type in Fig. 5.19, changing a signal flow graph in the left-hand side to a signal flow graph in the right-hand side is an equivalent transformation. It can be shown easily that for each type in the figure, changing a signal flow graph in the right-hand side to a signal flow graph in the left-hand side is also an equivalent transformation. This is why the symbol $\Leftrightarrow$ is employed in the figure indicating an equivalent transformation.

Using these equivalent transformations, let us change signal flow graphs. A signal flow graph in Fig. 5.18 is shown in Fig. 5.21(a). By the use of type III, we can delete node v as shown in Fig. 5.21(b). Then changing branch c to two branches in series, we have one shown in Fig. 5.21(c). In this signal flow graph, there are two branches having b as their weights. Using type IV, these two branches can be replaced by one branch as shown in Fig. 5.21(d) and so on. This shows that there are many signal flow graphs representing the same difference equation.

We will develop a basic structure of a signal flow graph for a transfer function $H(z)$. Consider a transfer function $H(z)$ of the form

$$H(z) = \frac{y(z)}{x(z)} = \frac{\sum_{k=0}^{M} b_k z^{-k}}{1 - \sum_{k=1}^{N} a_k x^{-k}} \tag{5.10}$$

Changing the equation in Eq. 5.10 as

$$y(z) = \sum_{k=1}^{N} a_k z^{-k} y(z) + \sum_{k=0}^{M} b_k z^{-k} x(z) \tag{5.11}$$

we can obtain a signal flow graph G as shown in Fig. 5.22(a). By Mason's formula, it is easily seen that this signal flow graph gives Equation 5.11. By using an equivalent transformation of type III, the graph G in Fig. 5.22(b) can be obtained. We change two branches entering at node t_0 whose weights are z^{-1} by one branch by type IV of equivalent transformations. In the resultant graph, we change two branches entering at the node t_1 whose weights are z^{-1} by one branch by type IV. Continuing this change by type IV of equivalent transformations, we will have a signal flow graph shown in

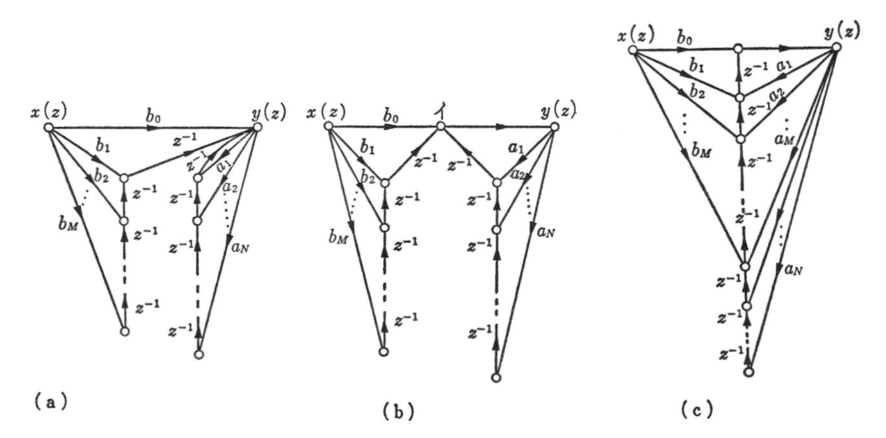

Figure 5.22 Signal Flow Graph. (a) Signal Flow Graph G_1; (b) Graph G_2; (c) Graph G_3.

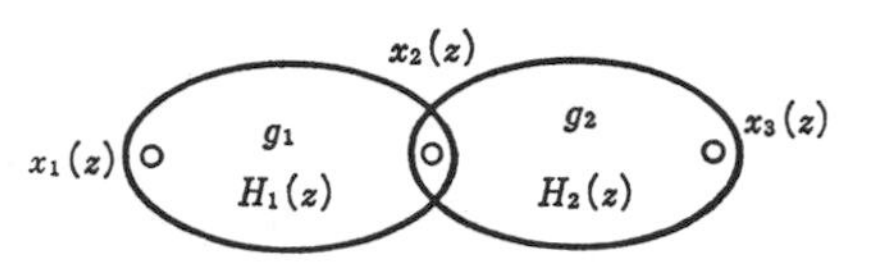

Figure 5.23 g_1 and g_2 in Series

Fig. 5.22(c) where $M < N$ is assumed. Even if $M > N$, however, a similar signal flow graph can be constructed. Notice that the number of z^{-1} in a signal flow graph G_1 is $M + N$ but that of a signal flow graph G_3 is M.

5.4 SERIES AND PARALLEL STRUCTURES OF DISCRETE TRANSFER FUNCTION

Suppose two signal flow graphs g_1 and g_2 are connected in a series as shown in Fig. 5.23. Let $H_1(z)$ be a transfer function $x_2(z)/x_1(z)$ of g_1 and $H_2(z)$ be a transfer function $x_3(z)/x_2(z)$ of g_2. Then a transfer function $x_3(z)/x_1(z)$ of the entire graph will be

$$\frac{x_3(z)}{x_1(z)} = H(z) = \frac{x_3(z)}{x_2(z)}\frac{x_2(z)}{x_1(z)} = H_2(z)H_1(z) \tag{5.12}$$

To generalize this, suppose graphs $g_1, g_2, \ldots,$ and g_m are in a series as shown in Fig. 5.24. Let $H_p(z)$ be a transfer function $x_{p+1}(z)/x_p(z)$ of g_p for $p = 1, 2, \ldots m$. Then a transfer function of a whole signal flow graph in the figure is

$$H(z) = \frac{x_{m+1}(z)}{x_1(z)} = \prod_{r=1}^{m} H_r(z) \tag{5.13}$$

Using this, we can obtain a signal flow graph of a transfer function $H(z)$ by expressing $H(z)$ as

$$H(z) = \frac{\prod_{k=1}^{M_1}(1 - a_k z^{-1}) \prod_{k=1}^{M_2}(1 - b_k z^{-1})(1 - \bar{b}_k z^{-1})}{\prod_{k=1}^{N_1}(1 - c_k z^{-1}) \prod_{k=1}^{N_2}(1 - d_k z^{-1})(1 - \bar{d}_k z^{-1})} \tag{5.14}$$

where a_k is a real zero, and b_k and $\bar{b}_k$ are a pair of complex conjugate zeros.

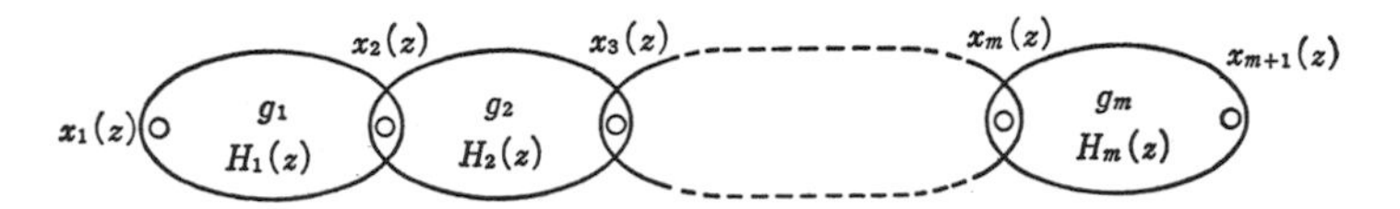

Figure 5.24 Graphs in Series

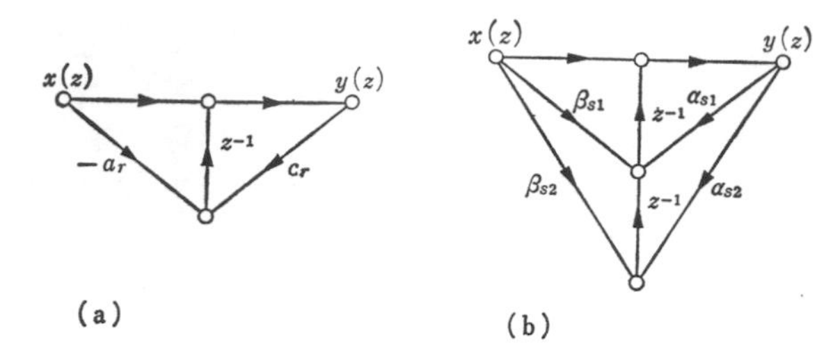

Figure 5.25 Subgraphs used in Series Structure. (a) Graph of $H_{1r}(z)$; (b) Graph of $H_{2s}(z)$.

Also c_k is a real pole, and d_k and $\bar{d}_k$ are a pair of complex conjugate poles. We can choose $H_r(z)$ to make the Eq. 5.14 in the form of Equation 5.13 as

$$\left.\begin{aligned} H_{11}(z) &= \frac{1 - a_1 z^{-1}}{1 - c_1 z^{-1}}, \quad H_{12}(z) = \frac{1 - a_2 z^{-1}}{1 - c_2 z^{-1}}, \ldots, \quad H_{1r}(z) = \frac{1 - a_r z^{-1}}{1 - c_r z^{-1}} \\ H_{21}(z) &= \frac{(1 - b_1 z^{-1})(1 - \bar{b}_1 z^{-1})}{(1 - d_1 z^{-1})(1 - \bar{d}_1 z^{-1})}, \ldots, \quad H_{2s}(z) = \frac{(1 - b_s z^{-1})(1 - \bar{b}_s z^{-1})}{(1 - d_s z^{-1})(1 - \bar{d}_s z^{-1})} \end{aligned}\right\} \tag{5.15}$$

When $M_1 = N_1$ and $M_2 = N_2$, we can consider $H_{1r}(z)$ and $H_{2s}(z)$ as transfer functions of subgraphs g_r and g_s in a signal flow graph in Fig. 5.26. Hence, we only need to show that there are signal flow graphs whose transfer functions are $H_{1r}(z)$ and $H_{2s}(z)$, respectively. A transfer function of a signal flow graph in Fig. 5.25(a) is $H_{1r}(z)$. Hence, we only need to obtain a signal flow graph whose transfer function is $H_{2s}(z)$. By expressing $H_{2s}(z)$ as

$$H_{2s}(z) = \frac{(1 - c_s x^{-1})(1 - c_s^* z^{-1})}{(1 - d_s z^{-1})(1 - d_s^* x^{-1})} = \frac{1 + \beta_{s1} x^{-1} + \beta_{s2} z^{-2}}{1 - \alpha_{s1} x^{-1} - \alpha_{s2} x^{-2}} \tag{5.16}$$

we can see that a signal flow graph in Fig. 5.25(b) has a transfer function equal to the Equation 5.16. Hence, a signal flow graph whose transfer function is $H(z)$ in Equation 5.14 can be one shown in Fig. 5.26.

By modifying the above procedure, we can obtain a signal flow graph whose structure is one shown in Fig. 5.26 even if $M_1 \neq N_1$ or $M_2 \neq N_2$. I will not discuss this point in this book, however.

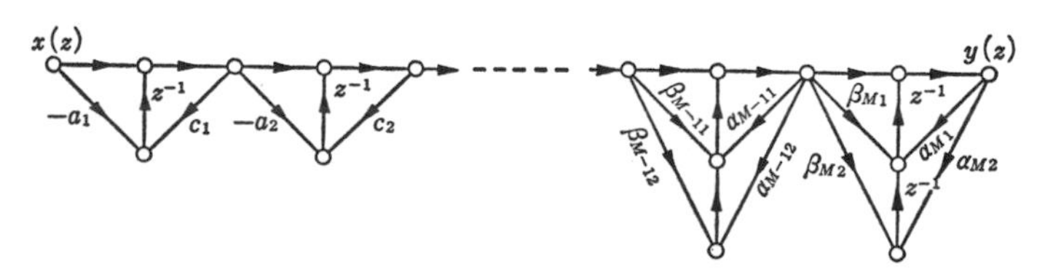

Figure 5.26 Series Structure

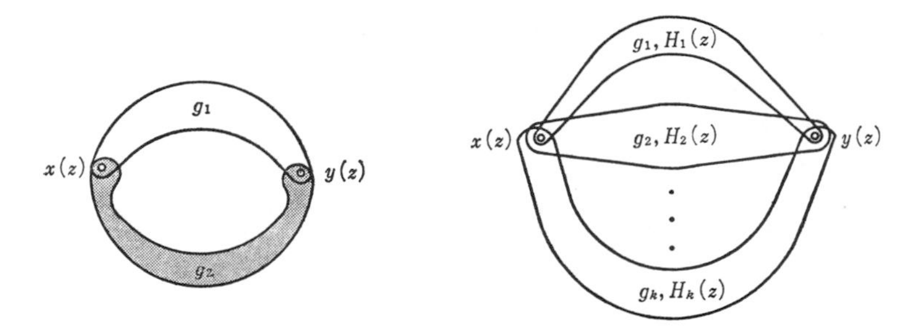

Figure 5.27 Two Graphs in Parallel

Figure 5.28 Parallel Structure

Consider that two signal flow graphs g_1 and g_2 are in parallel as shown in Fig. 5.27. Let $H_1(z)$ and $H_2(z)$ be transfer functions of g_1 and g_2, respectively. This means that when only g_1 is considered, $y(z)/x(z) = H_1(z)$. Also when only g_2 is considered, $y(z)/x(z) = H_2(z)$. Then a transfer function $H(z)$ of a whole signal flow graph in Fig. 5.27 is

$$\frac{y(z)}{x(z)} = H(z) = H_1(z) + H_2(z) \tag{5.17}$$

To generalize this, consider that k signal flow graphs $g_1, g_2, \ldots, g_r, \ldots$ and g_k are in series as shown in Fig. 5.28 where $H_r(z) = y(z)/x(z)$ is a transfer function of g_r when only g_r is present. Then the transfer function $H(z)$ of a whole signal flow graph in Fig. 5.28 is

$$H(z) = \sum_{r=1}^{k} H_r(z) \tag{5.18}$$

A signal flow graph in Fig. 5.28 whose transfer function is the one in Eq. 5.18 can be obtained by expressing the transfer function as

$$H(z) = \sum_{r=1}^{k_1} \frac{A_r}{1 - c_r z^{-1}} + \sum_{r=1}^{k_2} \frac{B_r(1 - e_r z^{-1})}{(1 - d_r z^{-1})(1 - d_r^* z^{-1})} + \sum_{r=1}^{k-k_1-k_2} h_r z^{-1} \tag{5.19}$$

Then we take each term in the right-hand side of the equation as a transfer function of a signal flow graph g_r and connect them in parallel. A signal flow graph in Fig. 5.29(a) is one whose transfer function is of the form $A_r/(1 - c_r z^{-1})$. A term $B_r(1 - e_r z^{-1})/[(1 - d_r z^{-1})(1 - d_r^* z^{-1})]$ can be expressed as

$$\frac{B_r(1 - e_r z^{-1})}{(1 - d_r z^{-1})(1 - d_r^* z^{-1})} = \frac{B_r(1 - e_r z^{-1})}{1 - \beta_{r1} z^{-1} - \beta_{r1} z^{-2}} \tag{5.20}$$

Hence, a signal flow graph in Fig. 5.29(b) is the one whose transfer function is of the form in Equation 5.20. Finally, $h_r x^{-1}$ can be a transfer function of a signal flow graph in Fig. 5.29(c). Thus, the parallel structure of a signal flow

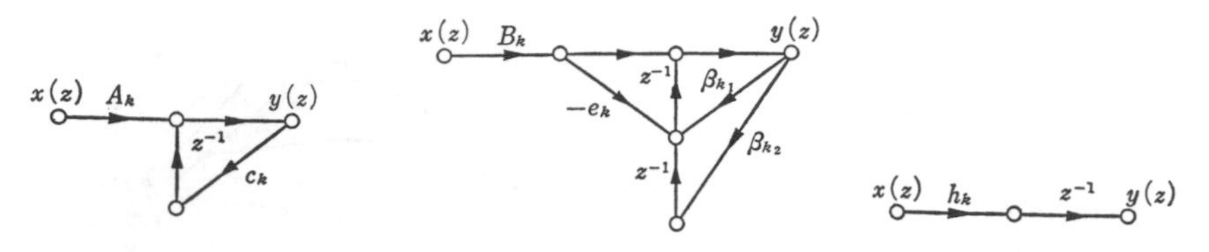

Figure 5.29 Subgraphs for Parallel Structure. (a) Graph of $A_r/(1 - c_r z^{-1})$; (b) Graph of $\dfrac{B_r(1 - c_r z^{-1})}{(1 - d_r z^{-1})(1 - d_r^* z^{-1})}$; (c) Graph of $h_r z^{-1}$.

graph in Fig. 5.30 having a transfer function $H(z)$ in Equation 5.19 can be constructed.

It must be noticed that when a node is $x(z)$ or $y(z)$, then z^{-1}, which is a weight of a branch, is a variable. However, when a node is $x(n)$ or $y(n)$, z^{-1} is a unit delay. For a system satisfying shift invariant and initial rest conditions, these two cases become identical. Hence we can change $x(z)$ and $y(z)$ to $z(n)$ and $y(n)$, and vice versa.

The symbol $d(v)$ indicates the number of branches entering at a node v. Suppose a computer is employed for calculating $y(n)$, and a signal flow graph is used to indicate the process of calculating $y(n)$. Then $d(v) - 1$ is the number of addition at that node v, the number of branches whose weights are not 1 is the number of multiplication, and the number of z^{-1} is the number of shift necessary for the calculation. Hence, for a given transfer function, a kind of a signal flow graph that should be chosen to reduce a computation time is an important problem. I will not discuss this problem here, however.

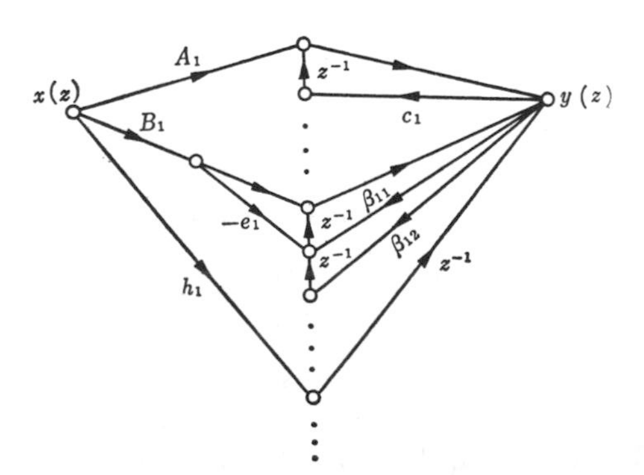

Figure 5.30 Parallel Structure

5.5 SUMMARY

1. A sink is a node in a signal flow graph such that all branches connected at the node are entering at the node.
2. A source is a node in a signal flow graph such that all branches connected at the node are leaving from the node.
3. A forward path from a node x_m to a node x_0 is a minimum set of branches such that starting from x_m we can reach x_0 by tracing each branch in the set only in the direction of its arrow. The node x_m is called the "initial node," and the node x_0 is called the "final node."
4. A closed loop is a forward path with the initial node and the final node being the same.
5. Mason's formula of a signal flow graph G is

$$\frac{x_0}{x_m} = \frac{\sum_{(k)} \Pi P_{m0_k}[1 - C_1 + C_2 - C_3 + \cdots \text{ of } G(\overline{P_{m0_k}})]}{1 - C_1 + C_2 - C_3 + \cdots + (-1)^r C_r + \cdots}$$

where x_m is a source, $C_r = \sum_{(k)} (\Pi L_{rk})$, L_{rk} is r closed loops, any two of which have no nodes in common, ΠL_{rk} is a product of the weight of all branches in L_{rk}, P_{m0_k} is a forward path from x_m to x_0, and $G(\overline{P_{m0_k}})$ is a graph obtained from G by deleting all nodes of P_{m0_k} and all branches connected at these nodes.
6. Two signal flow graphs are equivalent if the output signal y of these graphs is the same for a input signal x.
7. Some equivalent transformations for a signal flow graph are shown in Fig. 5.31.

5.6 PROBLEMS

1. Draw a signal flow graph from the following simultaneous equations:

a. $x_0 = 2x_1 + 3x_2$
$x_1 = x_0 + 2x_1 + 4x_2$

b. $x_0 = x_1 + 2x_2 + 3x_3$
$x_1 = x_0 + 2x_2 + x_3$
$x_2 = x_1 + 2x_2$

c. $x_0 = 2x_0 + 3x_1 + x_2 + x_3$
$x_1 = 2x_1 + 2x_2$
$x_2 = x_0 + x_1 + x_3$

d. $x_0 = x_2 + x_3$
$x_1 = x_0 + x_4$
$x_2 = x_1 + x_3 + x_4$
$x_3 = 2x_0 + 3x_1 + 2x_2 + 2x_3$

2. Obtain all forward paths from the node 1 to the node 2 of signal flow graphs in Fig. 5.32.
3. Obtain all closed loops of signal flow graphs in Fig. 5.33.

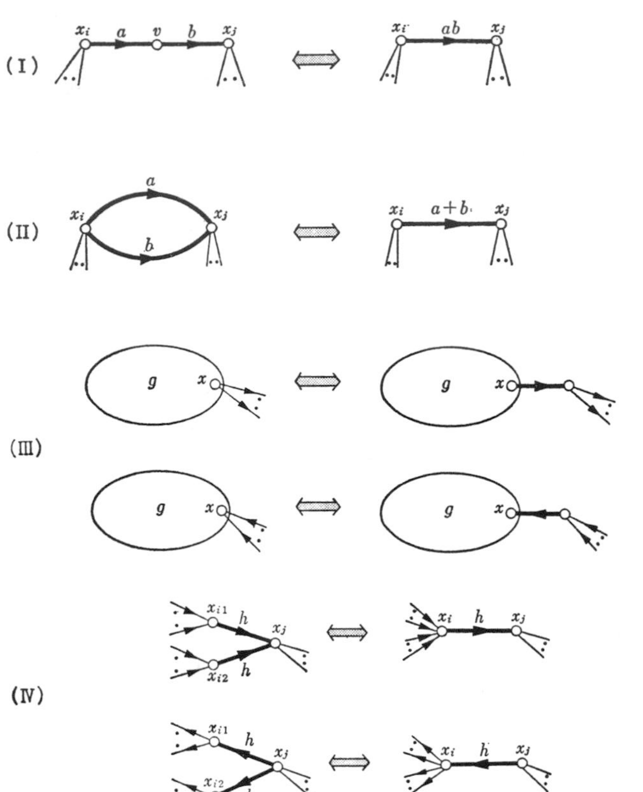

Where x is a source and y is a sink.

(V)

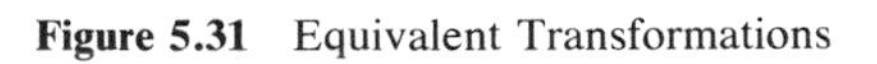

Figure 5.31 Equivalent Transformations

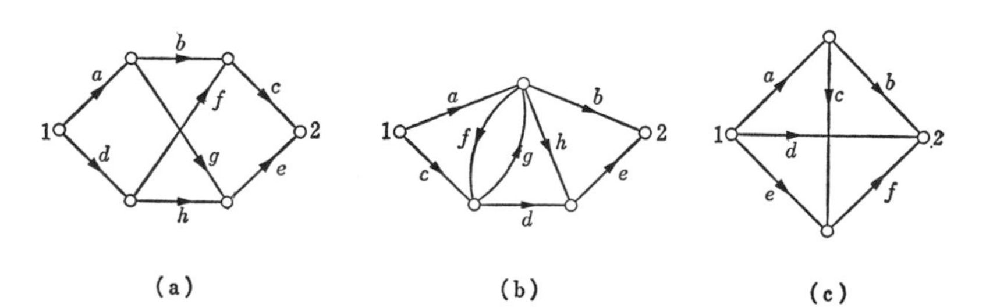

Figure 5.32 Graphs for Problem 2

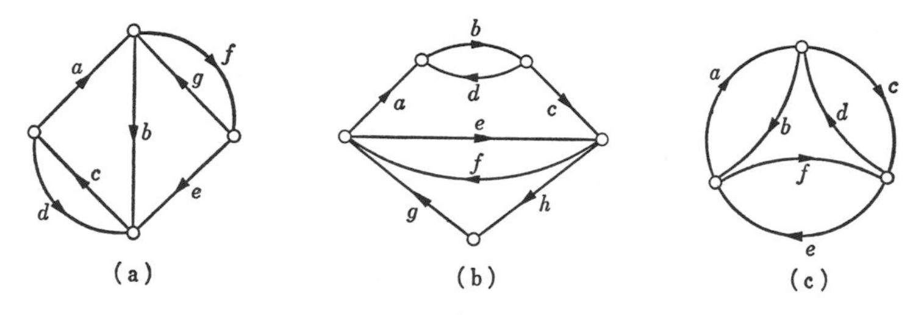

Figure 5.33 Graphs for Problem 3

4. Draw a signal flow graph of the following simultaneous equations and obtain x_0/x_3 by using the Mason's formula.

a.
$$\begin{aligned} x_0 &= x_1 + 2x_2 \\ x_1 &= 2x_1 + 2x_2 + 3x_3 \\ x_2 &= x_0 + x_1 + x_3 \end{aligned}$$

b.
$$\begin{aligned} x_0 &= x_1 + x_2 + x_3 \\ x_1 &= x_1 + x_2 + x_3 + x_4 \\ x_2 &= x_0 + 2x_2 + 3x_3 + x_4 \\ x_4 &= x_0 + x_2 + x_3 \end{aligned}$$

c.
$$\begin{aligned} x_0 &= x_1 + 3x_2 - x_3 \\ x_1 &= 2x_1 + x_2 - x_3 + x_4 \\ x_2 &= x_0 - 2x_2 + x_4 \\ x_4 &= x_1 + 2x_2 - x_3 \end{aligned}$$

d.
$$\begin{aligned} x_0 &= x_1 - x_2 + x_3 \\ x_1 &= x_2 + x_3 - 2x_4 \\ x_2 &= x_0 + 2x_2 - x_3 + x_4 \\ x_4 &= x_0 - 2x_2 - x_3 + 2x_4 \end{aligned}$$

5. Draw a signal flow graph for each of the following transfer functions;

a. $H(z) = \dfrac{1 - z^{-1} + 2z^{-2}}{1 - z^{-1} - z^{-2} - z^{-3}}$ **b.** $H(z) = \dfrac{1 - 2z^{-1} + z^{-2}}{1 - z^{-1} - z^{-2} - z^{-3}}$

c. $H(z) = \dfrac{1 - z^{-2}}{1 - 3z^{-1} + 3z^{-2} - z^{-3}}$ **d.** $H(z) = \dfrac{1 - z^{-3}}{1 - 3z^{-1} + 3z^{-2} - z^{-3}}$

6. Obtain a series structure of a signal flow graph for each of the following transfer functions:

a. $H(z) = \dfrac{1 + 2z^{-1} + z^{-2}}{1 - 3z^{-1} + 3z^{-1} - z^{-3}}$ **b.** $H(z) = \dfrac{1 - 2z^{-1} + z^{-2}}{1 - z^{-1} - z^{-2} + z^{-3}}$

c. $H(z) = \dfrac{1 - 2z^{-1} + z^{-2}}{1 - z^{-1} + z^{-2} - z^{-3}}$ **d.** $H(z) = \dfrac{1 - 2z^{-1} + z^{-2}}{1 - 2z^{-2} + z^{-4}}$

7. Obtain a parallel structure of a signal flow graph for each of the transfer functions given in the previous problem.

6
Discrete Fourier Transformation

6.1 DISCRETE FOURIER SERIES

In Chapter 3, we studied the discrete function Fourier transformation $X(e^{j\omega})$ of a discrete function $x(n)$ given by Equation 3.57 as

$$X(e^{j\infty}) = \sum_{k=-\infty}^{\infty} x(k)e^{-j\omega k}$$

Also we have found the $X(e^{j\omega})$ is a periodic function of a period 2π. Hence, we can express the inverse discrete function Fourier transformation as

$$x(n) = \frac{1}{2\pi}\int_{-\pi}^{\pi} X(e^{j\omega})e^{j\omega n}\,d\omega$$

Furthermore, the discrete function Fourier transformation of convolution of two functions $x(n)$ and $h(n)$ is

$$\sum_{k=-\infty}^{\infty} [x(k) * h(k)]e^{-j\omega k} = \sum_{k=-\infty}^{\infty}\left[\sum_{m=-\infty}^{\infty} x(m)h(k-m)\right] e^{-j\omega k} = X(e^{j\omega})H(e^{j\omega})$$

where $X(e^{j\omega})$ and $H(e^{j\omega})$ are the discrete function Fourier transformation of $x(n)$ and $h(n)$, respectively. This shows that convolution of $x(n)$ and $h(n)$

can be obtained by the inverse discrete function Fourier transformation of product of $X(e^{j\omega})$ and $H(e^{j\omega})$.

Because $X(e^{j\omega})$ and $H(e^{j\omega})$ are continuous functions of ω, we should change these to discrete functions before multiplying $X(e^{j\omega})$ and $H(e^{j\omega})$ by computers. If there is a transformation that transforms a discrete function into a discrete function, and the convolution of two discrete functions is the inverse transformation of product of the transformation of these two functions, computation of discrete functions by computers will become much simpler. Such a transformation is a "discrete Fourier transformation" or simply called a DFT.

First we will study a case when a discrete function is periodic. Because a periodic discrete function is a key for development of the DFT, we will use the symbol "o" to a function as $\mathring{x}(n)$ and $\mathring{y}(n)$ to indicate that these are periodic discrete functions.

A discrete function $\mathring{x}(n)$ has a period N means that for any nonzero integer k, $\mathring{x}(n)$ satisfies

$$\mathring{x}(n) = \mathring{x}(n + Nk)$$

We have learned already that the Fourier exponential series of an analog function $x(t)$ having a period T_0 can be expressed as

$$x(t) = \sum_{k=-\infty}^{\infty} X(k)\, e^{j2\pi k f_0 t} \tag{6.1}$$

By sampling this function $x(t)$ with a proper interval, we can change Equation 6.1 as

$$\mathring{x}(n) = \sum_{k=-\infty}^{\infty} X(k)\, e^{j(2\pi/N)nk} \tag{6.2}$$

Splitting this summation in the right-hand side by a period N, Equation 6.2 becomes

$$\mathring{x}(n) = \cdots + \sum_{k=0}^{N-1} X(k)\, e^{j(2\pi/N)nk} + \sum_{k=N}^{2N-1} X(k)\, e^{j(2\pi/N)nk} + \sum_{k=2N}^{3N-1} X(k)\, e^{j(2\pi/N)nk} + \cdots \tag{6.3}$$

Changing k in $e^{j(2\pi/N)nk}$ to $k + Nm$, we have

$$e^{j(2\pi/N)n(k+Nm)} = e^{j(2\pi/N)nk} e^{j2\pi nm}$$

Because $e^{j2\pi nm}$ is $\cos(2\pi nm) + j \sin(2\pi nm)$, $e^{j2\pi nm}$ is 1 for any value of integer nm. Hence,

$$e^{j(2\pi/N)n(k+Nm)} = e^{j(2\pi/N)nk}$$

This means that $e^{j(2\pi/N)nk}$ is a periodic function of a period N. Hence, Equation 6.3 can be expressed as

$$\mathring{x}(n) = \sum_{k=-\infty}^{\infty} (\cdots + X(k) + X(k+N) + X(k+2N) + \cdots) e^{j(2\pi/N)nk} \quad (6.4)$$

If $x(n)$ is finite, the right-hand side of Equation 6.4 must also be finite. So the sum of the right-hand side

$$X_t = \sum_{p=-\infty}^{\infty} X(k+PN)$$

should be finite. Because the summation is from $-\infty$ to ∞, in order for X to be finite, a periodic function $X(k+pN)$ must not contain non-zero values, which would be useless. Hence, this transformation is not a desired transformation. Suppose we use one term $\sum_{k=0}^{N-1} X(k)e^{j(2\pi/N)nk}$ in Equation 6.3 as a transformation of $\mathring{x}(n)$. That is,

$$\mathring{x}(n) = \frac{1}{N}\sum_{k=0}^{N-1} X(k)e^{j(2\pi/N)nk} \quad (6.5)$$

Is $X(k)$ a periodic discrete function? Multiplying $e^{-j(2\pi/N)nr}$ by both sides of Equation 6.5 and summing the result for $n = 0$ to $n = N-1$, we have

$$\sum_{n=0}^{N-1} \mathring{x}(n)e^{-j(2\pi/N)nr} = \sum_{n=0}^{N-1}\left(\frac{1}{N}\sum_{k=0}^{N-1} X(k)e^{j(2\pi/N)nk}\right) e^{-j(2\pi/N)nr} \quad (6.6)$$

The right-hand side of this equation becomes

$$\sum_{n=0}^{N-1}\left(\frac{1}{N}\sum_{k=0}^{N-1} X(k)e^{j(2\pi/N)nk}\right) e^{-j(2\pi/N)nr} = \sum_{k=0}^{N-1} X(k)\sum_{n=0}^{N-1}\frac{1}{N}e^{j(2\pi/N)n(k-r)}$$

This $\sum_{n=0}^{N-1}\frac{1}{N}e^{j(2\pi/N)n(k-r)}$ when $K = r$ is

$$\sum_{n=0}^{N-1}\frac{1}{N} = 1$$

However, when $k \neq r$

$$\sum_{n=0}^{N-1}\frac{1}{N}e^{j(2\pi/N)nq} = \frac{1}{N}(1 + e^{j(2\pi/N)q} + e^{j(2\pi/N)2q} + e^{j(2\pi/N)3q} + \cdots + e^{j(2\pi/N)(N-1)q} = \frac{1}{N}\frac{1-e^{j2\pi q}}{1-e^{j(2\pi/N)q}} \quad (6.7)$$

where an integer $q = k - r$. Both k and r are integers that are changing from 0 to $N - 1$. When $q = k - r$ is not 0, integer q will be changed from $-(N - 1)$ to $N - 1$. Hence, when $q \neq 0$, $e^{j(2\pi/N)q}$ will not take a value of 1. Thus, the denominator $1 - e^{j(2\pi/N)q}$ of Equation 6.7 will not become 0. Conversely, the numerator $e^{j2\pi q}$ will be 1 for any integer q. Hence, Equation 6.7 is always 0, or

$$\sum_{n=0}^{N-1} \frac{1}{N} e^{j(2\pi/N)n(k-r)} = \begin{cases} 1 & k = r \\ 0 & k \neq r \end{cases}$$

Thus, Equation 6.6 is

$$\sum_{n=0}^{N-1} \mathring{x}(n) e^{-j(2\pi/N)nr} = X(r) \tag{6.8}$$

Because $\mathring{x}(n)$ and $e^{-j(2\pi/N)nr}$ are periodic functions of a period N, $X(r)$ is a periodic function of a period N. Furthermore, because $\mathring{x}(n)$ and $e^{-j(2\pi/N)nr}$ are discrete functions, $X(r)$ will be a discrete function.

Using W_N, which is defined as

$$W_N = e^{-j(2\pi/N)} \tag{6.9}$$

Equation 6.17 can be written as

$$\mathring{X}(k) = \sum_{n=0}^{N-1} \mathring{x}(n) W_N^{nk} \tag{6.10}$$

Notice that we use $\mathring{X}(k)$ because it is a periodic function. Also Equation 6.5 can be rewritten as

$$\mathring{x}(n) = \frac{1}{N} \sum_{k=1}^{N-1} \mathring{X}(k) W_N^{-nk} \tag{6.11}$$

$\mathring{X}(k)$ in Equation 6.10 is called a DFS, and $\mathring{x}(n)$ in Equation 6.11 is called an inverse DFS. As an example, consider $\mathring{x}(n)$ shown in Fig. 6.1(a). Because the period N of $\mathring{x}(n)$ is 4, $e^{-j(2\pi/4)} = W_r$ is $(-j)$. From Equation 6.10, $\mathring{X}(k)$ are

$$\mathring{X}(0) = \sum_{m=0}^{3} \mathring{x}(n)(-j)^{n0} = 6$$

$$\mathring{X}(1) = \sum_{n=0}^{3} \mathring{x}(n)(-j)^{n} = -2$$

$$\mathring{X}(2) = \sum_{n=0}^{3} \mathring{x}(n)(-j)^{n2} = -2$$

$$\mathring{X}(3) = \sum_{n=0}^{3} \mathring{x}(n)(-j)^{n3} = -2$$

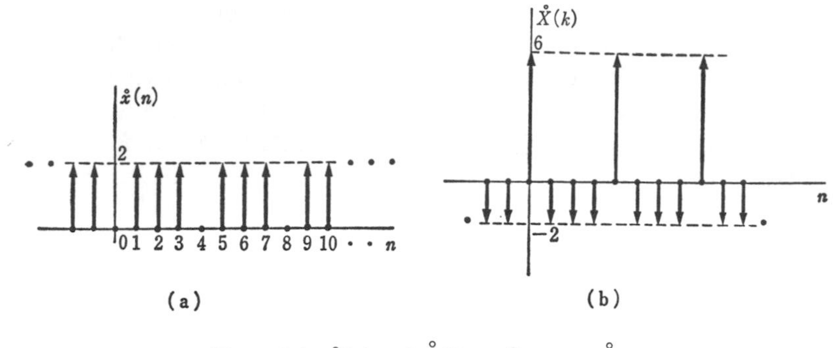

Figure 6.1 $\mathring{x}(n)$ and $\mathring{X}(k)$ (a) $\mathring{x}(n)$; (b) $\mathring{X}(k)$

These $\mathring{X}(k)$ are shown in Fig. 6.1(b). The reader should try to calculate $\mathring{x}(n)$ from these $\mathring{X}(k)$ by Equation 6.11.

For another example, consider $\mathring{x}(n)$ shown in Fig. 6.2(a). Because a period of this discrete function $\mathring{x}(n)$ is 8, $W_8 = e^{-j2\pi/8} = (1 - j1)/\sqrt{2}$. Because a DFS of $\mathring{x}(n)$ can be expressed by Equation 6.10 as

$$\mathring{X}(k) = \sum_{n=0}^{7} \mathring{x}(n) \left(\frac{1 - j1}{\sqrt{2}}\right)^{nk}$$

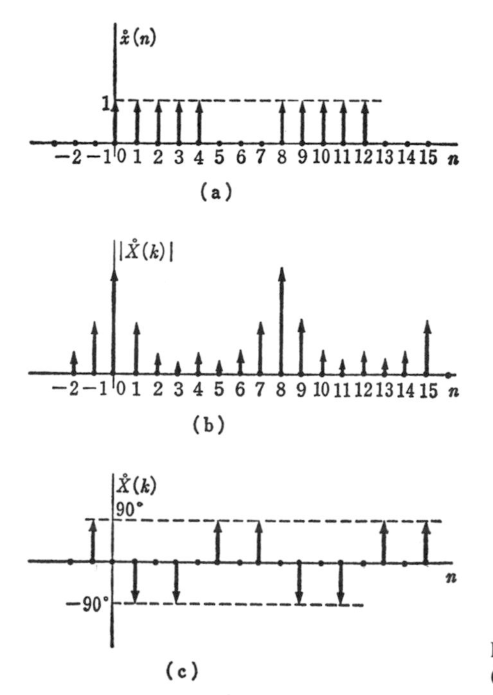

Figure 6.2 $\mathring{x}(n)$ and $\mathring{X}(k)$. (a) $\mathring{x}(n)$; (b) $|\mathring{X}(k)|$; (c) Phase of $\mathring{X}(k)$.

we have

$$\mathring{X}(0) = 5$$

$$\mathring{X}(1) = 1 + \frac{1 - j1}{\sqrt{2}} + \left(\frac{1 - j1}{\sqrt{2}}\right)^2 + \left(\frac{1 - j1}{\sqrt{2}}\right)^3 + \left(\frac{1 - j1}{\sqrt{2}}\right)^4 = -j(1 + \sqrt{2})$$

$$\mathring{X}(2) = 1 + (-j) + (-j)^2 + (-j)^3 + (-j)^4 = 1$$

$$\mathring{X}(3) = j(1 - \sqrt{2})$$

$$\mathring{X}(4) = 1$$

$$\mathring{X}(5) = -j(1 - \sqrt{2})$$

$$\mathring{X}(6) = 1$$

$$\mathring{X}(7) = j(1 + \sqrt{2})$$

The magnitude and the phase of these $\mathring{X}(k)$ are shown in Fig. 6.2(b) and (c).

Let us investigate the relationship between the DFS and the z-transformation. The z-transformation of $\mathring{x}(n)$ is (Equation 4.36)

$$\mathring{X}(z) = \sum_{n=-\infty}^{\infty} \mathring{x}(n) z^{-n} \tag{6.12}$$

To compare this and the DFS in Equation 6.10 is very difficult because one is sum from 0 to $N - 1$, and the other is sum from $-\infty$ to ∞. Hence, we will compare the z-transformation of one period of $\mathring{x}(n)$ and the DFS. Let $x(n)$ be one period of $\mathring{x}(n)$. That is, $x(n)$ is

$$x(n) = \begin{cases} \mathring{x}(n), & 0 \le n \le N - 1 \\ 0, & \text{Others} \end{cases} \tag{6.13}$$

The z-transformation of this $x(n)$ is

$$X(z) = \sum_{n=0}^{\infty} x(n) z^{-n} = \sum_{n=0}^{N-1} \mathring{x}(n) z^{-n} \tag{6.14}$$

Because a variable z is

$$z = re^{j\infty}$$

letting $r = 1$ and $\omega = (2\pi/N)k_1$ the Equation 6.14 becomes Equation 6.10. Hence,

$$\mathring{X}(k) = X(z)\big|_{z=e^{j(2\pi/N)k}} \tag{6.15}$$

This means that sampling $X(z)$ at a point $2\pi/N$ on a circumference C whose center is the origin and the radius is 1 becomes $\mathring{X}(k)$. As an example, consider $\mathring{x}(n)$ in Fig. 6.2(a). Let $x(n)$ be one period of $\mathring{x}(n)$. The z-transformation of $x(n)$ is

$$X(z) = \sum_{n=-\infty}^{\infty} x(n)z^{-n} = 1 + z^{-1} + z^{-2} + z^{-3} + z^{-4}$$

By changing z by $e^{j(2\pi/8)k}$, the preceding equation becomes

$$\mathring{X}(k) = 1 + e^{-j(\pi/4)k} + e^{-j(\pi/4)k2} + e^{-j(\pi/4)k3} + e^{-j(\pi/4)k4}$$

Taking $k = 0, 1, 2, \ldots$, we can obtain those $\mathring{X}(0)$, $\mathring{X}(1)$, $\mathring{X}(2)$, . . . which have been obtained previously.

6.2 PROPERTIES OF DFS

The properties of the DFS of discrete periodic functions are very similar to those of the z-transformation of discrete functions. First, let $\mathring{x}_1(n)$ and $\mathring{x}_2(n)$ be periodic functions having the same period N. Consider $\mathring{x}_3(n)$, which is

$$\mathring{x}_3(n) = a\mathring{x}_1(n) + b\mathring{x}_2(n)$$

By Equation 6.10, the DFS of $\mathring{x}_3(n)$ is

$$\mathring{X}_3(k) = \sum_{n=0}^{N-1} \mathring{x}_3(n)W_N^{kn} = \sum_{n=0}^{N-1} (a\mathring{x}_1(n) + b\mathring{x}_2(n))W_N^{kn} = a\mathring{X}_1(k) + b\mathring{X}_2(k)$$

which shows that the DFS has the linearity property.

Let $\mathring{X}(k)$ be the DFS of $\mathring{x}(n)$. Then the DFS of $\mathring{x}(n + m)$ is (by Equation 6.10)

$$\sum_{n=0}^{N-1} \mathring{x}(n + m)W_N^{kn} = \sum_{n'=0}^{N-1} \mathring{x}(n')W_N^{k(n'-m)} = W_N^{-km}\mathring{X}(k)$$

except when $m > N$, $m' = m \bmod N$ should be used. This equation is similar to $z^{-m}X(k)$ obtained by the z-transformation of $x(n + m)$.

Let $X_1(z)$ and $X_2(z)$ be the z-transformation of $x_1(n)$ and $x_2(n)$, respectively. We have seen already that product $X_1(z)X_2(z)$ is equal to the z-transformation of convolution of $x_1(n)$ and $x_2(n)$. Let $\mathring{X}_1(k)$ and $\mathring{X}_2(k)$ be the DFS of $\mathring{x}_1(n)$ and $\mathring{x}_2(n)$, respectively. Suppose $\mathring{X}_3(k)$ is product of $\mathring{X}_1(k)$ and $\mathring{X}_2(k)$. That is,

$$\mathring{X}_3(k) = \mathring{X}_1(k)\mathring{X}_2(k) = \sum_{n=0}^{N-1} \mathring{x}_1(n)W_N^{kn} \sum_{m=0}^{N-1} \mathring{x}_2(m)W_N^{km}$$

$$= \sum_{n=0}^{N-1} \sum_{m=0}^{N-1} \mathring{x}_1(n)\mathring{x}_2(m)W_N^{k(n+m)}$$

By Equation 6.11, the inverse DFS of $\mathring{X}_3(k)$ is

$$\mathring{x}_3(r) = \frac{1}{N}\sum_{k=0}^{N-1} \mathring{X}_3(k) W_N^{-rk} = \frac{1}{N}\sum_{k=0}^{N-1} \mathring{X}_1(k)\mathring{X}_2(k) W_N^{-rk}$$

$$= \frac{1}{N}\sum_{k=0}^{N-1}\sum_{n=0}^{N-1}\sum_{m=0}^{N-1} \mathring{x}_1(n)\mathring{x}_2(m) W_N^{k(n+m)} W_N^{-rk} \tag{6.16}$$

$$= \sum_{n=0}^{N-1} \mathring{x}_1(n) \sum_{m=0}^{N-1} \mathring{x}_2(m) \left[\frac{1}{N}\sum_{k=0}^{N-1} W_N^{-k(r-n-m)}\right]$$

When $r - n - m = 0$, $\frac{1}{N}\sum W_N^{-k(r-n-m)}$ in the right-hand side of this equation is

$$\frac{1}{N}\sum_{k=0}^{N-1} W_N^{-k(r-n-m)} = \frac{1}{N}\sum_{k=0}^{N-1} e^{j0} = 1$$

which is also true when $r - n - m = +N$. When $r - n - m \neq 0$ or $r - n - m \neq +N$, by letting $p = r - n - m$, $\frac{1}{N}\sum W_N^{-k(r-n-m)}$ can be expressed as

$$\frac{1}{N}\sum_{k=0}^{N-1} W_N^{-kp} = \frac{1}{N}[1 + e^{j(2\pi/N)p} + e^{j(2\pi/N)2p} + e^{j(2\pi/N)3p} + \cdots + e^{j(2\pi/N)(N-1)p}$$

$$= \frac{1}{N}\frac{1 - e^{j2\pi p}}{1 - e^{j(2\pi/N)p}}$$

which is always 0 because neither $p = 0$ nor $p = \pm N$. Hence, Equation 6.16 becomes

$$\mathring{x}_3(r) = \sum_{n=0}^{N-1} \mathring{x}_1(n)\mathring{x}_2(r-n) \tag{6.17}$$

Equation 6.17 looks like convolution but summation is from 0 to $N - 1$. Hence, we call this "circular convolution." We can conclude that the DFS of a circular convolution of $\mathring{x}_1(n)$ and $\mathring{x}_2(n)$ is equal to product $\mathring{X}_1(k)\mathring{X}_2(k)$ where $\mathring{X}_1(k)$ and $\mathring{X}_2(k)$ are the DFS of $\mathring{x}_1(n)$ and $\mathring{x}_2(n)$, respectively.

Let us use an example to exercise circular convolution. Consider $\mathring{x}_1(n)$ and $\mathring{x}_2(n)$ in Fig. 6.3. First let us obtain $\mathring{x}_3(n)$ where

$$\mathring{x}_3(0) = \sum_{n=0}^{N-1} \mathring{x}_1(n)\mathring{x}_2(-n)$$

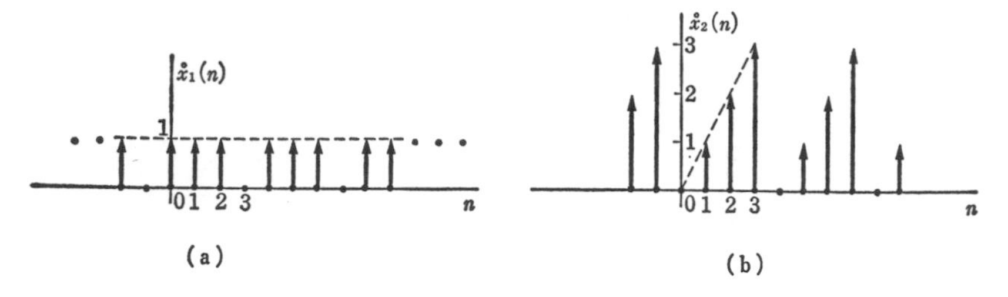

Figure 6.3 Discrete Periodic Functions $\mathring{x}_1(n)$ and $\mathring{x}_2(n)$. (a) $\mathring{x}_1(n)$; (b) $\mathring{x}_2(n)$.

Fig. 6.4(a) shows $\mathring{x}_2(-n)$. Multiplication of this $\mathring{x}_2(-n)$ and $\mathring{x}_1(n)$ gives the result as shown in Fig. 6.4(b). Because sum of this result in one period is $\mathring{x}_3(0)$, $\mathring{x}_3(0) = 5$.

For $\mathring{x}_3(1)$, Equation 6.17 becomes

$$\mathring{x}_3(1) = \sum_{n=0}^{3} \mathring{x}_1(n)\mathring{x}_2(1 - n)$$

To obtain $\mathring{x}_2(1 - n)$, we shift $\mathring{x}_2(-n)$ in Fig. 6.4(a) 1 unit to the right as shown in Fig. 6.5(a). Multiplying this and $\mathring{x}_1(n)$ gives the result in Fig. 6.5(b). The sum of this result in one period gives $\mathring{x}_3(1) = 4$. $\mathring{x}_3(3)$ is

$\mathring{x}_3(2)$ is

$$\mathring{x}_3(2) = \sum_{n=0}^{3} \mathring{x}_1(n)\mathring{x}_2(2 - n)$$

To obtain $\mathring{x}_2(2 - n)$, we shift $\mathring{x}_2(1 - n)$ in Fig. 6.5(a) 1 unit to the right as shown in Fig. 6.6(a). Multiplying this and $\mathring{x}_1(n)$ gives the result in Fig. 6.6(b). Summing the result in one period gives $\mathring{x}_3(2) = 3$.

Finally, $\mathring{x}_3(3)$ is

$$\mathring{x}_3(3) = \sum_{n=0}^{3} \mathring{x}_1(n)\mathring{x}_2(3 - n)$$

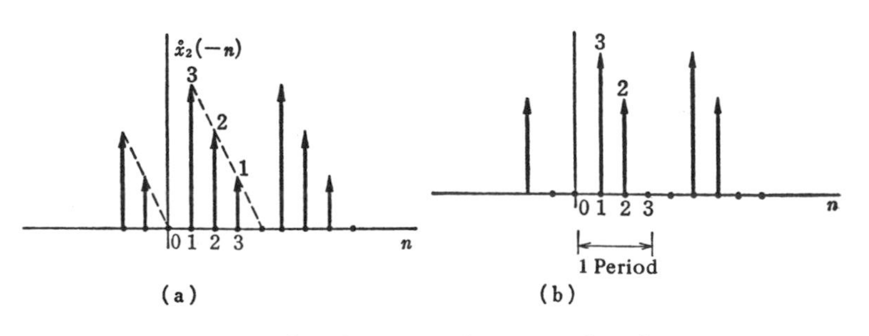

Figure 6.4 $\mathring{x}_1(n)\mathring{x}_2(-n)$. (a) $\mathring{x}_2(-n)$; (b) $\mathring{x}_1(n)\mathring{x}_2(-n)$.

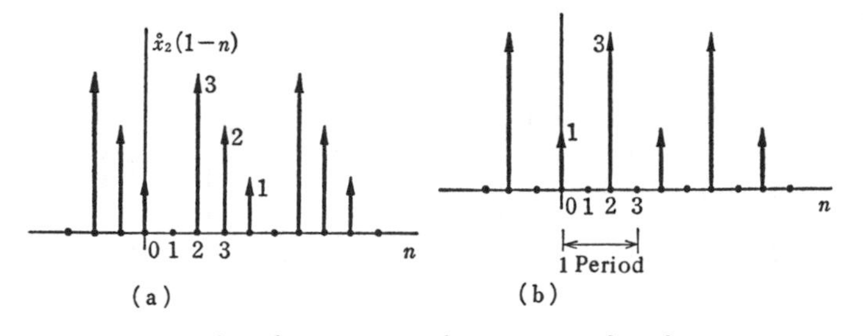

Figure 6.5 $\mathring{x}_1(n)\mathring{x}_2(1-n)$. (a) $\mathring{x}_2(1-n)$; (b) $\mathring{x}_1(n)\mathring{x}_2(1-n)$.

Shifting $\mathring{x}_2(2-n)$ in Fig. 6.6(a) 1 unit to the right, $\mathring{x}_2(3-n)$ in Fig. 6.7(a) can be obtained. Multiplying this and $\mathring{x}_1(n)$, we have the result in Fig. 6.7(b). Summing the result in one period produces $\mathring{x}_3(3) = 6$. By this result, we have $\mathring{x}_3(n)$ as shown in Fig. 6.8. Because the period of this $\mathring{x}_3(n)$ is 4, $W_4 = e^{-j(2\pi/4)} = -j$. Hence, the DFS $\mathring{X}_3(k)$ for $k = 0, 1, 2, 3$ of this $\mathring{x}_3(n)$ are

$$\mathring{X}_3(0) = \sum_{k=0}^{3} \mathring{x}_3(k)(-j)^0 = 18$$

$$\mathring{X}_3(1) = \sum_{k=0}^{3} \mathring{x}_3(k)(-j)^k = 5 + 4(-j) + 3(-1) + 6(j) = 2 + j2$$

$$\mathring{X}_3(2) = \sum_{k=0}^{3} \mathring{x}_3(k)(-j)^{2k} = 5 + 4(-1) + 3 + 6(-1) = -2$$

$$\mathring{X}_3(3) = \sum_{k=0}^{3} \mathring{x}_3(k)(-j)^{3k} = 5 + 4(j) + 3(-1) + 6(-j) = 2 - j2$$

Because $\mathring{x}_3(n)$ is obtained by circular convolution of $\mathring{x}_1(n)$ and $\mathring{x}_2(n)$, $\mathring{X}_3(k)$ must be equal to product of $\mathring{X}_1(k)$ and $\mathring{X}_2(k)$ where $\mathring{X}_1(k)$ and $\mathring{X}_2(k)$ are the

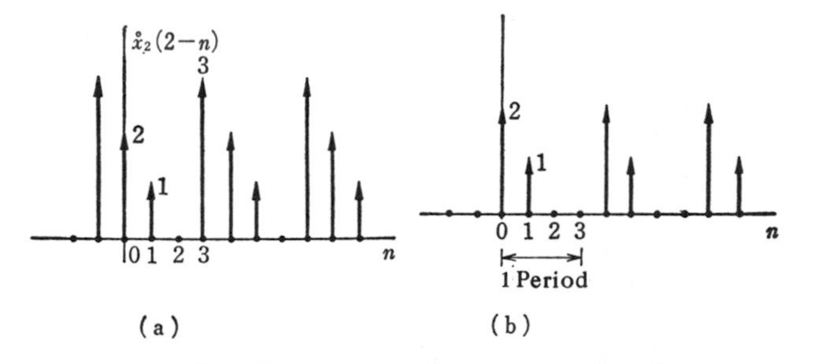

Figure 6.6 $\mathring{x}_1(n)\mathring{x}_2(2-n)$. (a) $\mathring{x}_2(2-n)$; (b) $\mathring{x}_1(n)\mathring{x}_2(2-n)$.

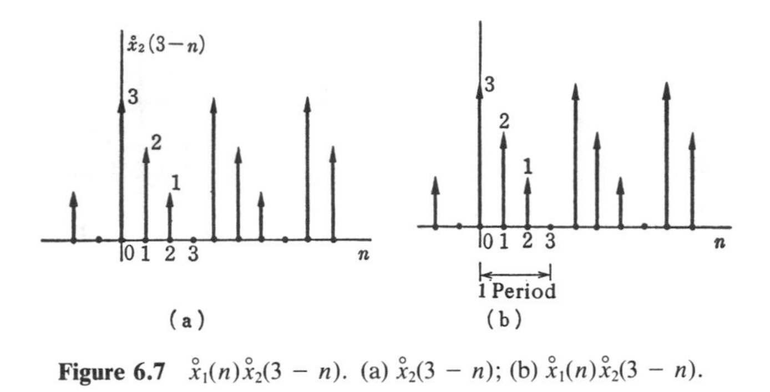

Figure 6.7 $\mathring{x}_1(n)\mathring{x}_2(3-n)$. (a) $\mathring{x}_2(3-n)$; (b) $\mathring{x}_1(n)\mathring{x}_2(3-n)$.

DFS of $\mathring{x}_1(n)$ and $\mathring{x}_2(n)$, respectively. Let us check this fact. First, we need the DFS $\mathring{X}_1(k)$ of $\mathring{x}_1(n)$ in Fig. 6.3(a), which can be calculated as

$$\mathring{X}_1(0) = \sum_{n=0}^{3} \mathring{x}_1(n) W_4^0 = 3$$

$$\mathring{X}_1(0) = \sum_{n=0}^{3} \mathring{x}_1(n) W_4^n = 1 + (-j) + (-1) = -j$$

$$\mathring{X}_1(2) = \sum_{n=0}^{3} \mathring{x}_1(n) W_4^{3n} = 1 + (-1) + 1 = 1$$

$$\mathring{X}_1(3) = \sum_{n=0}^{3} \mathring{x}_1(n) W_4^{3n} = 1 + (j) + (-1) = j$$

Next the DFS $\mathring{X}_2(k)$ of $\mathring{x}_2(n)$ in Fig. 6.3(b) are

$$\mathring{X}_2(0) = \sum_{n=0}^{3} \mathring{x}_2(n) W_4^0 = 6$$

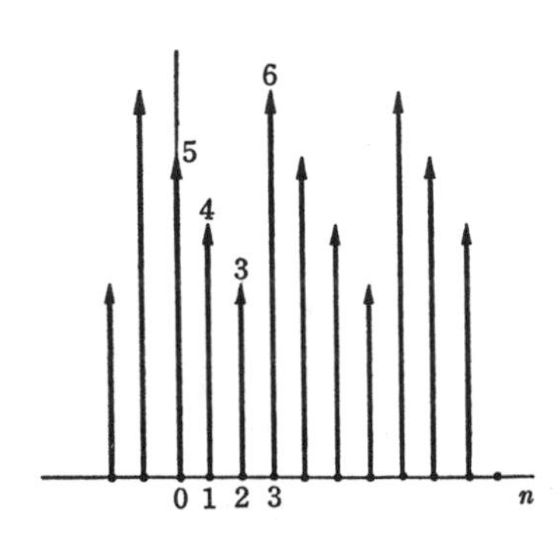

Figure 6.8 $\mathring{x}_3(k) = \sum_{n=0}^{3} \mathring{x}_1(n)\mathring{x}_2(k-n)$

$$\mathring{X}_2(1) = \sum_{n=0}^{3} \mathring{x}_2(n) W_4^n = (-j) + 2(-1) + 3(j) = -2 + j2$$

$$\mathring{X}_2(2) = \sum_{n=0}^{3} \mathring{x}_2(n) W_4^{2n} = (-1) + 2(1) + 3(-1) = -2$$

$$\mathring{X}_2(3) = \sum_{n=0}^{3} \mathring{x}_2(n) W_4^{3n} = (j) + 2(-1) + 3(-j) = -2 - j2$$

From these, we can find that product of $\mathring{X}_1(k)$ and $\mathring{X}_2(k)$ is equal to $\mathring{X}_3(k)$.

We have found that the DFS of circular convolution of two discrete periodic functions $\mathring{x}_1(n)$ and $\mathring{x}_2(n)$ is equal to product of $\mathring{X}_1(k)$ and $\mathring{X}_2(k)$ where $\mathring{X}_1(k)$ and $\mathring{X}_2(k)$ are the DFS of $\mathring{x}_1(n)$ and $\mathring{x}_2(n)$, respectively. What is the DFS of product of $\mathring{x}_1(n)$ and $\mathring{x}_2(n)$ equal to? By Equation 6.10, the DFS of $\mathring{x}_1(n)\mathring{x}_2(n)$ is $\sum_{n=0}^{N-1} \mathring{x}_1(n)\mathring{x}_2(n) W_N^{nk}$. Changing $\mathring{x}_1(n)$ and $\mathring{x}_2(n)$ by Equation 6.11, we have

$$\sum_{n=0}^{N-1} \mathring{x}_1(n)\mathring{x}_2(n) W_N^{nk} = \sum_{n=0}^{N-1} \left(\frac{1}{N} \sum_{q=0}^{N-1} \mathring{X}_1(q) W_N^{-qn}\right) \left(\frac{1}{N} \sum_{r=0}^{N-1} \mathring{X}_2(r) W_N^{-rn}\right) W_N^{nk} \tag{6.18}$$

$$= \frac{1}{N^2} \sum_{q=0}^{N-1} \sum_{r=0}^{N-1} \mathring{X}_1(q)\mathring{X}_2(r) \sum_{n=0}^{N-1} W_N^{n(k-q-r)}$$

The term $\sum_{n=0}^{N-1} W_N^{n(k-q-r)}$ in the right-hand side of this equation is

$$\sum_{n=0}^{N-1} W_N^{n(k-q-r)} = \begin{cases} N, & k - q - r = 0 \text{ or } \pm N \\ 0, & k - q - r \neq 0 \text{ nor } \pm N \end{cases}$$

Hence, Equation 6.18 can be written as

$$\sum_{n=0}^{N-1} \mathring{x}_1(n)\mathring{x}_2(n) W_N^{nk} = \frac{1}{N} \sum_{q=0}^{N-1} \mathring{X}_1(q)\mathring{X}_2(k - q) \tag{6.19}$$

This equation indicates that the DFS of product $\mathring{x}_1(n)\mathring{x}_2(n)$ is equal to $1/N$ times circular convolution of $\mathring{X}_1(k)$ and $\mathring{X}_2(k)$ where $\mathring{X}_1(k)$ and $\mathring{X}_2(k)$ are the DFS of $\mathring{x}_1(n)$ and $\mathring{x}_2(n)$, respectively.

6.3 DISCRETE FOURIER TRANSFORMATION (DFT)

To obtain the DFS of a discrete function $x(n)$, we must be sure that $x(n)$ is periodic. Those discrete functions that appear in practice are usually nonperiodic, however. What kind of practical discrete functions must we consider? When we employ computers to treat discrete functions, these functions should not have nonzero values for $n \to \infty$. Because otherwise, any computer cannot handle such functions in a finite time. In other words, discrete functions that can be handled by computers are those whose nonzero values appear within a finite interval. Such a discrete function is called a "finite duration function" or a "finite duration signal." For example,

$$x(n) = \begin{cases} 1, & 0 \le n \le N-1 \\ 0, & \text{Otherwise} \end{cases} \tag{6.20}$$

is a finite duration function. In other words, a finite duration of a discrete function being K means that the function is 0 for all $n \ge K$. By Equation 3.57, the discrete function Fourier transformation of a finite duration function $x(n)$ in Equation 6.20 is

$$X(e^{j\omega}) = \sum_{n=-\infty}^{\infty} x(n)e^{-j\omega n} = \sum_{n=0}^{N-1} e^{-j\omega n} = \frac{1 - e^{-j\omega N}}{1 - e^{-j\omega}}$$

which is clearly not a discrete function. Conversely, if a discrete function is periodic, then we know that the DFS of the function is discrete. Hence, by considering a finite duration function $x(n)$ as one period of a periodic function $\mathring{x}(n)$, that is

$$x(n) = \begin{cases} \mathring{x}(n), & 0 \le n \le N-1 \\ 0, & \text{Otherwise} \end{cases} \tag{6.21}$$

and taking one period of the DFS $\mathring{X}(k)$ of $\mathring{x}(n)$ as $X(k)$ of $x(n)$, we can have a discrete function $X(k)$ for a discrete function $x(n)$. This $X(k)$ is called the DFT of a discrete function $x(n)$. Whenever it is convenient, the DFT $X(k)$ of $x(n)$ is indicated by the symbol $\mathcal{D}[x(n)]$. In other words, the DFT of a finite duration function $x(n)$, which can be expressed by Equation 6.39, is

$$\mathcal{D}[x(n)] = X(k) = \begin{cases} \displaystyle\sum_{n=0}^{N-1} x(n)W_N^{kn}, & 0 \le k \le N-1 \\ 0, & \text{Otherwise} \end{cases} \tag{6.22}$$

and the inverse DFT of the preceding function $X(k)$ is

$$x(n) = \begin{cases} \displaystyle\frac{1}{N}\sum_{k=0}^{N-1} X(k)W_N^{-kn}, & 0 \le n \le N-1 \\ 0, & \text{Otherwise} \end{cases} \tag{6.23}$$

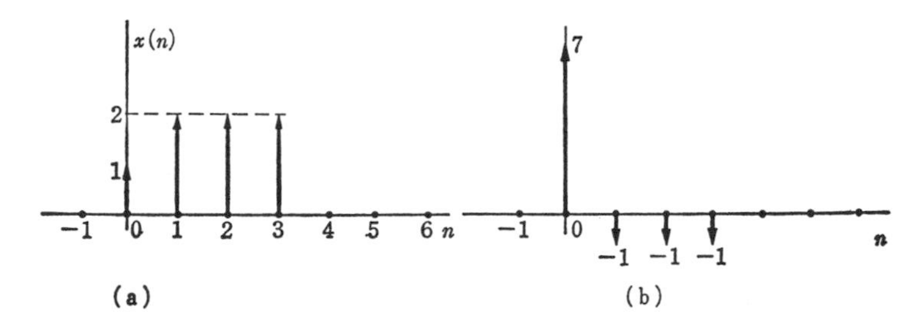

Figure 6.9 $x(n)$ and DFT $X(k)$. (a) Finite Duration Function $x(n)$; (b) DFT $X(k)$

which can be indicated by the symbol $\mathcal{D}^{-1}[X(k)]$ whenever it is convenient.

As an example, consider a finite duration function $x(n)$ in Fig. 6.9(a). The finite duration of this $x(n)$ is from 0 to 3. Hence, $N = 4$. Thus, the DFT of this $x(n)$ is

$$\mathcal{D}[x(n)] = X(k) = \sum_{n=0}^{3} x(n)W_4^{kn}$$

Because $W_4^{kn} = e^{-j(2\pi/4)kn} = (-j)^{kn}$, $X(k)$ for $k = 0, 1, 2, 3$ are

$$X(0) = 1 + 2(1) + 2(1)^2 + 2(1)^3 = 7$$

$$X(1) = 1 + 2(-j) + 2(-j)^2 + 2(-j)^3 = -1$$

$$X(2) = 1 + 2(-1) + 2(1) + 2(-1) = -1$$

$$X(3) = 1 + 2(j) + 2(-1) + 2(-j) = -1$$

by which $X(k)$ in Fig. 6.9(b) is obtained.

Because a finite duration function $x(n)$ is considered as one period of a periodic function $\mathring{x}(n)$, delaying $x(n)$ m unit means delaying $\mathring{x}(n)$ m unit to make $\mathring{x}(n - m)$ and taking the same one period of $\mathring{x}(n - m)$ to define $x(n - m)$. For example, consider a finite duration function $x(n)$ in Fig. 6.10(a). Let us obtain $x(n - 3)$, which indicated delaying $x(n)$ 3 unit. First we form $\mathring{x}(n - 3)$ by delaying $\mathring{x}(n)$ in Fig. 6.10(b) 3 unit as shown in Fig. 6.10(c). Then we take one period of $\mathring{x}(n - 3)$ as $x(n - 3)$, which is shown in Fig. 6.10(d). To show this clearly, we define the symbol $R_N(n)$ as

$$R_N(n) = \begin{cases} 1, & 0 \le n \le N - 1 \\ 0, & \text{Otherwise} \end{cases} \tag{6.24}$$

By using $R_N(n)$, $x(n)$ can be expressed as

$$x(n) = \mathring{x}(n)R_N(n) \tag{6.25}$$

where $\mathring{x}(n)$ is a periodic function, and $x(n)$ is one period from $n = 0$ to $n =$

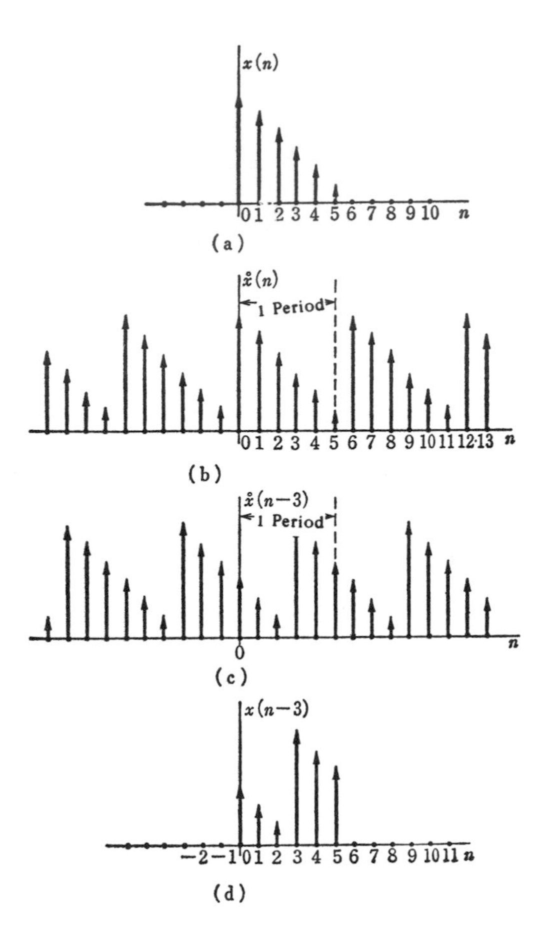

Figure 6.10 $x(n)$ and $x(n-3)$. (a) Finite Duration Function $x(n)$; (b) Periodic Function $\mathring{x}(n)$ for $x(n)$; (c) $\mathring{x}(n-3)$; (d) $x(n-3)$.

$N-1$ of $\mathring{x}(n)$. By Equation 6.25, delaying m unit of $x(n)$ can be expressed as

$$x(n-m) = \mathring{x}(n-m)R_N(n)$$

Similarly, $X(k)$ is one period from $n = 0$ to $n = N - 1$ of a periodic function $\mathring{X}(k)$, which can be expressed as

$$X(k) = \mathring{X}(k)R_N(k)$$

The DFT $\mathscr{D}[x(n-m)]$ of $x(n-m)$, which is delaying m unit of a finite duration function $x(n)$, is

$$\mathscr{D}[x(n-m)] = \left(\sum_{n=0}^{N-1} \mathring{x}(n-m)W_N^{kn}\right) R_N(k) \tag{6.26}$$

Because W_N^{kn} is

$$W_N^{kn} = e^{-j(2\pi/N)kn} = e^{-j(2\pi/N)k(n-m)}e^{-j(2\pi/N)km} = W_N^{k(n-m)}W_N^{km}$$

Equation 6.26 can be changed as

$$\mathcal{D}[x(n-m)] = W_N^{km}\left(\sum_{n=0}^{N-1} \mathring{x}(n-m)W_N^{k(n-m)}\right) R_N(k)$$

$$= W_N^{km}\mathcal{D}[x(n)]$$

Finally, the linearity property of the DFT can be seen by

$$\mathcal{D}[x_1(n) + ax_2(n)] = \left(\sum_{n=0}^{N-1} (\mathring{x}_1(n) + a\mathring{x}_2(n))W_N^{kn}\right) R_N(k)$$

$$= \left(\sum_{n=0}^{N-1} \mathring{x}_1(n)W_N^{kn}\right) R_N(k) + a\left(\sum_{n=0}^{N-1} \mathring{x}_2(n)W_N^{kn}\right) R_N(k)$$

$$= \mathcal{D}[x_1(n)] + a\mathcal{D}[x_2(n)]$$

6.4 CIRCULAR CONVOLUTION

We have found that circular convolution of periodic functions $\mathring{x}_1(n)$ and $\mathring{x}_2(n)$ is equal to the inverse DFS of product $\mathring{X}_1(k)\mathring{X}_2(k)$ where $\mathring{X}_1(k)$ and $\mathring{X}_2(k)$ are the DFS of $\mathring{x}_1(n)$ and $\mathring{x}_2(n)$. We will see here that this is also true for the DFT of finite duration functions.

Let $X_1(k)$ and $X_2(k)$ be the DFT of finite duration functions $x_1(n)$ and $x_2(n)$, respectively. Then the inverse DFT $\mathcal{D}^{-1}[X_1(k)X_2(k)]$ of product $X_1(k)X_2(k)$ can be expressed by Equations 6.23 and 6.24 as

$$\mathcal{D}^{-1}[X_1(k)X_2(k)] = \left(\frac{1}{N}\sum_{k=0}^{N-1} X_1(k)X_2(k)\,W_N^{-kn}\right) R_N(n) \tag{6.27}$$

Because $X_1(k)$ and $X_2(k)$ can be expressed as

$$X_1(k) = \sum_{m=0}^{N-1} \mathring{x}_1(m)W_N^{km}R_N(k)$$

and

$$X_2(k) = \sum_{p=0}^{N-1} \mathring{x}_2(p)W_N^{kp}R_N(k)$$

Equation 6.27 can be written as

$$\mathcal{D}^{-1}[X_1(k)X_2(k)] = \frac{1}{N}\sum_{k=0}^{N-1}\left(\sum_{m=0}^{N-1} \mathring{x}_1(m) W_N^{km} R_N(k)\right)$$

$$\times \left(\sum_{p=0}^{N-1} \mathring{x}_2(p) W_N^{kp} R_N(k)\right) W_N^{-kn} R_N(n) \qquad (6.28)$$

$$= \sum_{m=0}^{N-1}\sum_{p=0}^{N-1} \mathring{x}_2(m)\mathring{x}_1(p) R_N(n) \sum_{k=0}^{N-1}\left(\frac{1}{N} W_N^{-k(n-m-p)} R_N(k)\right)$$

It can be seen that $\sum_{k=0}^{N-1}\left(\frac{1}{N} W_N^{-k(n-m-p)} R_N(k)\right)$ is 1 when $n - m - p$ is either 0 or $\pm N$ and 0 for integer $n - m - p$ is neither 0 nor $\pm N$. Hence, Equation 6.28 becomes

$$\mathcal{D}^{-1}[X_1(k)X_2(k)] = \sum_{m=0}^{N-1} \mathring{x}_1(m)\mathring{x}_2(n-m) R_N(n) \qquad (6.29)$$

The right-hand side of this equation is the circular convolution of finite duration functions $x_1(m)$ and $x_2(m)$, which will be indicated by $x_1(n)*x_2(n)$. Hence, Equation 6.29 can be expressed as

$$\mathcal{D}^{-1}[X_1(k)X_2(k)] = x_1(n)*x_2(n) \qquad (6.30)$$

We will see this circular convolution by using an example. Let us investigate steps to obtain $x_3(n) = \sum_{m=0}^{N-1} \mathring{x}_1(m)\mathring{x}_2(n-m) R_N(n)$ for given $x_1(n)$ and $x_2(n)$ shown in Fig. 6.11.

Let us change the m axis for $x_1(m)$ in a circle having N point as shown in Fig. 6.12(a). Similarly, let us change the m axis for $x_2(m)$ to a circle having N point as shown in Fig. 6.12(b). $x_2(-m)$ can be obtained from $x_2(m)$ by reversing its direction as shown in Fig. 6.12(c). Multiplying $x_2(-m)$ and

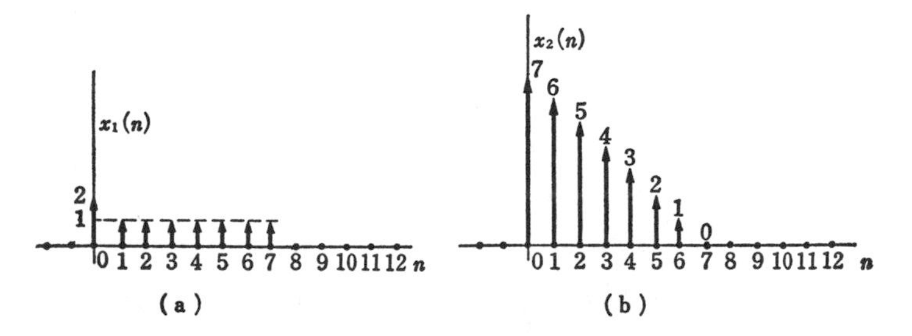

Figure 6.11 Finite Duration Function $x_1(n)$ and $x_2(n)$. (a) $x_1(n)$; (b) $x_2(n)$

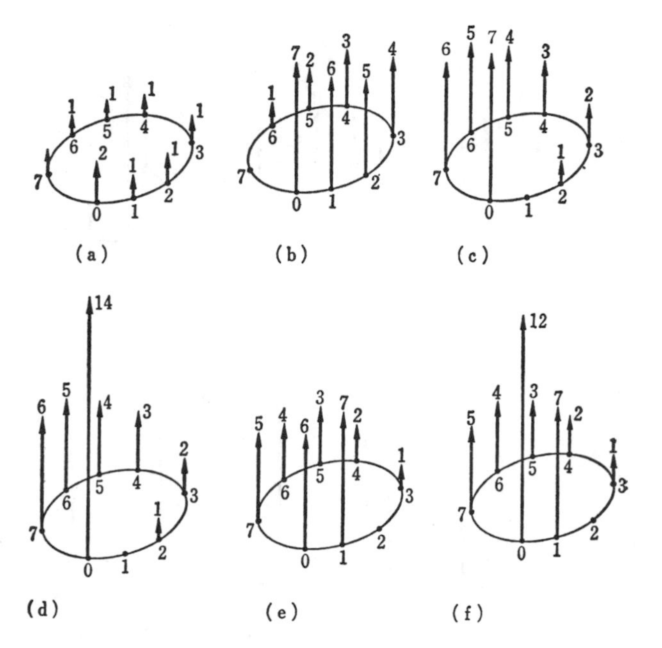

Figure 6.12 Circular Convolution. (a) $x_1(m)$; (b) $x_2(m)$; (c) $x_2(-m)$; (d) $x_1(m)x_2(-m)$; (e) $x_2(1 - m)$; (f) $x_1(m)x_2(1 - m)$.

$x_1(m)$ gives the result shown in Fig. 6.12(d), which is $x_1(m)x_2(-m)$. From $x_3(0) = \sum_{m=0}^{7} \mathring{x}_1(m)\mathring{x}_2(-m)$, summing all values in this figure gives $x_3(0)$. Thus, $x_3(0) = 35$.

Next, because $x_3(1)$ is $\sum_{m=0}^{7} \mathring{x}_1(m)\mathring{x}_2(1 - m)$, it is necessary to have $x_2(1 - m)$, which can be obtained from $x_2(-m)$ by shifting $x_2(-m)$ in Fig. 6.12(c) 1 unit counterclockwise as shown in Fig. 6.12(e). Multiplying $x_2(1 - m)$ and $x_1(m)$ gives $x_1(m)x_2(1 - m)$ as shown in Fig. 6.12(f). Because $x_3(1) = \sum_{m=0}^{7} \mathring{x}_1(m)\mathring{x}_2(1 - m)$, summation of values in the figure gives $x_3(1) = 34$.

Because $x_3(2)$ is $x_3(2) = \sum_{m=0}^{7} \mathring{x}_1(m)\mathring{x}_2(2 - m)$, we shift $x_2(1 - m)$ in Fig. 6.12(e) 1 unit counterclockwise to obtain $x_2(2 - m)$ as shown in Fig. 6.13(a). Multiplying $x_2(2 - m)$ and $x_1(m)$ gives the result shown in Fig. 6.13(b). The sum of the values in this figure, which is 33, is $x_3(2)$.

Next, shifting $x_2(2 - m)$ in Fig. 6.13(a) 1 unit counterclockwise gives $x_2(3 - m)$. Multiplying this with $x_1(m)$ gives $x_1(m)x_2(3 - m)$ as shown in

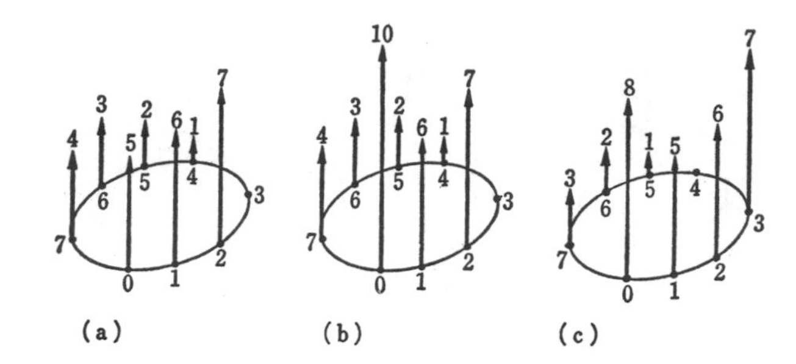

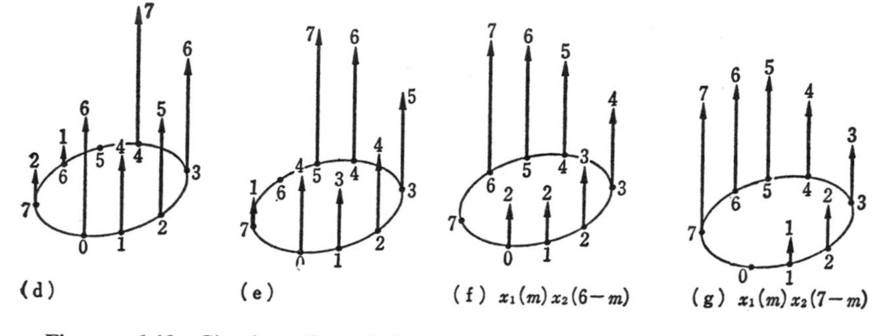

Figure 6.13 Circular Convolution. (a) $x_2(2-m)$; (b) $x_1(m)x_2(2-m)$; (c) $x_1(m)x_2(3-m)$; (d) $x_1(m)x_2(4-m)$; (e) $x_1(m)x_2(5-m)$; (f) $x_1(m)x_2(6-m)$; (g) $x_1(m)x_2(7-m)$.

Fig. 6.13(c), from which we can obtain $x_3(3) = 32$. $x_3(4)$ is the sum of values in Fig. 6.13(d) obtained by multiplying $x_1(m)$ and $x_2(4-m)$, and $x_3(5)$ is the sum of values in Fig. 6.13(e), which is the product of $x_1(m)$ and $x_2(5-m)$, and so on; all values of $x_3(n)$ as shown in Fig. 6.14 can be obtained. It should be clear from this example that circular convolution is convolution associated with circular axis.

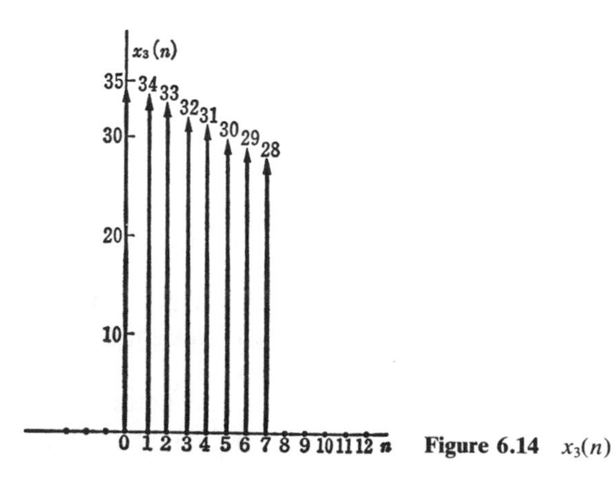

Figure 6.14 $x_3(n)$

Using this example, let us check Equation 6.29. By Equation 6.22, the DFT of $x_1(n)$ is

$$X_1(k) = \sum_{n=0}^{N-1} x_1(n) W_N^{kn} = \sum_{n=0}^{7} x_1(n) W_8^{kn}, \quad 0 \le k \le 7$$

where W_8 is

$$W_8 = e^{-j(2\pi/8)} = e^{-j(\pi/4)}$$

Hence, we can calculate $x_1(0)$, $X_1(1)$, $X_1(2)$, . . . as

$$X_1(0) = \sum_{n=0}^{7} x_1(n) = 9$$

$$X_1(1) = \sum_{n=0}^{7} x_1(n)e^{-j(\pi/4)n} = 2 + \frac{1}{\sqrt{2}}(1 - j1) + (-j)$$
$$+ \frac{1}{\sqrt{2}}(-1 - j1) + (-1) + \frac{1}{\sqrt{2}}(-1 + j1) + (j) + \frac{1}{\sqrt{2}}(1 + j1) = 1$$

$$X_1(2) = \sum_{n=0}^{7} x_1(n)e^{-j(\pi/2)n} = 2 + (-j) + (-1) + (j)$$
$$+ (1) + (-j) + (-1) + (j) = 1$$

Continuing this calculation, we have $X_1(3) = 1$, $X_1(4) = 1$, $X_1(5) = 1$, $X_1(6) = 1$ and $X_1(7) = 1$. Similarly, the DFT of $x_2(n)$ is

$$X_1(0) = \sum_{n=0}^{7} x_2(n) = 28$$

$$X_2(1) = \sum_{n=0}^{7} x_2(n)e^{-j(\pi/4)n} = 7 + \frac{6}{\sqrt{2}}(1 - j1) + (-j5) + \frac{4}{\sqrt{2}}(-1 - j1)$$
$$+ (-3) + \frac{2}{\sqrt{2}}(-1 + j1) + (j) + 0 = 4 - j4(\sqrt{2} + 1)$$

$$X_2(2) = \sum_{n=0}^{7} x_2(n)e^{-j(\pi/2)n} = 7 + (-j6) + (-5) + (j4)$$
$$+ (3) + (-j2) + (-1) + 0 = 4 - j4$$

Continuing this, we will have $X_2(3) = 4 - j4(\sqrt{2} - 1)$, $X_2(4) = 4$, $X_2(5) = 4 + j4(\sqrt{2} - 1)$, $X_2(6) = 4 + j4$, and $X_2(7) = 4 + j4(\sqrt{2} + 1)$.

Because $X_3(k) = X_1(k)X_2(k)$, we have

$$
\begin{aligned}
X_3(0) &= X_1(0)X_2(0) = 252 \\
X_3(1) &= X_1(1)X_2(1) = 4 - j4(\sqrt{2} + 1) \\
X_3(2) &= X_1(2)X_2(2) = 4 - j4 \\
X_3(3) &= 4 - j4(\sqrt{2} - 1) \\
X_3(4) &= 4 \\
X_3(5) &= 4 + j4(\sqrt{2} - 1) \\
X_3(6) &= 4 + j4
\end{aligned}
$$

and $$X_3(7) = 4 + j4(\sqrt{2} + 1)$$

The inverse DFT of these $X_3(k)$ should give those $x_3(n)$ obtained by circular convolution, which we have exercised previously. By Equation 6.23, the inverse DFT of X_3 is

$$x_3(n) = \frac{1}{8}\sum_{n=0}^{7} X_3(k)\, W_8^{-kn}, \quad 0 \leq n \leq 7$$

Hence, $x_3(0)$, $x_3(1)$, $x_3(2)$, . . . are

$$x_3(0) = \frac{1}{8}\sum_{n=0}^{7} X_3(k) = 35$$

$$
\begin{aligned}
x_3(1) &= \frac{1}{8}\sum_{n=0}^{7} X_3(k)\, e^{j(\pi/4)k} \\
&= \frac{1}{8}\{252 + [4(1 + \sqrt{2}) - j4] + (4 + j4) + [4(1 - \sqrt{2}) + j4] \\
&\quad + (-4) + [4(1 - \sqrt{2}) - j4] + (4 - j4) + [4(1 + \sqrt{2}) + j4]\} \\
&= 34
\end{aligned}
$$

Continuing this, we can obtain $x_3(2) = 33$, $x_3(3) = 32$, and so on, which are same values as those we have obtained by circular convolution. Thus, Equation 6.29 is satisfied for this example.

As another example, consider finite duration functions $x_1(n)$ and $x_2(n)$ as shown in Fig. 6.15. Let $x_3(n)$ be the circular convolution of $x_1(n)$ and

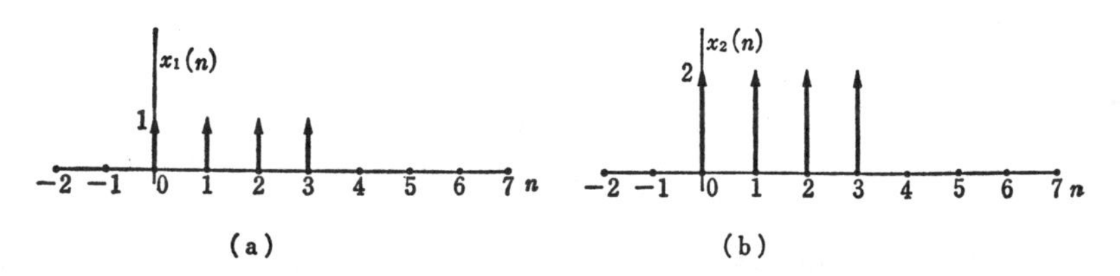

Figure 6.15 Finite Duration Functions $x_1(n)$ and $x_2(n)$. (a) $x_1(n)$; (b) $x_2(n)$.

$x_2(n)$. That is, $x_3(n) = x_1(n)*x_2(n)$. To obtain the DFT $X_3(k)$ of $x_3(n)$, first we will obtain the DFT of $x_1(n)$. By choosing $N = 4$, W_4 is $e^{-j(\pi/2)}$. Hence, from $x_1(n)$ in Fig. 6.15(a), $X_1(k)$ is

$$X_1(0) = \sum_{n=0}^{3} x_1(n) = \sum_{n=0}^{3} 1 = 4$$

$$X_1(1) = \sum_{n=0}^{3} x_1(n)e^{-j(\pi/2)n} = \sum_{n=0}^{3} e^{-j(\pi/2)n} = 1 + (-j) + (-1) + (j) = 0$$

$$X_1(2) = \sum_{n=0}^{3} e^{-j\pi n} = 1 + (-1) + (1) + (-1) = 0$$

and $X_1(3) = \sum_{n=0}^{3} e^{-j(3\pi/2)n} = 1 + (j) + (-1) + (-j) = 0$

Also, from Fig. 6.15(b), $x_2(n)$ is $2x_1(n)$. Hence, $X_2(k)$ is

$$X_2(0) = 8,\ X_2(1) = 0,\ X_2(2) = 0,\ X_2(3) = 0$$

Thus, $X_3(k) = X_1(k)X_2(k)$ is

$$X_3(0) = 32,\ X_3(1) = 0,\ X_3(2) = 0,\ X_3(3) = 0$$

or $X_3(k) = 4X_2(k)$. Thus, the inverse DFT of $X_3(k)$ is $x_3(n) = 4x_2(n)$ or $x_3(n) = 8R_4(n)$. Let us express $x_1(m)$ and $x_2(m)$ on circular axis with $N = 4$ points as shown in Fig. 6.16(a) and (b). $x_2(-m)$ can be obtained from $x_2(m)$ by reversing its direction. We can see that the result will be the same figure as that of $x_2(m)$. Hence, $x_3(0) = \sum_{m=0}^{3} \mathring{x}_1(m)\mathring{x}_2(-m)$, which is the sum of the value in the result of the multiplication of $x_1(m)$ and $x_2(-m)$, which is 8. For $x_3(1) = \sum_{m=0}^{3} \mathring{x}_1(m)\mathring{x}_2(1 - m)$, we need $x_2(1 - m)$, which can be obtained from $x_2(-m)$ by shifting $x_2(-m)$ 1 unit counterclockwise. We can see, however, that shifting 1 unit counterclockwise will not change the

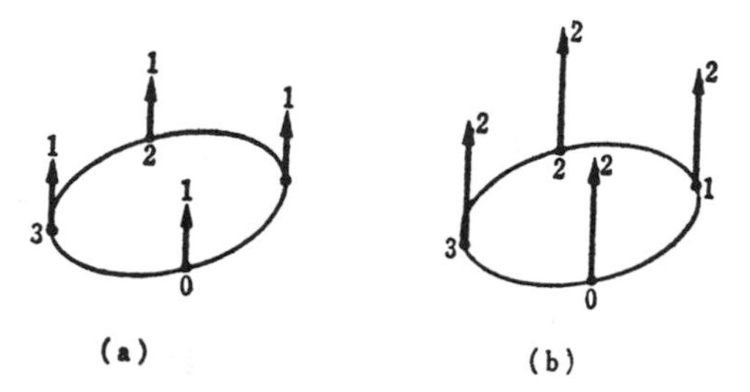

Figure 6.16 $x_1(m)$ and $x_2(m)$. (a) $x_1(m)$; (b) $x_2(m)$.

figure. Hence, $x_2(1 - m)$ is the same as $x_2(-m)$. Thus, $x_3(1)$, which is the sum of the values in the result of multiplication of $x_1(m)$ and $x_2(1 - m)$, will be the *same* as $x_3(0)$, which is 8. Similarly, we have $x_3(2) = 8$ and $x_3(3) = 8$. Thus, $x_3(n) = 4x_2(n)$ shows that Equation 6.29 is true for this case.

6.5 PERIOD OF CIRCULAR CONVOLUTION

We have studied that a finite duration function is considered as one period of a periodic function to obtain the DFT. Many periodic functions exist whose one period is equal to a given finite duration function. Will choice of periodic function make any difference when we obtain the DFT of a given finite duration function? To see an effect of choice of periodic functions on the DFT, we will use a period N to be 8 rather than 4 this time for the same finite duration functions $x_1(n)$ and $x_2(n)$ in Fig. 6.15.

Because a circular axis has 8 points, representing $x_1(m)$ and $x_2(m)$ on a circular axis becomes as shown in Fig. 6.17(a) and (b). From this $x_2(m)$,

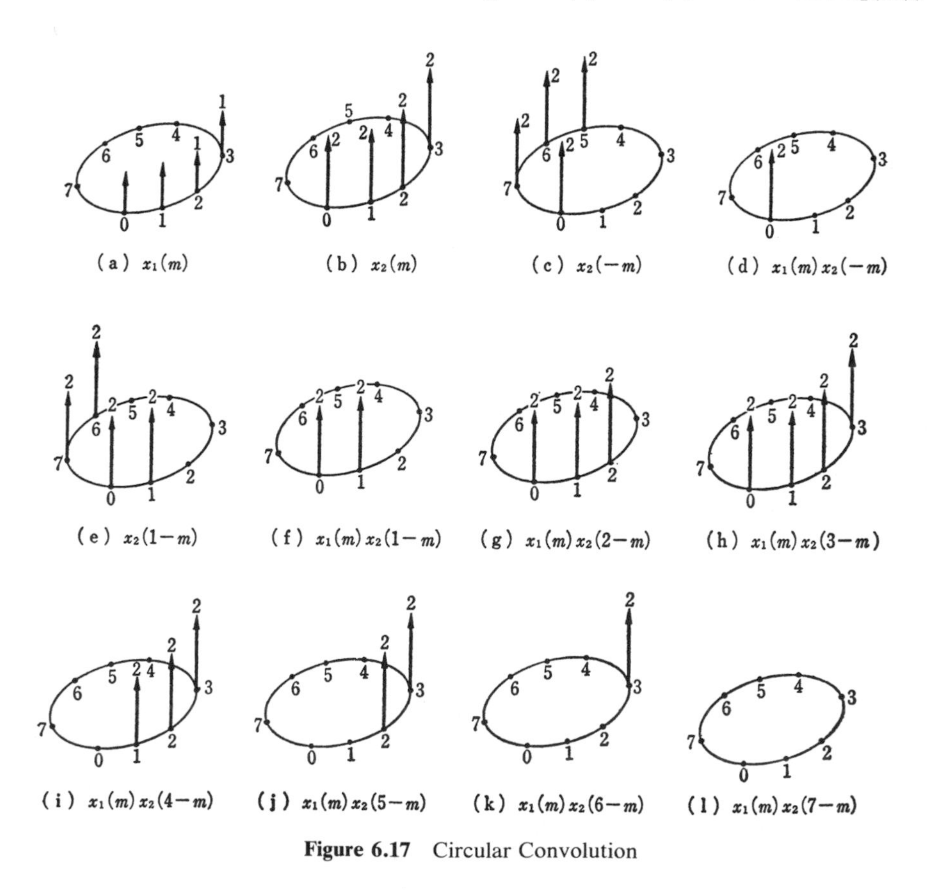

Figure 6.17 Circular Convolution

$x_2(-m)$ will be the result shown in Fig. 6.17(c). Hence, $x_1(m)x_2(-m)$ is the result shown in Fig. 6.17(d). Thus, $x_3(0) = 2$ is different from the result obtained previously. Why is the value of $x_3(0)$ different from that in the previous case? Because we take $N = 8$, the circular convolution of $x_1(n)$ and $x_2(n)$ is

$$x_3(n) = x_1(n)*x_2(n) = \sum_{n=0}^{7} x_1(m)x_2(n - m)$$

Hence, $x_3(n)$ will have the different value compared with one obtained previously. With this circular convolution does Equation 6.29 satisfy? To check this, we will obtain $x_3(n)$ for $n = 0, 1, 2, \ldots, 8$, first. Because we already have obtained $x_3(0)$, let us obtain $x_3(1)$. For $x_3(1)$, $x_2(1 - m)$ is obtained by shifting 1 unit of $x_2(-m)$ counterclockwise as shown in Fig. 6.17(e). Then multiplying $x_1(m)$ and $x_2(1 - m)$ gives the result in Fig. 6.17(f), from which $x_3(1) = 4$ is obtained.

Next, we shift $x_2(1 - m)$ 1 unit counterclockwise to obtain $x_2(2 - m)$, and calculate $x_1(m)x_2(2 - m)$ as shown in Fig. 6.17(g). From this result, we can obtain $x_3(2) = 6$. Similarly, we can obtain $x_1(m)x_2(3 - m)$ in Fig. 6.17(h), $x_1(m)x_2(4 - m)$ in Fig. 6.17(i), $x_1(m)x_2(5 - m)$ in Fig. 6.17(j), $x_1(m)x_2(6 - m)$ in Fig. 6.17(k), and $x_1(m)x_2(7 - m)$ in Fig. 6.17(l). From these, we have $x_3(3) = 8$, $x_3(4) = 6$, $x_3(5) = 4$, $x_3(6) = 2$, and $x_3(7) = 0$, which are shown in Fig. 6.18. This $x_3(n)$ is the same as the convolution of $x_1(n)$ and $x_2(n)$, which is

$$x_3(n) = \sum_{m=-\infty}^{\infty} \mathring{x}_1(m)\mathring{x}_2(n - m)R_N(n)$$

This means that if we take a period N large enough, the result of circular convolution becomes the same as that of convolution. A question is whether Equation 6.29 can satisfy when $N = 8$. The DFT $X_1(k)$ of $x_1(n)$ when $N = 8$ is

$$X_1(k) = \sum_{n=0}^{7} x_1(n) W_8^{kn}$$

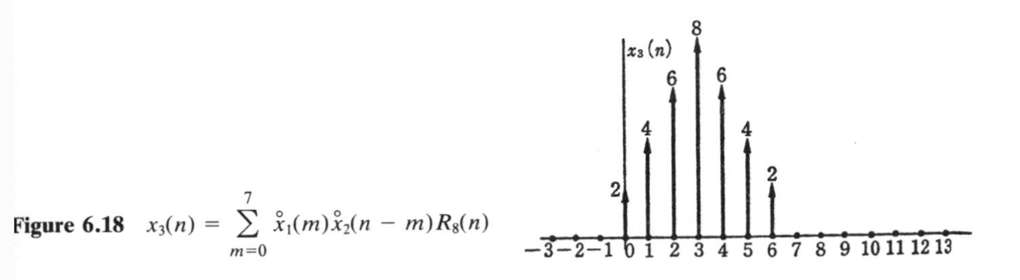

Figure 6.18 $x_3(n) = \sum_{m=0}^{7} \mathring{x}_1(m)\mathring{x}_2(n - m)R_8(n)$

and W_8 is

$$W_8 = e^{-i(2\pi/8)} = e^{-j(\pi/4)}$$

Hence,

$$X_1(0) = \sum_{n=0}^{7} x_1(n) = 4$$

$$X_1(1) = \sum_{n=0}^{7} x_1(n)e^{-j(\pi/4)n} = 1 + \frac{1}{\sqrt{2}}(1 - j1) + (-j) + \frac{1}{\sqrt{2}}(-1 - j1) = 1 - j(1 + \sqrt{2})$$

$$X_1(2) = \sum_{n=0}^{7} x_1(n)e^{-j(\pi/2)n} = 1 + (-j) + (-1) + (j) = 0$$

Continuing this to obtain $X_1(k)$ for $k = 3, 4, 5, 6, 7$, we have $X_1(3) = 1 + j(1 - \sqrt{2})$, $X_1(4) = 0$, $X_1(5) = 1 - j(1 - \sqrt{2})$, $X_1(6) = 0$, and $X_1(7) = 1 + j(1 + \sqrt{2})$.

Because $x_2(n) = 2x_1(n)$, the DFT $X_2(k)$ of $x_2(n)$ for $N = 8$ is

$$X_2(k) = \sum_{n=0}^{7} x_2(n) W_8^{kn} = 2 \sum_{n=0}^{7} x_1(n) W_8^{kn} = 2X_1(k)$$

Hence, from the $X_1(k)$ obtained previously, we have $X_2(0) = 8$, $X_2(1) = 2[1 - j(1 + \sqrt{2})]$, $X_2(2) = 0$, $X_2(3) = 2[1 + j(1 - \sqrt{2})]$, $X_2(4) = 0$, $X_2(5) = 2[1 - j(1 - \sqrt{2})]$, $X_2(6) = 0$, $X_2(7) = 2[1 + j(1 + \sqrt{2})]$. To obtain $x_3(n)$, by the use of Equation 6.29 we will calculate $X_3(k) = X_1(k)X_2(k)$, which is

$$X_3(0) = 32,\ X_3(1) = 4(1 + \sqrt{2})(-1 - j1),\ X_3(2) = 0$$
$$X_3(3) = 4(1 - \sqrt{2})(-1 + j1),\ X_3(4) = 0,\ X_3(5) = 4(1 - \sqrt{2})(-1 - j1)$$
$$X_3(6) = 0,\ X_3(7) = 4(1 + \sqrt{2})(-1 + j1)$$

Hence, by Equation 6.23, the inverse DFT of $X_3(k)$ is

$$x_3(n) = \frac{1}{8} \sum_{k=0}^{7} X_3(k) W_8^{-kn}, \qquad 0 \le n \le N - 1$$

Thus,

$$x_3(0) = \frac{1}{8} \sum_{k=0}^{7} X_3(k) = \frac{1}{8}[32 + 4(1 + \sqrt{2})(-1 - j1) + 4(1 - \sqrt{2})(-1 + j1) + 4(1 - \sqrt{2})(-1 - j1) + 4(1 + \sqrt{2})(-1 + j1)] = 2$$

$$x_3(1) = \frac{1}{8}\sum_{k=0}^{7} X_3(k)e^{j(\pi/4)k} = \frac{1}{8}[32 + 4(1 + \sqrt{2})(-j\sqrt{2})$$
$$+ 4(1 - \sqrt{2})(-j\sqrt{2}) + 4(1 - \sqrt{2})(j\sqrt{2}) + 4(1 + \sqrt{2})(j\sqrt{2})]$$
$$= 4$$

$$x_3(2) = \frac{1}{8}\sum_{k=0}^{7} X_3(k)e^{j(\pi/2)k} = \frac{1}{8}[32 + 4(1 + \sqrt{2})(1 - j1)$$
$$+ 4(1 - \sqrt{2})(1 + j1) + 4(1 - \sqrt{2})(1 - j1) + 4(1 + \sqrt{2})(1 + j1)]$$
$$= 6$$

By continuing this process, we can calculate, $x_3(n)$ for $n = 3, 4, 5, 6, 7$ as $x_3(3) = 8, x_3(4) = 6, x_3(5) = 4, x_3(6) = 2$, and $x_3(7) = 0$. It is clear that these results are the same as those obtained by circular convolution of $x_1(n)$ and $x_2(n)$ calculated previously. The reader should make sure that choosing the large value of a period N will not influence the development of Equation 6.29.

6.6 SUMMARY

1. The DFS of a periodic function $\mathring{x}(n)$ is

$$\mathring{X}(k) = \sum_{n=0}^{N-1} \mathring{x}(n)W_N^{kn}$$

and its inverse DFS is

$$\mathring{x}(n) = \frac{1}{N}\sum_{k=0}^{N-1} \mathring{X}(k)W_N^{-kn}$$

where N is a period and $W_N = e^{-j(2\pi/N)}$

2. Let $\mathring{X}_r(k)$ be the DFS of a periodic function $\mathring{x}_r(n)$. Then,

a. the DFS $\mathring{X}_3(n)$ of $\mathring{x}_3(n) = a\mathring{x}_1(n) + b\mathring{x}_2(n)$ is

$$\mathring{X}_3(k) = a\mathring{X}_1(k) + b\mathring{X}_2(k)$$

b. the DFS $\mathring{X}(p)$ of $\mathring{x}(n + m)$ is

$$\mathring{X}(p) = W_N^{-km}\mathring{X}(k)$$

where $\mathring{X}(k)$ is the DFS of $\mathring{x}(n)$, and

c. the DFS $\mathring{X}_3(k)$ of $\mathring{x}_3(n)$, which is the circular convolution of $\mathring{x}_1(n)$ and $\mathring{x}_2(n)$, that is,

$$\mathring{x}_3(n) = \sum_{m=0}^{N-1} \mathring{x}_1(m)\mathring{x}_2(n-m)$$

is

$$\mathring{X}_3(k) = \mathring{X}_1(k)\mathring{X}_2(k)$$

3. A finite duration (interval) of a finite duration function $x(n)$ is from 0 to $N - 1$ means $x(n) = 0$ for all n satisfying $n < 0$ or $n > N$.

4. A periodic function $\mathring{x}(n)$ is a corresponding function of a finite duration function $x(n)$ if one period of $\mathring{x}(n)$ is equal to $x(n)$.

5. The DFT $X(k)$ of a finite duration function $x(n)$ is

$$X(k) = \begin{cases} \sum_{n=0}^{N-1} x(n)W_N^{kn}, & 0 \le k \le N-1 \\ 0, & \text{Otherwise} \end{cases}$$

and its inverse DFT is

$$x(n) = \begin{cases} \dfrac{1}{N}\sum_{k=0}^{N-1} X(k)W_N^{-kn}, & 0 \le n \le N-1 \\ 0, & \text{Otherwise} \end{cases}$$

where the finite duration pf $x(n)$ is from 0 to $N - 1$.

6. The symbol $R_N(p)$ indicates a finite duration function defined by

$$R_N(p) = \begin{cases} 1 & 0 \le p \le N-1 \\ 0 & \text{Otherwise} \end{cases}$$

7. A relationship between a finite duration function $x(n)$ and a corresponding periodic function $\mathring{x}(n)$ is

$$x(n) = \mathring{x}(n)R_N(n)$$

Also a relationship between the DFT $X(k)$ of $x(n)$ and the DFT $\mathring{X}(k)$ is $\mathring{x}(n)$ is

$$X(k) = \mathring{X}(k)R_N(k)$$

8. Let $\mathfrak{D}[x_1(n)]$ and $\mathfrak{D}[x_2(n)]$ be the DFT of $x_1(n)$ and $x_2(n)$, respectively. Then

a. The DFT $\mathfrak{D}[ax_1(n) + bx_2(n)]$ of $ax_1(n) + bx_2(n)$ is

$$\mathfrak{D}[ax_1(n) + bx_2(n)] = a\mathfrak{D}[x_1(n)] + b\mathfrak{D}[x_2(n)]$$

b. the DFT $\mathfrak{D}[x_1(n + m)]$ of $x_1(n + m)$ is defined as

$$\mathfrak{D}[x_1(n + m)] = W_N^{-km}\mathfrak{D}[x_1(n)]$$

and

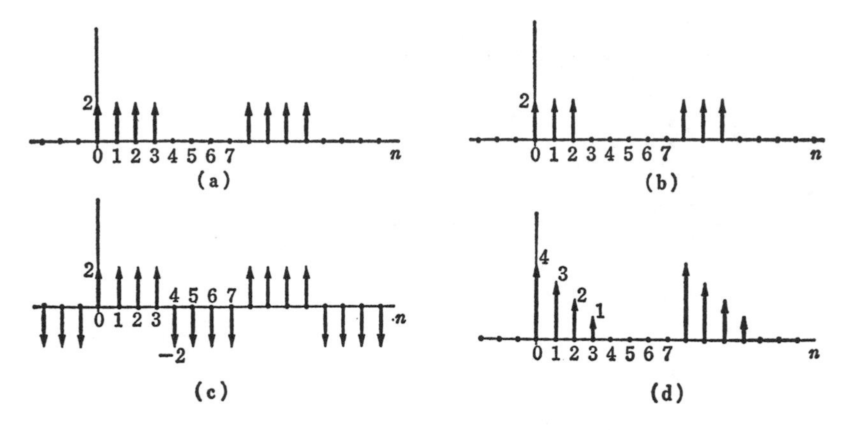

Figure 6.19 Periodic Functions

c. the DFT $\mathscr{D}[x_1(n) * x_2(n)]$ of circular convolution $x_1(n) * x_2(n)$ is

$$\mathscr{D}[x_1(n) * x_2(n)] = \mathscr{D}[x_1(n)]\mathscr{D}[x_2(n)]$$

where $x_1(n) * x_2(n) = \sum_{m=0}^{N-1} \mathring{x}_1(m)\mathring{x}_2(n - m)R_N(n)$

6.7 PROBLEMS

1. Obtain the DFS of periodic functions shown in Fig. 6.19.
2. Obtain the DFS of functions obtained from periodic functions shown in Fig. 6.19 by delaying k unit where (a) $k = 4$ and (b) $k = 8$.

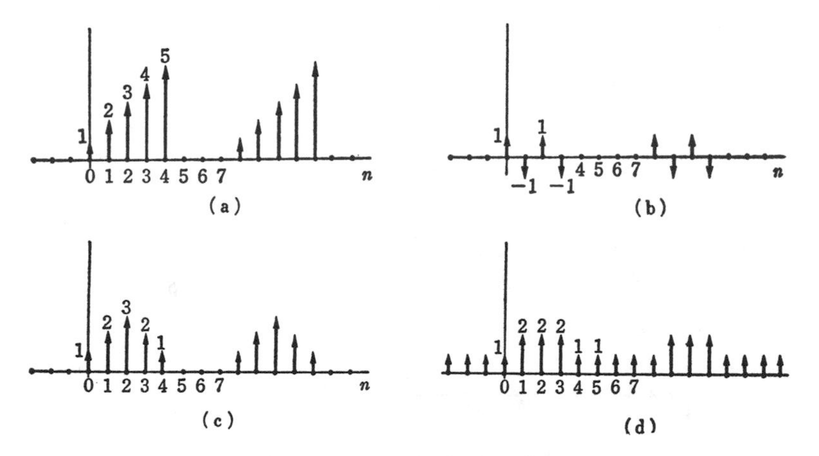

Figure 6.20 Periodic Functions

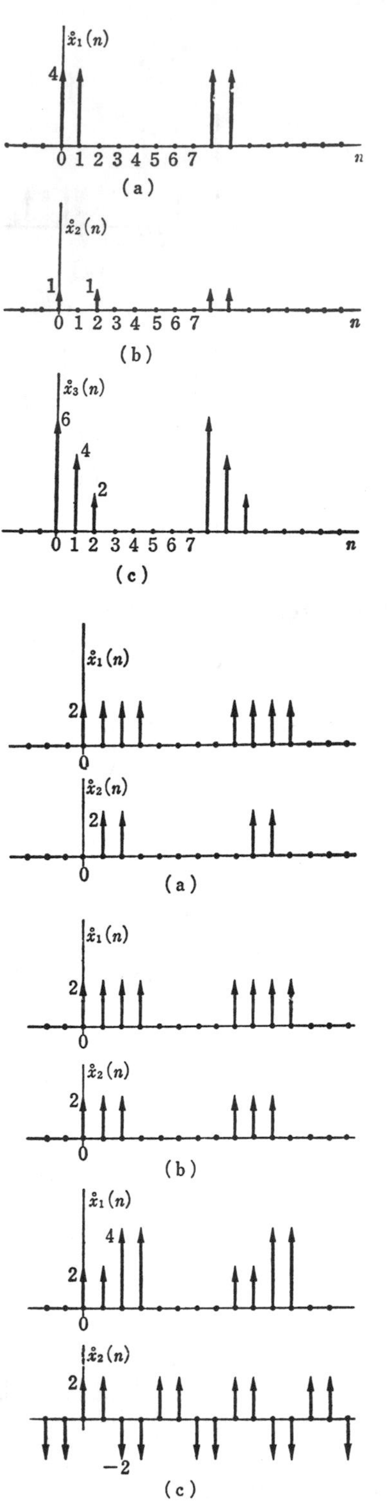

Figure 6.21 $\mathring{x}_1(n)$, $\mathring{x}_2(n)$, and $\mathring{x}_3(n)$. (a) $\mathring{x}_1(n)$; (b) $\mathring{x}_2(n)$; (c) $\mathring{x}_3(n)$.

Figure 6.22 $\mathring{x}_1(n)$ and $\mathring{x}_2(n)$

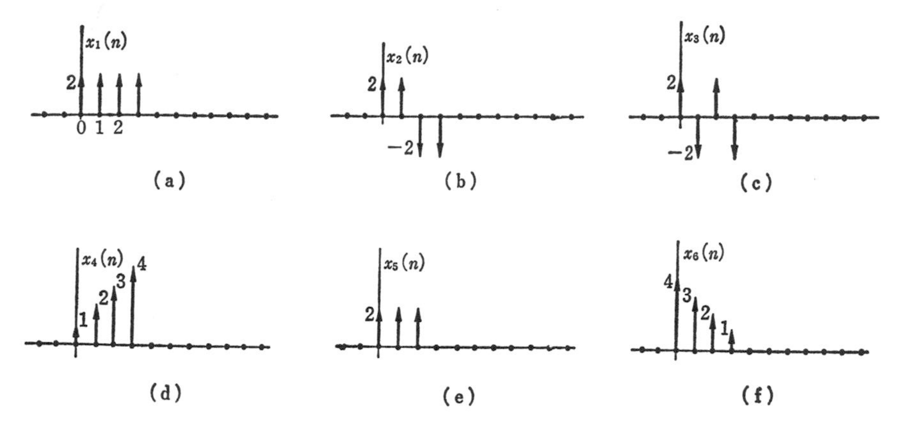

Figure 6.23 Finite Duration Functions

3. Let N be the period of a periodic function $\mathring{x}(n)$ and $\mathring{X}_1(k)$ be the DFS of $\mathring{x}(n)$ by using N as its period. Let $\mathring{X}_2(k)$ be the DFS of $\mathring{x}(n)$ by taking $2N$ as a period of $\mathring{x}(n)$. Explain the relationship between $\mathring{X}_1(k)$ and $\mathring{X}_2(k)$.
4. Obtain the DFS of periodic functions in Fig. 6.19 by using the z-transformation.
5. Obtain the DFS of periodic functions in Fig. 6.20.
6. Obtain the DFS of $\mathring{x}_1(n)$, $\mathring{x}_2(n)$, and $\mathring{x}_3(n)$ in Fig. 6.21.
7. Calculate the circular convolution $\sum_{m=0}^{N-1} \mathring{x}_1(m)\mathring{x}_2(n-m)R_N(n)$ of $\mathring{x}_1(n)$ and $\mathring{x}_2(n)$ in Fig. 6.22.
8. Calculate the DFT of finite duration functions in Fig. 6.23 with $N = 4$.
9. Obtain the DFT of finite duration functions in Fig. 6.23 with $N = 8$.
10. Obtain the DFT of (1) $x_1(n) + 2x_2(n)$, (2) $x_1(n) + x_3(n)$, (3) $x_1(n) - x_3(n)$, (4) $x_2(n) + x_4(n)$, (5) $x_1(n) - x_4(n)$, (6) $x_4(n) - x_6(n)$ with $N = 4$ where $x_1(n)$, $x_2(n)$, $x_3(n)$, $x_4(n)$, $x_5(n)$, and $x_6(n)$ are shown in Fig. 6.23.
11. Calculate the circular convolution of the following finite durations functions

    a. $x_1(n)$ and $x_2(n)$ **b.** $x_1(n)$ and $x_3(n)$ **c.** $x_1(n)$ and $x_4(n)$
    d. $x_1(n)$ and $x_5(n)$ **e.** $x_1(n)$ and $x_6(n)$ **f.** $x_2(n)$ and $x_3(n)$
    g. $x_2(n)$ and $x_4(n)$ **h.** $x_2(n)$ and $x_5(n)$ **i.** $x_3(n)$ and $x_4(n)$
    j. $x_3(n)$ and $x_5(n)$ **k.** $x_4(n)$ and $x_5(n)$ **l.** $x_4(n)$ and $x_6(n)$
    m. $x_5(n)$ and $x_6(n)$

    where each finite duration function is shown in Fig. 6.23.

7
Fast Digital Fourier Transformation

7.1 REDUCTION OF STEPS FOR CALCULATING *X(k)* WHEN $N = 2^q$

When a computer is employed for digital signal processing, an important factor is the number of steps necessary for calculation. Consider the number of steps required to obtain the DFT of a finite duration function. The DFT of a finite duration function $x(n)$, discussed in Chapter 6 is

$$X(k) = \sum_{n=0}^{N-1} x(n) W_N^{kn} \qquad k = 0, 1, \cdots, N-1 \tag{7.1}$$

where W_N in the right-hand side is

$$W_N = e^{-j(2\pi/N)} \tag{7.2}$$

By employing the symbol Re F for expressing the real part and the symbol Im F for indicating the imaginary part of the function F, Equation 7.1 becomes

$$X(k) = \sum_{n=0}^{N-1} [(\text{Re } x(n) \text{Re } W_N^{kn} - \text{Im } x(n) \text{Im } W_N^{kn}) + j(\text{Re } x(n) \text{Im } W_N^{kn} + \text{Im } x(n) \text{Re } W_N^{kn})]$$

This equation shows that to calculate $X(k)$ for one value of k, it requires $4N$ multiplications and $4N - 2$ additions. Because k changes from 0 to $N - 1$, to obtain $X(k)$ for all k, it requires $4N^2$ multiplications and $(4N - 2)N$ additions. If a computer can handle each arithmetic operation of complex numbers by one step, these number of steps will reduce to N^2 multiplications and $(n - 1)N$ additions. One effective way of reducing these numbers of steps for calculating the DFT of a finite duration function is the FFT. The FFT employs two techniques for reduction of steps, one of which is to combine the calculation of small parts to obtain $X(k)$, and the other is to use properties of W_N^{kn} effectively.

To explain the steps of FFT, we will first consider the case when $N = 2^q$ for an integer q. Because N is an even number 2^q, the DFT in Equation 7.1 can be expressed as

$$X(k) = \sum_{n=0}^{N-1} x(n)W_N^{kn} = \sum_{n \text{ even}} x(n)W_N^{kn} + \sum_{n \text{ odd}} x(n)W_N^{kn} \tag{7.3}$$

Those n that are of even number can be expressed as $2r$, and those n that are of odd number can be expressed as $2r + 1$. Hence, Equation 7.3 can be changed as

$$\begin{aligned} X(k) &= \sum_{r=0}^{(N/2)-1} x(2r)W_N^{k(2r)} + \sum_{r=0}^{(N/2)-1} x(2r+1)W_N^{k(2r+1)} \\ &= \sum_{r=0}^{(N/2)-1} x(2r)(W_N^2)^{kr} + W_N^k \sum_{r=0}^{(N/2)-1} x(2r+1)(W_N^2)^{kr} \end{aligned} \tag{7.4}$$

Because W_N^2 in the second term in the right-hand side can be expressed as

$$W_N^2 = e^{-j(2\pi/N)2} = e^{-j(2\pi)(N/2)} = W_{N/2}$$

Equation 7.4 can further be changed as

$$X(k) = \sum_{r=0}^{(N/2)-1} x(2r)W_{N/2}^{kr} + W_N^k \sum_{r=0}^{(N/2)-1} x(2r+1)W_{N/2}^{kr} \tag{7.5}$$

We can express terms in the right-hand side of Equation 7.5 as

$$X(k) = B_0(k) + W_N^k B_1(k) \tag{7.6}$$

which shows that $X(k)$ can be obtained by first calculating $B_0(k)$ and $B_1(k)$, and then summing the results as shown by a signal flow graph in Fig. 7.1.

From Equations 7.5 and 7.6, we can see that $B_0(k)$ and $B_1(k)$ are the DFT of finite duration functions, whose period is, respectively, $N/2$. For convenience, the DFT of a finite duration function of $N = k$ is called the "k-point DFT." With this terminology, these $B_0(k)$ and $B_1(k)$ are $N/2$-point DFTs.

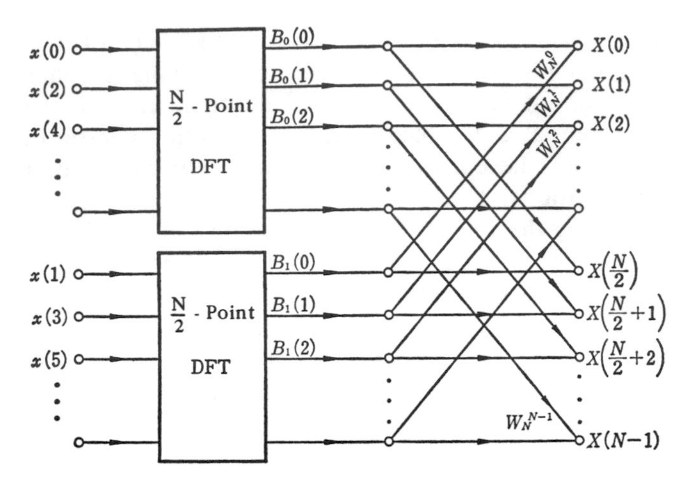

Figure 7.1 Combination of Two $N/2$-Point DFTs

We can see that the number of steps necessary to compute $B_0(0)$, $B_0(1)$, . . . and $B_0(N/2 - 1)$ is $N/2\,(N/2 - 1)$ additions and $(N/2)^2$ multiplications. Similarly, the number of steps necessary to calculate $B_1(0)$, $B_1(1)$, . . . and $B_1(N/2 - 1)$ is $N/2\,(N/2 - 1)$ additions and $(N/2)^2$ multiplications. Because of $B_0(k) + W_N^k B_1(k)$ in the right-hand side of Equation 8.8, calculation of $X(k)$ requires N multiplications and N additions to these steps to obtain $B_0(k)$ and $B_1(k)$. Hence, the number of steps necessary to obtain $X(k)$ by Equation 7.6 is

$$2\frac{N}{2}\left(\frac{N}{2} - 1\right) + N = N\left(\frac{N}{2} - 1\right) + N$$

additions and

$$2\left(\frac{N}{2}\right)^2 + N = \frac{N^2}{2} + N$$

multiplications, which is fewer than those necessary to obtain $X(k)$ compared with that by using Equation 7.1.

Because $N = 2^q$, $N/2$ is an even number. Hence, each $B_0(k)$ and $B_1(k)$ can be split again to reduce the number of steps for calculation.

From Equations 7.5 and 7.6, $B_0(k)$ is

$$B_0(k) = \sum_{r=0}^{(N/2)-1} x(2r)W_{N/2}^{kr}$$

By r being an even or odd number, the right-hand side of this equation can be changed as

$$B_0(k) = \sum_{\text{even } r} x(2r)W_{N/2}^{kr} + \sum_{\text{odd } r} x(2r)W_{N/2}^{kr}$$

Those r that are of even number can be expressed as $2s$, and those r that are of odd number can be expressed as $2s + 1$. Hence, the preceding equation can be written as

$$\begin{aligned} B_0(k) &= \sum_{s=0}^{(N/4)-1} x(4s)W_{N/2}^{k(2s)} + \sum_{s=0}^{(N/4)-1} x(4s+2)W_{N/2}^{k(2s+1)} \\ &= \sum_{s=0}^{(N/4)-1} x(4s)W_{N/4}^{ks} + W_{N/2}^{k} \sum_{s=0}^{(N/4)-1} x(4s+2)W_{N/4}^{ks} \end{aligned} \tag{7.7}$$

Let us use the symbols $B_{00}(k)$ and $B_{01}(k)$ to indicate the right-hand side of Equation 7.7 as

$$B_0(k) = B_{00}(k) + W_{N/2}^{k}B_{01}(k) \tag{7.8}$$

Similarly, $B_1(k)$ can be expressed as

$$B_1(k) = \sum_{r=0}^{(N/2)-1} x(2r+1)W_{N/2}^{kr} \tag{7.9}$$

By splitting the right-hand side of this equation according to r being an even or odd number, we can express Equation 7.13 as

$$B_1(k) = \sum_{s=0}^{(N/4)-1} x(4s+1)W_{N/4}^{ks} + W_{N/2}^{k} \sum_{s=0}^{(N/4)-1} x(4s+3)W_{N/4}^{ks}$$

By using the symbols $B_{10}(k)$ and $B_{11}(k)$, this equation can be expressed as

$$B_1(k) = B_{10}(k) + W_{N/2}^{k}B_{11}(k) \tag{7.10}$$

By Equations 7.8 and 7.10, the DFTs $B_0(k)$ and $B_1(k)$ indicated by two boxes in Fig. 7.1 can be changed to four $n/4$-points DFTs $B_{00}(k)$, $B_{01}(k)$, $B_{10}(k)$, and $B_{11}(k)$ as shown in Fig. 7.2.

Because $N/4$ is an even number, we can split each of $N/4$-point DFTs $B_{00}(k)$, $B_{01}(k)$, $B_{10}(k)$ and $B_{11}(k)$ in this figure by the same technique as those used previously. This splitting of each DFT can be continued until we have the 2-point DFTs shown in Fig. 7.3(a).

The equation for the 2-point DFT is

$$u(k) = \sum_{n=0}^{1} x(v(n))W_2^{kn} \qquad k = 0, 1$$

By expanding this, we have

$$u(k) = x(v(0))W_2^{0k} + x(v(1))W_2^{1k} \tag{7.11}$$

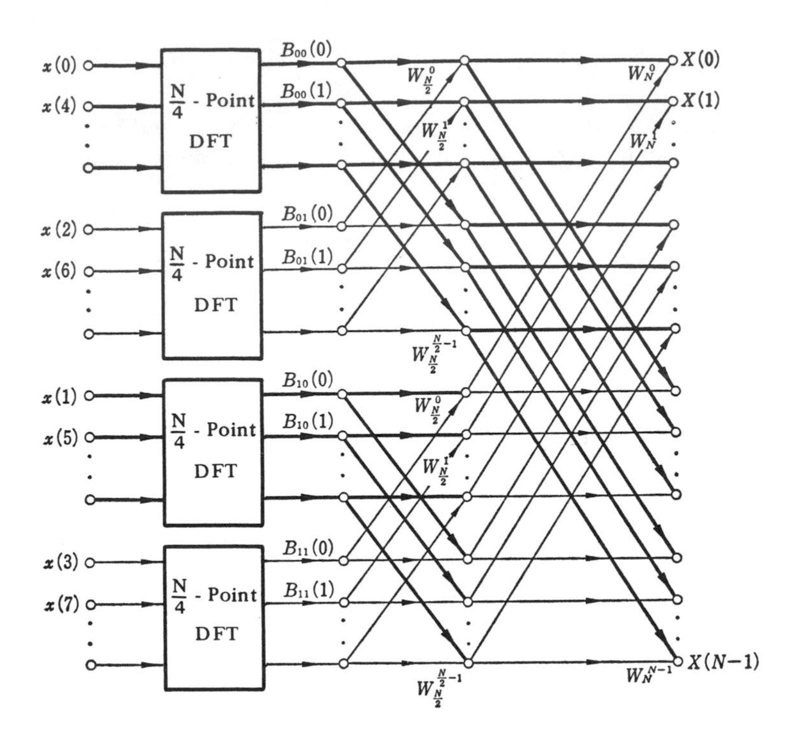

Figure 7.2 Combination of Four $N/4$-Point DFTs

where W_2^{0k} is

$$W_2^{0k} = e^{-j(2\pi/2)0k} = 1$$

and W_2^{1k} is

$$W_2^{1k} = e^{-j(2\pi/2)1k} = \begin{cases} 1, & k = 0 \\ -1, & k = 1 \end{cases}$$

Hence, Equation 7.17 can be expressed as

$$\left.\begin{aligned} u(0) &= x(v(0)) + x(v(1)) \\ u(1) &= x(v(0)) - x(v(1)) \end{aligned}\right\} \qquad (7.12)$$

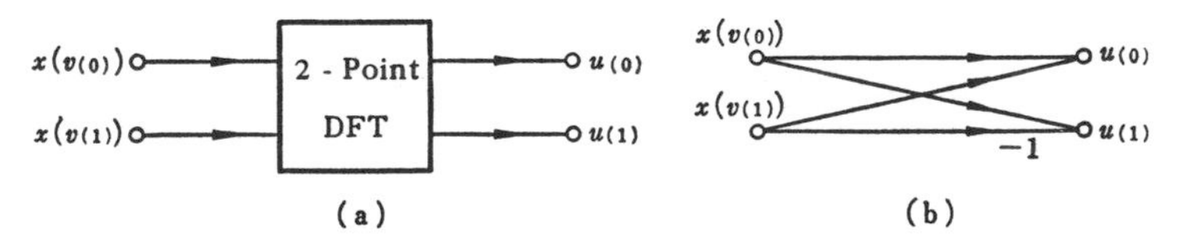

Figure 7.3 2-Point DFT. (a) 2-Point DFT; (b) Simplified 2-Point DFT.

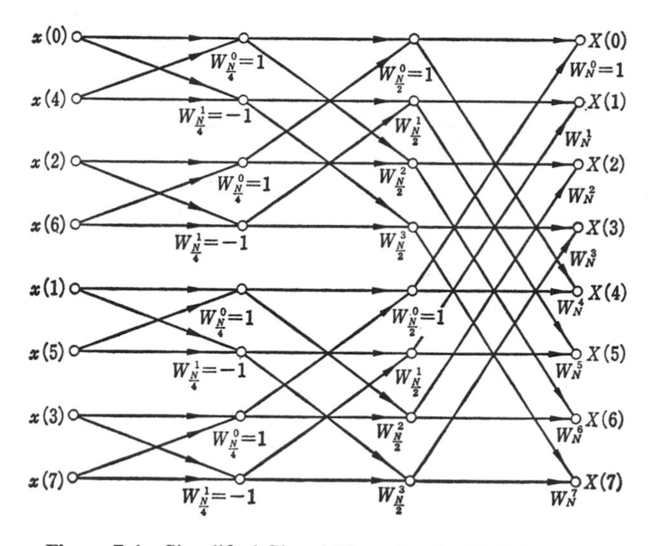

Figure 7.4 Simplified Signal Flow Graph of 8-Point DFT

Thus, this 2-point DFT can become a simple signal flow graph as shown in Fig. 7.3(b). With these simplifications, the number of steps necessary to calculate the N-point DFT, which is the DFT of a finite duration function with a period N, is $N \log_2 N$ multiplications and additions of complex numbers. As an example, a simplified flow graph of the DFT of a finite duration function with $N = 8$ is shown in Fig. 7.4.

7.2 DECIMATION-IN-TIME FFT AND DECIMATION-IN-FREQUENCY FFT

To reduce the number of steps further, consider connections at $X(r)$ in Fig. 7.1. At $X(0)$, there is a branch from $B_0(0)$ and a branch from $B_1(0)$. From $B_0(0)$ and $B_1(0)$ there are branches to $x(N/2)$ as shown in Fig. 7.5(a). In general, at $X(p)$ $(0 \leq p \leq (N/2) - 1)$ there are two branches from $B_0(p)$ and $B_1(p)$. Also from $B_0(p)$ and $B_1(p)$ there are branches to $X(N/2 + p)$ as shown in Fig. 7.5(b).

From $B_0(p)$, the weight of a branch to $X(p)$ is W_N^p and that of a branch to $X(N/2 + p)$ is $W_N^{(N/2+p)}$, where W_N^p is

$$W_N^p = e^{-j(2\pi/N)p}$$

and $W_N^{(N/2+p)}$ *is*

$$W_N^{(N/2+p)} = e^{-j(2\pi/N)[(N/2)+p]} = -e^{-j(2\pi/N)p} = -W_N^p$$

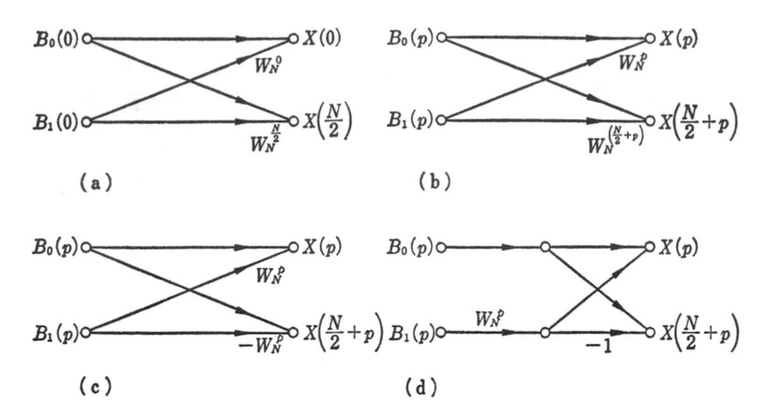

Figure 7.5 Simplifying Subgraph. (a) Branches from $B_0(0)$ and $B_1(0)$; (b) Branches from $B_0(p)$ and $B_1(p)$; (c) Changing $W_N^{(N/2+p)}$ to $-W_N^p$; (d) Simplified Graph.

Hence, $W_N^{(N/2+p)}$ in Fig. 7.5(b) can be changed to $-W_N^p$ as shown in Fig. 7.5(c). Because by an equivalent transformation of types I and IV, the signal flow graph in Fig. 7.5(c) can be changed to the one in Fig. 7.5(d). Using this modified signal flow graph, the number of steps will be reduced by one. Hence, using this modification for all possible places in Fig. 7.1, we will have the signal flow graph shown in Fig. 7.6. Because W_N^0 is

$$W_N^0 = e^{-j(2\pi/N)0} = 1$$

the branches whose weight is W_N^0 will not indicate its weight.

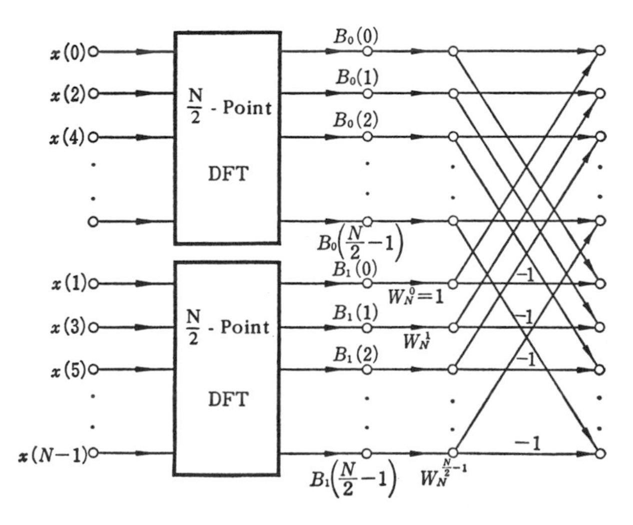

Figure 7.6 Simplified Subgraphs of Graph in Fig. 8.1.

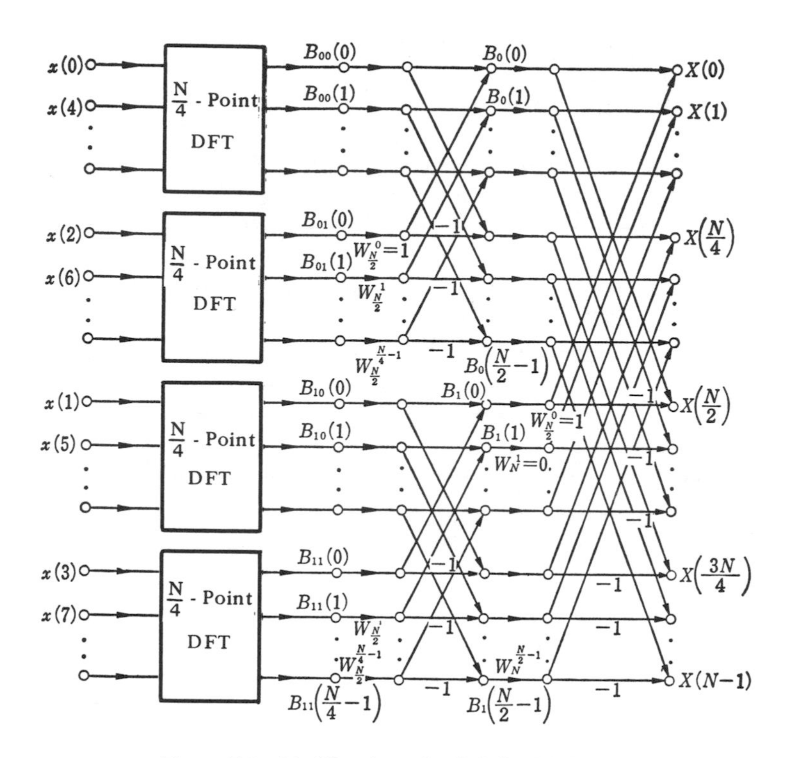

Figure 7.7 Modification of $N/2$-Point DFTs.

Similar simplification can be applied to those $N/2$-point DFTs. Hence, a signal flow graph in Fig. 7.2 can be modified as the one shown in Fig. 7.7. A signal flow graph obtained by applying this modification to all r-point DFTs for $r > 2$ is an algorithm called the "decimation-in-time FFT." It is clear that the number of steps necessary to compute the DFT of a finite duration function is reduced tremendously by the use of a decimation-in-time algorithm. The reader should compare a decimation-in-time algorithm for $N = 8$ as shown in Fig. 7.8.

We can see from a decimation-in-time FFT algorithm in Fig. 7.8 that the input signal $x(n)$ is splitting into the even part $x(2r)$ and the odd part $x(2r + 1)$ to make computation simpler. The so-called decimation-in frequency FFT algorithm makes computation simpler by splitting the output signal $X(k)$ into the even part $X(2r)$ and the odd part $X(2r + 1)$. Consider a DFT of a finite duration function $x(n)$ in Equation 7.1, which is

$$X(k) = \sum_{n=0}^{N-1} x(n) W_N^{kn}, \qquad k = 0, 1, \ldots, N - 1$$

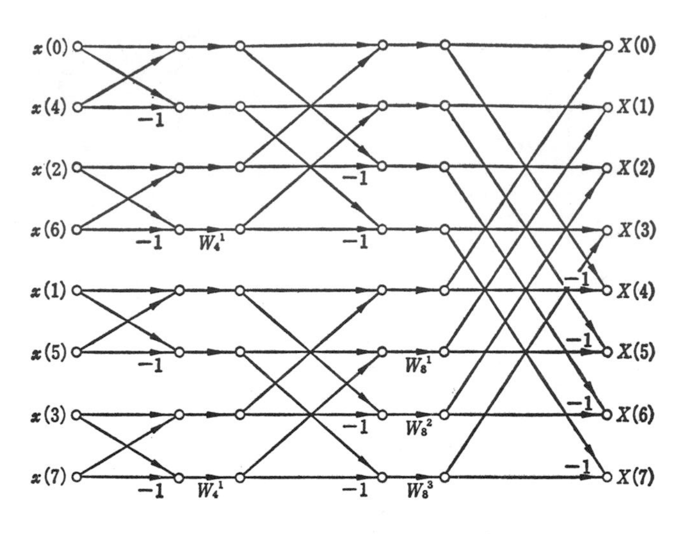

Figure 7.8 Decimation-in-time FFT for $N = 8$.

Expanding the right-hand side into two summations, one of which is from 0 to $N/2 - 1$, and the other is from $N/2$ to $N - 1$, as

$$X(k) = \sum_{n=0}^{(N/2)-1} x(n)\, W_N + \sum_{n=(N/2)}^{N-1} x(n)\, W_N^{kn}$$

$$= \sum_{n=0}^{(N/2)-1} x(n)\, W_N^{kn} + W_N^{k(N/2)} \sum_{n=0}^{(N/2)-1} x\left(n + \frac{N}{2}\right) W_N^{kn}$$

Since $W_N^{k(N/2)} = e^{-j(2\pi/N)k(N/2)} = (-1)^k$, the preceding equation can be expressed as

$$X(k) = \sum_{n}^{(N/2)-1} \left[x(n) + (-1)^k x\left(n + \frac{N}{2}\right)\right] W^{kn} \tag{7.13}$$

Now splitting $k = 0, 1, 2, \ldots, N - 1$ of $X(k)$ into the even number $2r$ and the odd number $2r + 1$ so that we have $X(2r)$ and $X(2r + 1)$ as

$$X(2r) = \sum_{n}^{(N/2)-1} \left[x(n) + x\left(n + \frac{N}{2}\right)\right] W_N^{2rn} \tag{7.14}$$

and

$$x(2r + 1) = \sum_{n=0}^{(N/2)-1} \left[x(n) - x\left(n + \frac{N}{2}\right)\right] W_N^{(2r+1)n} \tag{7.15}$$

Let $b_0(t)$, $b_1(t)$, $B_0(u)$, and $B_1(u)$ be

$$b_0(t) = x(n) + x\left(n + \frac{N}{2}\right)$$

$$b_1(t) = \left[x(n) - x\left(n + \frac{N}{2}\right)\right] W_N^n$$

$$B_0(u) = X(2r)$$

and

$$B_1(u) = X(2r + 1)$$

Then Equation 7.14 can be expressed as

$$B_0(u) = \sum_{t=0}^{(N/2)-1} b_0(t) W_N^{ut} \tag{7.16}$$

and Equation 7.15 can be expressed as

$$B_1(u) = \sum_{t=0}^{(N/2)-1} b_1(t) W_N^{ut} \tag{7.17}$$

Hence, process of calculation of these $X(2r)$ and $X(2r + 1)$ can be seen by a signal flow graph in Fig. 7.9(a). Notice that this is similar to that in Fig. 7.6 except that the output signal $X(k)$ is arranged as $X(0)$, $X(2)$, $X(4)$, . . . then $X(1)$, $X(3)$, ... These $N/2$-point DFT B_0 and B_1 in the figure can be split as shown in Fig. 7.9(b) because Equations 7.16 and 7.17 are the same form as Equation 7.13. Continuing this process of splitting until there are no DFT that can be split any more, we will have a signal flow graph representing an algorithm called a "decimation-in-frequency FFT." Fig. 7.9(c) is a decimation-in-frequency FFT algorithm for $N = 8$.

As an example, let us calculate $X(k)$ of a finite duration function $x(n)$ in Fig. 7.10(a) by a decimation-in-time FFT algorithm. Choosing $N = 4$ makes

$$W_4^1 = e^{-j(2\pi/4)} = -j$$

Hence, a signal flow graph of a decimation-in-time FFT algorithm becomes the one shown in Fig. 7.10(b). The values at each step in this algorithm are shown at each node in Fig. 7.10(c). It is clear that using this signal flow graph is much easier than using Equation 7.1 for obtaining $X(k)$.

As another example, consider the finite duration function $x(n)$ given in Fig. 7.11(a). By choosing $N = 8$, W_4^1 is $-j$, and all other necessary W_8^k are

$$W_8^1 = e^{-j(2\pi/8)} = \frac{1}{\sqrt{2}}(1 - j1)$$

$$W_8^2 = e^{-j(2\pi/8)2} = -j$$

and

$$W_8^3 = e^{-j(2\pi/8)3} = \frac{1}{\sqrt{2}}(-1 - j1)$$

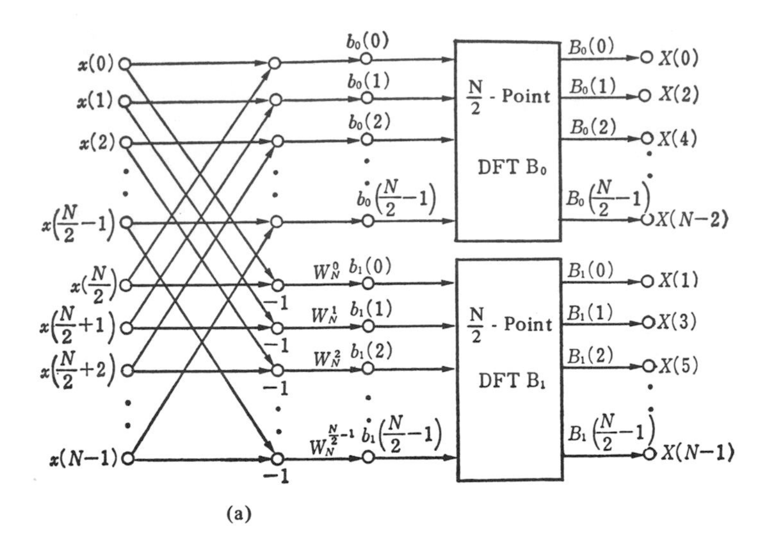

(a)

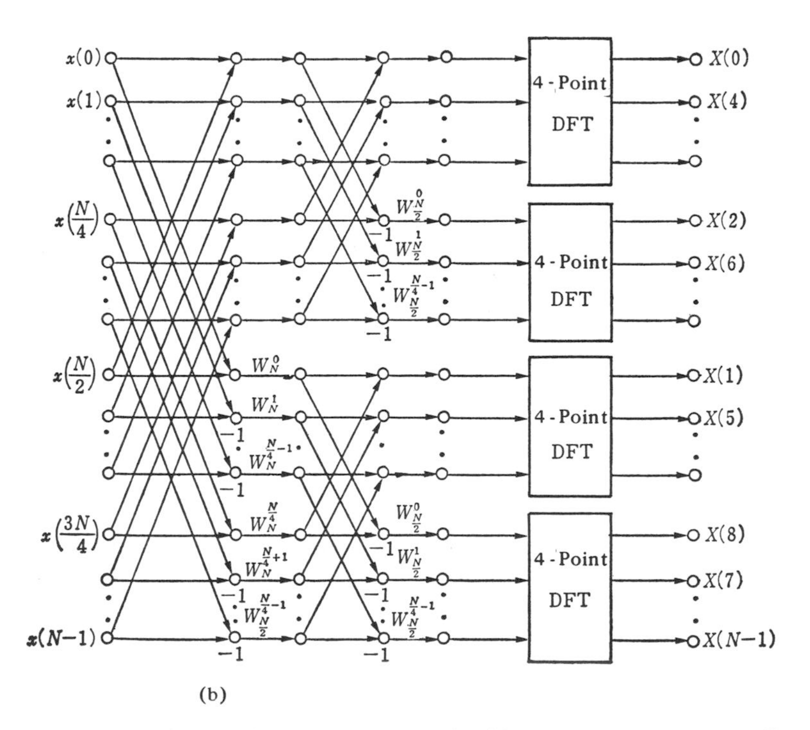

(b)

Figure 7.9 Decimation-in-Frequency FFT Algorithm. (a) Connecting Two $N/2$-Point DFTs; (b) Connecting four $N/4$-Point DFTs; (c) Decimation-in-Frequency FFT Algorithm with $N = 8$.

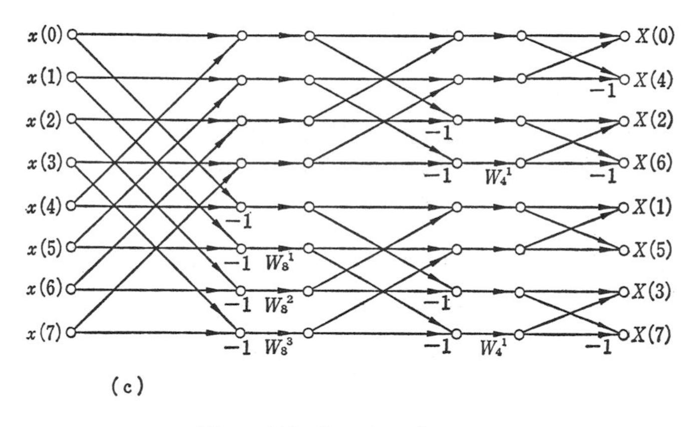

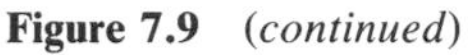

Figure 7.9 *(continued)*

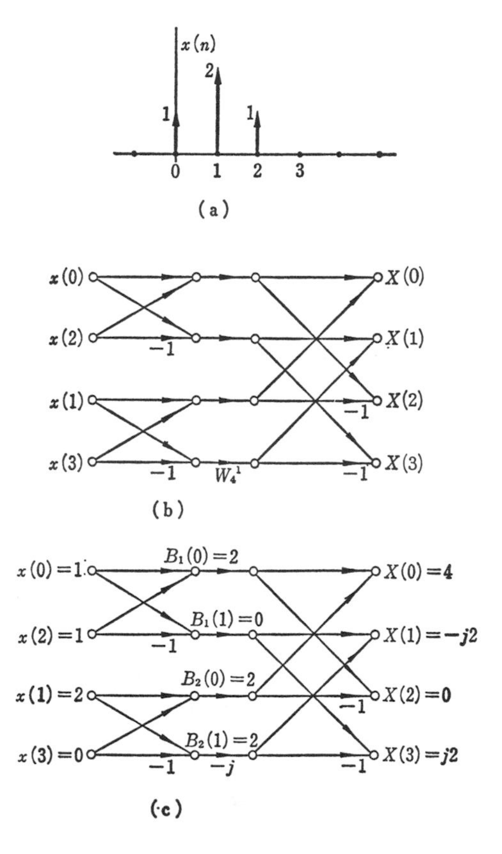

Figure 7.10 FFT with $N = 4$. (a) $x(n)$; (b) Decimation-in-Time FFT; (c) Calculation of $X(k)$.

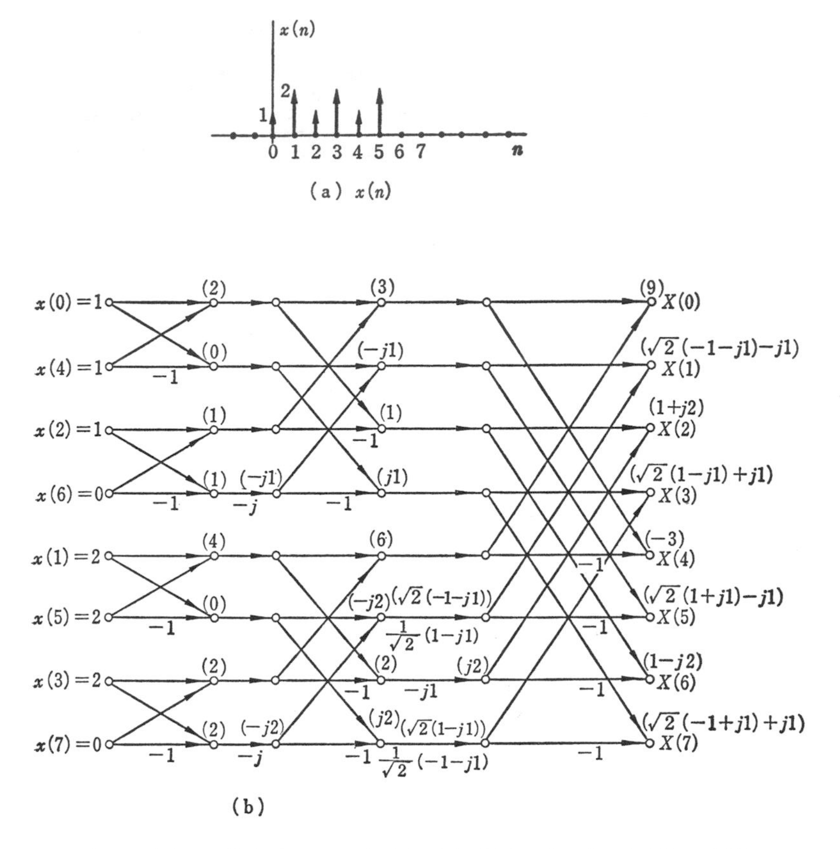

Figure 7.11 FFT with $N = 8$. (a) $x(n)$; (b) Decimation-in-Time FFT.

Hence, a signal flow graph for a decimation-in-time FFT algorithm becomes the one shown in Fig. 7.11(b) in which the value in the parentheses at each node is the result of calculation at that step. This example also shows that obtaining $X(k)$ by the use of a decimation-in-time algorithm is extremely effective.

7.3 FFT WITH $N \neq 2^q$

We have studied the construction of signal flow graphs of a decimation-in-time FFT and a decimation-in-frequency when N is 2^q. We can always choose N of a given finite duration function to be 2^q. The smallest period N we can choose for a finite duration function $x(n)$ in Fig. 7.10(a) is 3, but we choose N to be 4 to use the FFT with $N = 4$ in Fig. 7.10(b). Also, the smallest period N of $x(n)$ in Fig. 7.11(a) is 6, but we choose $N = 2^3$ so that the FFT with $N = 8$ in Fig. 7.11(b) can be used for the calculation of $X(k)$. Suppose by some reason it is impossible to choose N to be 2^q. Is it possible to obtain a signal flow graph similar to that of the FFT with $N = 2^q$?

Suppose N is a product of prime numbers $p_1, p_2, \ldots$ and p_t. Let q be

$$q = p_2 p_3 \cdots p_t$$

Then N is equal to

$$N = p_1 q$$

To split the right-hand side of the DFT of a finite duration function $x(n)$

$$X(k) = \sum_{n=0}^{N-1} x(n) W_N^{kn} \tag{7.18}$$

we will put n into groups $s_0, s_1, s_2, \ldots$ and s_{p_1-1} as

$$\begin{aligned}
s_0 &= \{0, p_1, 2p_1, \cdots, (q-1)p_1\} \\
s_1 &= \{1, p_1 + 1, 2p_1 + 1, \cdots, (q-1)p_1 + 1\} \\
&\ \ \vdots \\
s_i &= \{i, p_1 + i, 2p_1 + i, \cdots, (q-1)p_1 + i\} \\
&\ \ \vdots \\
s_{p_1-1} &= \{p_1 - 1, p_1 + (p_1 - 1), \cdots, (q-1)p_1 + (p_1 - 1)\}
\end{aligned}$$

Then expand the right-hand side of Equation 7.18 into p_1 summations as

$$X(k) = \sum_{n \varepsilon s_0} x(n) W_N^{kn} + \sum_{n \varepsilon s_1} x(n) W_N^{kn} + \cdots + \sum_{n \varepsilon s_i} x(n) W_N^{kn} + \cdots + \sum_{n \varepsilon s_{p_1-1}} x(n) W_N^{kn}$$

Using the symbol $B_i(k)$, this equation can be expressed as

$$X(k) = B_0(k) + W_N^k B_1(k) + \cdots + W_N^{ki} B_i(k) + \cdots + W_N^{k(p_1-1)} B_{p_1-1}(k) \tag{7.19}$$

A signal flow graph corresponding to Equation 7.19 is shown in Fig. 7.12. By the use of the preceding procedure, each q-point DFT can be changed into smaller DFTs. Continuing this modification will give a signal flow graph corresponding to a decimation-in-time FFT algorithm. As an example, when $p_1 = 3$ and $q = 2$, a signal flow graph of a decimation-in-time FFT with $N = 6$ is shown in Fig. 7.13. As in the case of $N = 2^q$, a signal flow graph may be simplified further by the use of equivalent transformation of signal flow graphs. A signal flow graph with $N \neq 2^q$, however, will have a very

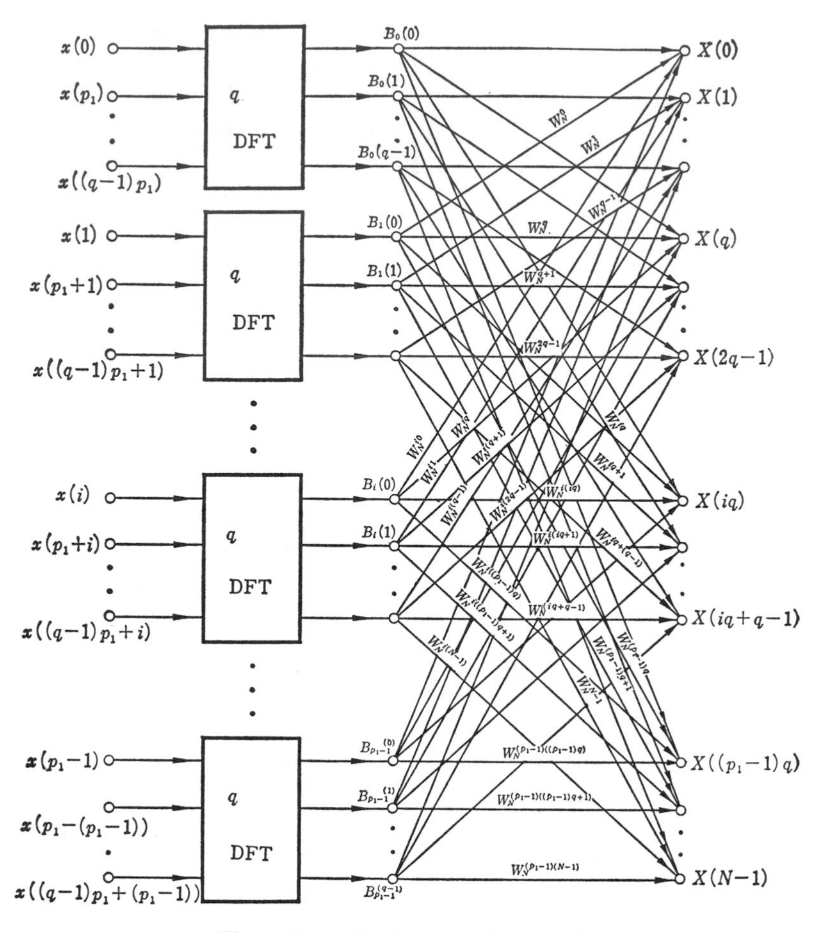

Figure 7.12 Combine p_1 q-Point DFTs.

complicated structure as shown in Fig. 7.12; one must be very careful to use equivalent transformations.

7.4 INVERSE FFT

The inverse DFT of $X(k)$ is given by Equation 6.23, which is

$$x(n) = \frac{1}{N} \sum_{k=0}^{N-1} X(k)\, W_N^{-kn} \tag{7.20}$$

Compared with Equation 7.1, we can see that multiplying $1/N$ with the right-hand side of Equation 7.1 and replacing W_N^{kn} with W_N^{-kN} makes the same form

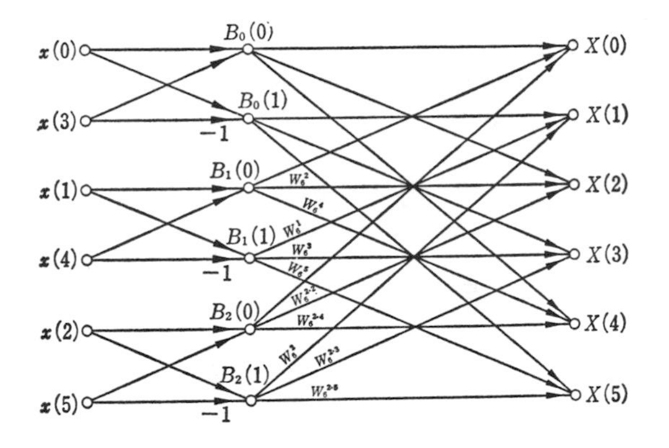

Figure 7.13 Decimation-in-Time FFT with $N = 6$.

as this equation. Hence, it may be possible to obtain a signal flow graph for the inverse FFT from a signal flow graph for the FFT by just minor modification. We can take care of $1/N$ easily by considering an input signal is $X(k)/N$ rather than $X(k)$. A problem is W_N^{-kn}. Suppose $N = 2^q$. Equation 7.20 can be expressed as

$$x(n) = \sum_{k=0}^{N-1} \frac{X(k)}{N} W_N^{-kn} = \sum_{k \text{ even}} \frac{X(k)}{N} W_N^{-kn} + \sum_{k \text{ odd}} \frac{X(k)}{N} W_N^{-kn}$$

which can be written as

$$x(n) = \sum_{r=0}^{(N/2)-1} \frac{X(2r)}{N} W_N^{-k(2r)} + \sum_{r=0}^{(N/t)-1} \frac{X(2r+1)}{N} W_N^{-k(2r+1)}$$

Modifying this equation, we have

$$x(n) = \sum_{r=0}^{(N/2)-1} \frac{X(2r)}{N} (W_N^2)^{-kr} + W_N^{-k} \sum_{r=0}^{(N/2)-1} \frac{X(2r+1)}{N} (W_N^2)^{-kr}$$

$$= \sum_{r=0}^{(N/2)-1} \frac{X(2r)}{N} W_{N/2}^{-kr} + W_N^{-k} \sum_{r=0}^{(N/2)-1} \frac{X(2r+1)}{N} W_{N/2}^{-kr}$$

It can be seen that the first term in the right-hand side of this equation is a form of Equation 7.20, and the second term is W_N^{-k} time, a form of Equation 7.20. Hence, by the use of symbols $C_0(k)$ and $C_1(k)$, this equation can be expressed as

$$x(n) = C_0(k) + W_N^{-k} C_1(k)$$

Comparing this equation with Equation 7.6, we can see that W_N^{kn} in Equation 7.6 is changed to W_N^{-kn} in this equation. $C_0(k)$ and $C_1(k)$ are called $N/2$-point inverse DFTs. The same result can be obtained when we split $C_0(k)$ and

$C_1(k)$ into $N/4$-point inverse DFTs. We can continue the process of splitting k-point inverse DFTs until each inverse DFT is a 2-point inverse DFT. An output $u(k)$ of such a 2-point inverse DFT can be expressed as

$$u(k) = \sum_{i=0}^{1} \frac{X(v(i))}{N} W_2^{-ik}, \qquad k = 0, 1$$

By expanding this, we have

$$u(k) = \frac{X(v(0))}{N} W_2^{-0k} + \frac{X(v(1))}{N} W_2^{-1k}, \qquad k = 0, 1$$

Because W_2^{-0k} is 1 and W_2^{-1k} is

$$W_2^{-1k} = e^{j(2\pi/2)k} = \begin{cases} 1, & k = 0 \\ -1, & k = 1 \end{cases}$$

$u(k)$ for $k = 0$ and $k = 1$ are

$$u(0) = \frac{X(v(0))}{N} + \frac{X(v(1))}{N}$$

$$u(1) = \frac{X(v(0))}{N} - \frac{X(v(1))}{N}$$

which are exactly the same form as Equation 7.12.

Finally, W_N^p and $W_N^{(N/2+p)}$ in Fig. 7.5b must be W_N^{-p} and $W_N^{-(N/2+1)}$ for the inverse DFT. Since

$$W_N^{-N/2} = e^{j(2\pi/N)(N/2)} = -1$$

changing W_N^p to W_N^{-p} in Fig. 7.5d gives a simplified signal flow graph for the inverse DFT. Thus, a signal flow graph for the inverse FFT algorithm can be obtained from a signal flow graph of the FFT algorithm by changing W_N^p to W_N^{-p}, input $x(n)$ to $X(n)/N$, and output $X(k)$ to $x(k)$. For example, a signal flow graph for the decimation-in-time inverse FFT with $N = 8$ can be obtained easily from a graph in Fig. 7.8 by changing (1) input $x(n)$ to $X(n)/N$, (2) output $X(k)$ to $x(k)$, and (3) W_4^1 to W_4^{-1} and W_8^p to W_8^{-p} as shown in Fig. 7.14. For example, consider $X(k)$ of FFT given in Fig. 7.11 where

$$X(0) = 8, X(1) = 0, \ X(2) = -j4, X(3) = 0$$

$$X(4) = 0, X(5) = 0, \ X(6) = j4, \quad X(7) = 0$$

To obtain $x(k)$ from these $X(k)$ by the use of the inverse FFT with $N = 8$, we will use the decimation-in-time inverse FFT as shown in Fig. 7.15 in which

$$W_4^{-1} = e^{j(2\pi/8)} = j1$$

$$W_8^{-1} = e^{j(2\pi/8)} = \frac{1}{\sqrt{2}}(1 + j1)$$

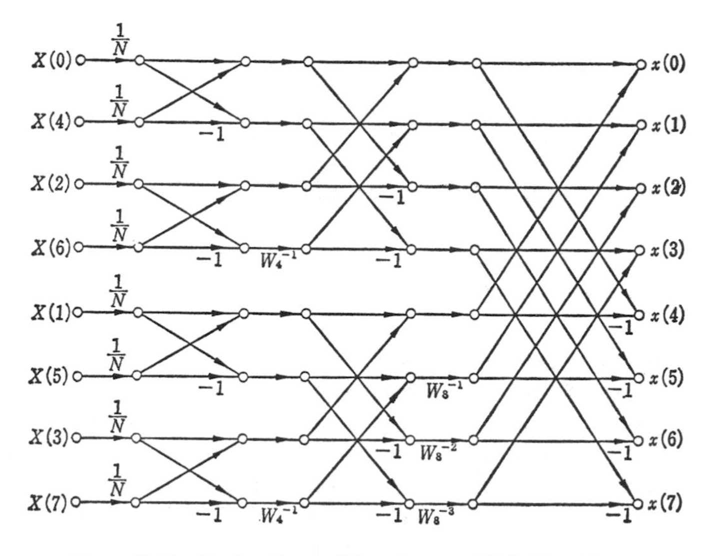

Figure 7.14 Decimation-in-Time Inverse FFT Algorithm.

$$W_8^{-2} = e^{j(2\pi/8)2} = j1$$

and

$$W_8^{-3} = e^{j(2\pi/8)3} = \frac{1}{\sqrt{2}}(-1 + j1).$$

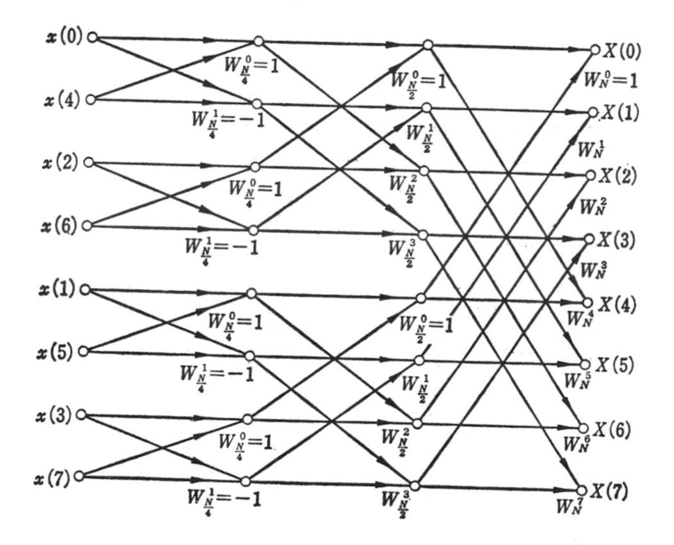

Figure 7.15 Decimation-in-Time Inverse FFT Algorithm.

The results shown in the figure are

$$x(0) = 1,\ x(1) = 2,\ x(2) = 1,\ x(3) = 0,$$
$$x(4) = 1,\ x(5) = 2,\ x(6) = 1,\ x(7) = 0$$

which is a set of given $x(k)$ to obtain $X(k)$ by FFT in Fig. 7.11.

7.5 APPLICATION OF FFT

Naturally, an FFT algorithm has been applied to obtaining DFTs, calculation of power spectrums, and circular convolutions. Here we will introduce particular ways of applying FFT. One is to use FFT for calculation of regular convolution of two discrete functions. Here we use the terminology "regular convolution" instead of using a familiar terminology, which is "convolution to distinguish it from circular convolution." Consider two finite duration functions $x(n)$ and $h(n)$ as shown in Fig. 7.16. As we have studied in Chapter 6, if we take N, which is the largest finite duration (interval) of $x(n)$ and $h(n)$, and calculate circular convolution, the result is different from regular convolution of these two functions. If we need to obtain regular convolution by using circular convolution, we must choose N much larger than that for calculating circular convolution. For example, to obtain circular convolution of $x(n)$ and $h(n)$ in Fig. 7.16, it is sufficient to choose $N = 6$. If we need to compute regular convolution of these functions by using circular convolution, however, we must choose N equal to at least 10. Fig. 7.17(a) is the result of circular convolution of these $x(n)$ and $h(n)$ with $N = 6$, whereas Fig. 7.17(b) is the result of circular convolution of these $x(n)$ and $h(n)$ with $N = 10$, which is the same result as regular convolution of $x(n)$ and $h(n)$.

Let us investigate whether employing a large N to use circular convolution for calculation of regular convolution is effective. When we calculate circular convolution directly, it requires N^2 multiplications and N^2 additions. A result $y(n)$ of circular convolution $y(n) = x(n) * h(n)$ is equal to

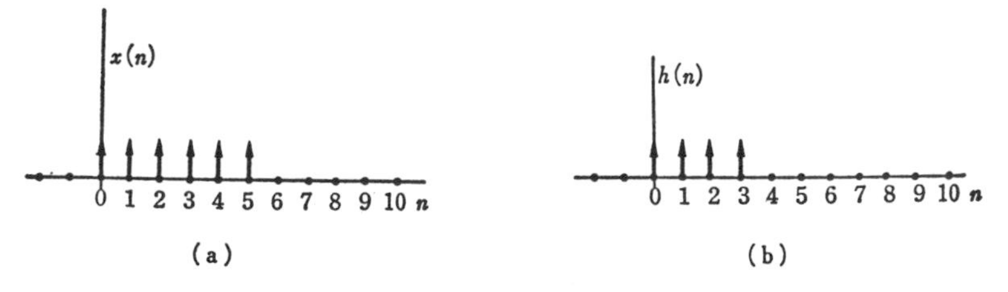

Figure 7.16 $x(n)$ and $h(n)$. (a) $x(n)$; (b) $h(n)$.

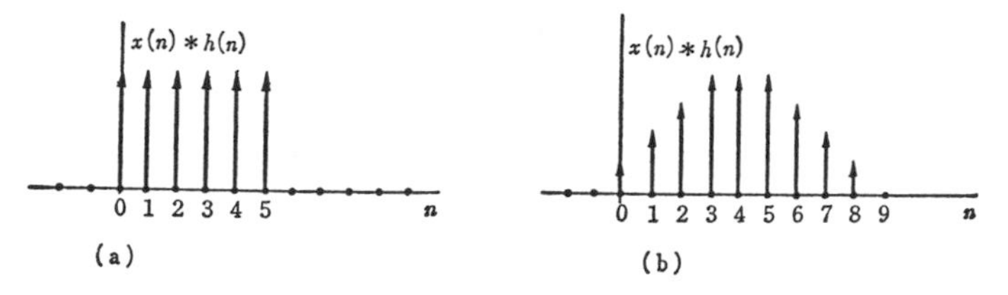

Figure 7.17 Circular Convolution. (a) $x(n)*h(n)$ with $N = 6$; (b) $x(n)*h(n)$ with $N = 10$.

$$y(n) = \sum_{k=0}^{N-1} X(k)H(k)W_N^{-kn}$$

where $X(k)$ is the DFT of $x(n)$, and $H(k)$ is the DFT of $h(n)$. Hence, if we obtain $y(n)$ by first calculating the DFTs $X(k)$ and $H(k)$ of $x(n)$ and $h(n)$, then obtain the product $X(k)H(k)$, and finally calculate the inverse DFT of $X(k)H(k)$, we need no more than $3N(\log_2 N) + N$ multiplications and no more than $3N(\log_2 N)$ additions. E. O. Brigham[1] stated that when $N = 4096$, using the FFT and the inverse FFT to obtain $y(n) = x(n) * h(n)$ takes 1/26 computer time compared with the calculation of circular convolution $x(n) * y(n)$ directly.

Let us investigate the DFT of a complex conjugate of $x(n)$. Suppose the symbol Re F indicates a real part of F, and the symbol Im F indicates an imaginary part of F. Then $x(n)W_N^{kn}$ is

$$x(n)W_N^{kn} = (\text{Re } x(n) \text{ Re } W_N^{kn} - \text{Im } x(n) \text{ Im } W_N^{kn}) + j(\text{Re } x(n) \text{ Im } W_N^{kn} + \text{Im } x(n) \text{ Re } W_N^{kn}) \tag{7.21}$$

Now replacing $x(n)$ and W_N^{kn} by their complex conjugates $\overline{x(n)}$ and $\overline{W_N^{kn}}$, we have

$$\text{Re } \overline{x(n)} = \text{Re } x(n), \qquad \text{Im } \overline{x(n)} = -\text{Im } x(n)$$

$$\text{Re } \overline{W_N^{kn}} = \text{Re } W_N^{kn}, \qquad \text{Im } \overline{W_N^{kn}} = -\text{Im } W_N^{kn}$$

Hence,

$$\overline{x(n)}\ \overline{W_N^{kn}} = (\text{Re } x(n) \text{ Re } W_N^{kn} - \text{Im } x(n) \text{ Im } W_N^{kn}) - j(\text{Re } x(n) \text{ Im } W_N^{kn} + \text{Im } x(n) \text{ Re } W_N^{kn}) \tag{7.22}$$

Because $\overline{x(n)W_N^{kn}}$ is equal to Equation 7.21, except that the sign of the imaginary part is different, we have

[1] E. O. Brigham, *The Fast Fourier Transform* (Englewood Cliffs, N.J.: Prentice Hall, 1974).

$$\overline{x(n)W_N^{kn}} = (\text{Re}\, x(n)\, \text{Re}\, W_N^{kn} - \text{Im}\, x(n)\, \text{Im}\, W_N^{kn}) \\ -j(\text{Re}\, x(n)\, \text{Im}\, W_N^{kn} + \text{Im}\, x(n)\, \text{Re}\, W_N^{kn})$$

We can see that this equation is equal to Equation 7.22. Hence,

$$\overline{x(n)W_N^{kn}} = \overline{x(n)}\ \overline{W_N^{kn}} \tag{7.23}$$

Because W_N^{kn} is

$$W_N^{kn} = e^{-j(2\pi/N)kn}$$

$\overline{W_N^{kn}}$ is

$$\overline{W_N^{kn}} = e^{j(2\pi/N)kn} = W_N^{-kn}$$

and by Equation 7.23, $\overline{x(n)W_N^{kn}}$ can be expressed as

$$\overline{x(n)W_N^{kn}} = \overline{x(n)}W_N^{-kn}$$

Using this result, $\overline{X(k)}$ is

$$\overline{X(k)} = \overline{\sum_{n=0}^{N-1} x(n)W_N^{kn}} = \sum_{n=0}^{N-1} \overline{x(n)}W_N^{-kn} \tag{7.24}$$

This relationship is very useful. For example, consider a finite duration function $y(n)$, which is

$$y(n) = \sum_{m=0}^{N-1} \mathring{x}(m)\mathring{h}(n+m)\, R_N(n) \tag{7.25}$$

The right-hand side of this equation is known as "circular correlation."

Let $\mathring{X}(k)$ be the DFT of $\mathring{x}(n)$ and $\mathring{H}(k)$ be the DFT of $\mathring{h}(n)$. Then the circular correlation in Equation 7.25 can be expressed as

$$y(n) = \sum_{m=0}^{N-1} \left(\frac{1}{N}\sum_{k=0}^{N-1} \mathring{X}(k)W_N^{-km}\right)\left(\frac{1}{N}\sum_{k=0}^{N-1} \mathring{H}(r)W_N^{-r(n+m)}\right) R_N(n) \tag{7.26}$$

By Equation 7.26, the term $\sum_{k=0}^{N-1} \mathring{X}(k)W_N^{-km}$ in the right-hand side of the equation can be expressed as

$$\sum_{k=0}^{N-1} \mathring{X}(k)W_N^{-km} = \overline{\overline{\sum_{k=0}^{N-1} \mathring{X}(k)W_N^{-km}}} = \overline{\sum_{k=0}^{N-1} \overline{\mathring{X}(k)}W_N^{km}}$$

Hence, Equation 7.26 becomes

$$y(n) = \sum_{m=0}^{N-1} \frac{1}{N}\overline{\sum_{k=0}^{N-1} \overline{\mathring{X}(k)}W_N^{km}}\left(\frac{1}{N}\sum_{r=0}^{N-1} \mathring{H}(r)W_N^{-r(n+m)}\right) R_N(n) \tag{7.27}$$

Suppose $x(n)$ is real. Then

$$x(n) = \frac{1}{N}\overline{\sum_{k=0}^{N-1} \mathring{X}(k) W_N^{km}} \text{ is } x(n) = \frac{1}{N}\sum_{k=0}^{N-1} \overline{\mathring{X}(k)} W_N^{km}.$$

Hence, Equation 7.27 can be written as

$$y(n) = \frac{1}{N}\sum_{k=0}^{N-1} \overline{\mathring{X}(k)} \sum_{r=0}^{N-1} \mathring{H}(r) W_N^{-rn} \left(\frac{1}{N}\sum_{m=0}^{N-1} W_N^{(k-r)m}\right) R_N(n) \tag{7.28}$$

The parenthetical portion in the right-hand side is

$$\frac{1}{N}\sum_{m=0}^{N-1} W_N^{(k-r)m} = \begin{cases} 1, & k = r \\ 0, & k \neq r \end{cases}$$

Hence, Equation 7.28 becomes

$$y(n) = \frac{1}{N}\sum_{k=0}^{N-1} \overline{\mathring{X}(k)} \mathring{H}(k) W_N^{-kn} \tag{7.29}$$

This indicates that for a real $x(n)$, circular correlation of $x(n)$ and $h(n)$ is by Equation 7.29

$$\sum_{m=0}^{N-1} x(m)h(n+m) = \frac{1}{N}\sum_{k=0}^{N-1} \overline{\mathring{X}(k)} \mathring{H}(k) W_N^{-kn} \tag{7.30}$$

which is the inverse DFT of product of conjugate of the DFT of $x(n)$ and the DFT of $h(n)$. By using the FFT and the inverse FFT, we can easily compute circular correlation of $x(n)$ and $h(n)$ when $x(n)$ is real.

7.6 SUMMARY

1. For $N = pq$, n will be grouped into p set $s_i = \{i, p+i, 2p+i, \cdots, (q-1)p+i\}$. With these s_i, the DFT $X(k)$ of $x(n)$ will be expressed as

$$X(k) = \sum_{n=0}^{N-1} x(n) W_N^{kn} = \sum_{n \in s_0} x(n) W_N^{kn} + \sum_{n \in s_1} x(n) W_N^{kn} + \cdots + \sum_{n \in s_i} x(n) W_N^{kn} + \cdots$$

Using this, the calculation of $X(k)$ becomes simpler because it becomes a combination of p q-point DFTs. Continuing this process of simplification, calculation of $X(k)$ becomes a combination of 2-point DFTs.

Then simplifying each 2-point DDT and others produces a decimation-in-time FFT algorithm. Fig. 7.8 is a decimation-in-time FFT with $N = 8$.

2. For $N = pq$, n will be grouped into p set $s_0 = \{0, p, \cdots, (q-1)p\}$, $s_1 = \{1, p+1, \cdots, (q-1)p+1\}, \cdots, s_i = \{i, p+i, \cdots, (q-1)p+i\}, \cdots$. With these s_i, the DFT $X(k)$ of $x(n)$ will be expressed as

$$X(k) = \sum_{n=0}^{N-1} x(n) W_N^{kn} = \sum_{n \in s_0} x(n) W_N^{kn} + \sum_{n \in s_1} x(n) W_N^{kn} + \cdots$$

Using this relationship, calculation of $X(k)$ becomes simpler because it becomes a combination of p q-point DFTs. Continuing this process of simplification, the signal flow graph becomes a combination of 2-point DFTs. Then simplifying each 2-point DFT and others produces a decimation-in-frequency FFT algorithm. Fig. 7.9(c) shows a decimation-in frequency FFT with $N = 8$.

3. Using an equation

$$X(k) = \frac{1}{N} \sum_{k=0}^{N-1} x(n) W_N^{-kn}$$

and simplifying calculation by the same process as that in part 1 will produce a decimation-in-time inverse FFT algorithm. Fig. 7.14 shows a decimation-in-time inverse FFT with $N = 8$.

4. Using an equation

$$X(k) = \frac{1}{N} \sum_{k=0}^{N-1} x(n) W_N^{-kn}$$

and simplifying calculation by the same process as that in part 2 will produce a decimation-in-frequency inverse FFT algorithm.

5. The circular correlation of $x(n)$ and $h(n)$ is

$$\sum_{m=-\infty}^{\infty} x(m) h(n+m)$$

When $x(n)$ is real, the equation becomes

$$\sum_{m=-\infty}^{\infty} x(m) h(n+m) = \frac{1}{N} \sum_{k=0}^{N-1} \overline{X(k)} H(k) W_N^{-kn}$$

where $X(k)$ is the DFT of $x(n)$, and $H(k)$ is the DFT of $h(n)$.

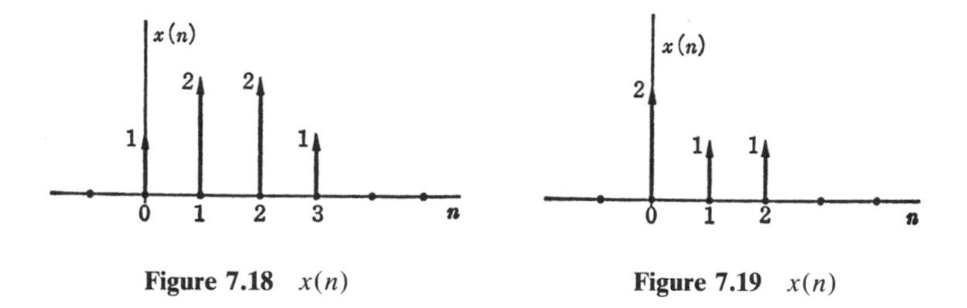

Figure 7.18 $x(n)$

Figure 7.19 $x(n)$

7.7 PROBLEMS

1. Draw a signal flow graph for a decimation-in-time FFT with $N = 4$ and a signal flow graph for a decimation-in-frequency FFT with $N = 4$.
2. Draw a signal flow graph for a decimation-in-time FFT with $N = 8$.
3. Obtain the DFT $X(k)$ of $x(n)$ in Fig. 7.18 by using a signal flow graph for a decimation-in-time FFT algorithm.
4. Obtain the DFT $X(k)$ of $x(n)$ in Fig. 7.19 by using a signal flow graph for a decimation-in-time FFT algorithm.
5. Compute the DFT $X(k)$ of the following $x(n)$ by using a signal flow graph for a decimation-in-time FFT algorithm.

$$x(n) = \begin{cases} 1 + j1, & n = 0 \\ 1 - j1, & n = 1 \\ 1 + j1, & n = 2 \\ 1 - j1, & n = 3 \\ 0 \quad , & \text{Otherwise} \end{cases}$$

6. For $x(n)$ in Fig. 7.20, obtain the DFT $X(k)$ by a signal flow graph for a decimation-in-time FFT algorithm.
7. Draw a signal flow graph for a decimation-in-time FFT with $N = 9$.
8. Draw a signal flow graph for a decimation-in-time FFT with $N = 6$.

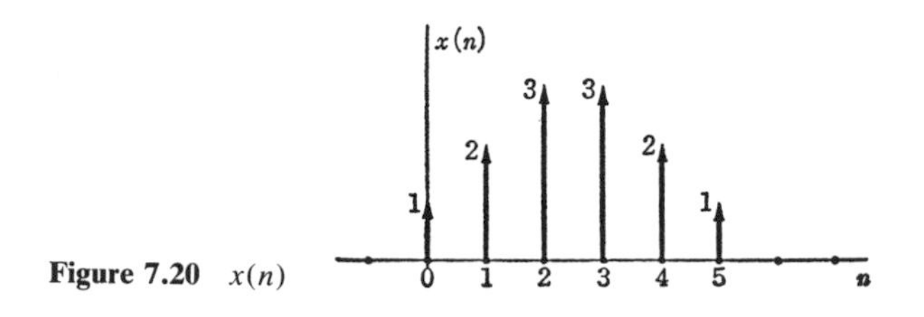

Figure 7.20 $x(n)$

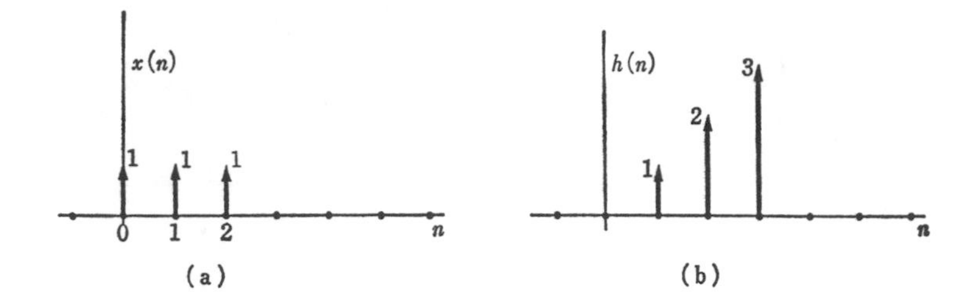

Figure 7.21 $x(n)$ and $h(n)$. (a) $x(n)$; (b) $h(n)$.

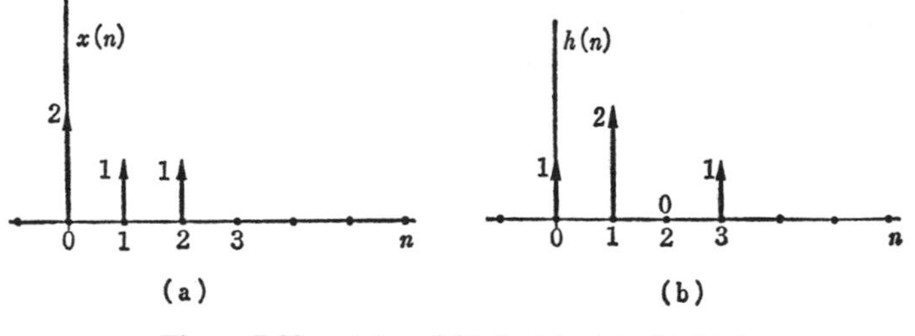

Figure 7.22 $x(n)$ and $h(n)$. (a) $x(n)$; (b) $h(n)$.

9. Calculate regular convolution of $x(n)$ and $h(n)$ in Fig. 7.21 by using the FFT and the inverse FFT.

10. Obtain regular convolution of $x(n)$ and $h(n)$ in Fig. 7.22 by using the FFT and the inverse FFT.

8 Digital Filter

8.1 APPROXIMATION OF INFINITE IMPULSE RESPONSE FUNCTION BY WINDOW

The purpose of a digital filter is the same as that of an analog filter, which is a system that gives a desired output signal for a given input signal. An output signal $y(n)$, which should be composed of only low-frequency components of an input signal, is a good example of filters.

A unit sample response $h(n)$ indicates that a digital filter can be divided into two types. It has been studied that a unit sample response $h(n)$ is a response when an input signal is a unit sample. When $h(n)$ is a finite duration function, then $h(n)$ is called a "finite impulse response function," or simply a "FIR function"; a system whose $h(n)$ is a FIR function is called a "finite impulse response system," or simply a "FIR system." If $h(n)$ is not a finite impulse response, then it is called an "infinite impulse response function," or simply an "IIR function"; a system whose $h(n)$ is an IIR function is called an "infinite impulse response system," or simply an "IIR system."

In other words, when a unit sample $u_I(n)$ is entered at the input of a system, and the output signal $y(n)$ becomes 0 after a finite n, then the system is a FIR system. Conversely, when a unit sample $u_I(n)$ is entered at the

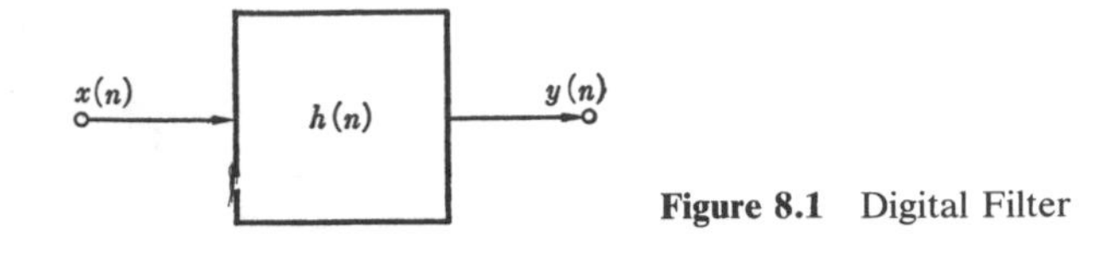

Figure 8.1 Digital Filter

input of a system, and the output signal $y(n)$ will have a value other than 0 for n larger than any integer N, then the system is an IIR system.

A system whose unit sample response is a FIR function is called a FIR digital filter, and a system whose unit sample response is an IIR function is called an IIR digital filter. For example, the system in Fig. 8.1 whose unit sample response $h(n)$ is

$$h(n) = \begin{cases} (-1)^n a^n, & 0 \leq n \leq N \\ 0, & \text{Otherwise} \end{cases} \tag{8.1}$$

is a FIR digital filter.

A characteristic of a digital filter, whose unit sample response is $h(n)$, is usually indicated by a frequency response $H(e^{j\omega})$, which is the discrete function Fourier transformation of $h(n)$ in Equation 3.54, that is,

$$H(e^{j\omega}) = \sum_{n=-\infty}^{\infty} h(n)e^{-j\omega n} \tag{8.2}$$

where $h(n)$ is the inverse DFT of $H(e^{j\omega})$ in Equation 3.56, that is,

$$h(n) = \frac{1}{2\pi}\int_{-\pi}^{\pi} H(e^{j\omega})e^{j\omega n}\, d\omega \tag{8.3}$$

In many cases, $h(n)$ is an IIR function. For example, suppose a desired frequency response $H_d(e^{j\omega})$ of a digital filter is the one shown in Fig. 8.2(a). Then a unit sample response $h_d(n)$ is

$$h_d(n) = \frac{1}{2\pi}\int_{-\omega_c}^{\omega_c} e^{j\omega n}\, d\omega = \frac{\sin w_c n}{\pi n}$$

which is shown in Fig. 8.2(b). This figure indicates that $h_\alpha(n)$ is an IIR function.

One way to change an IIR function to a FIR function is to set the value of the IIR function $h_d(n)$ to 0 for all n larger than an integer N. Suppose $h_f(n)$ is obtained from a IIR function $h_d(n)$ by changing the value of $h_d(n)$ at n for all $n > N$ to 0. A question is how a frequency response of $h_f(n)$ is different from that of $h_d(n)$. It is easily seen from Equation 8.2 that if an IIR

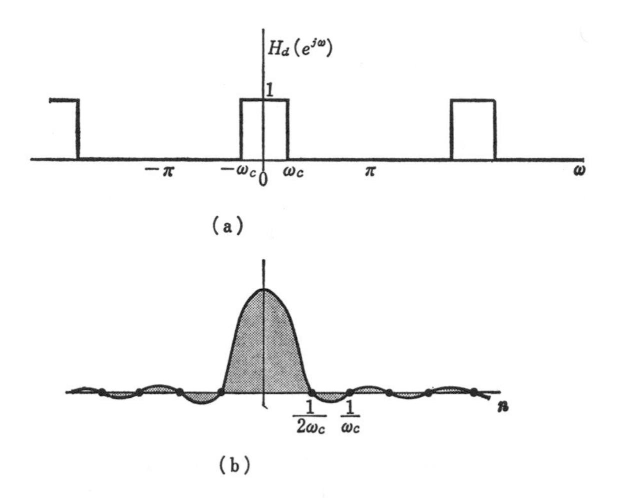

Figure 8.2 $H_d(e^{j\omega})$ and $h_d(n)$. (a) $H_d(e^{j\omega})$; (b) $h_d(n)$.

function $h_d(n)$ has the property that the value is almost 0 for large n, then by choosing a reasonably large N to obtain $h_f(n)$ as

$$h_f(n) = \begin{cases} h_d(n), & 0 \le n \le N-1 \\ 0, & \text{Otherwise} \end{cases} \tag{8.4}$$

the frequency response of $h_f(n)$ will be almost identical with a frequency response of an original IIR function $h_d(n)$. Let $w(n)$ be

$$w(n) = \begin{cases} 1, & 0 \le n \le N-1 \\ 0, & \text{Otherwise} \end{cases} \tag{8.5}$$

which is called a "rectangular window." With this $w(n)$, we can express FIR function $h_f(n)$ obtained from an IIR function $h_d(n)$ by changing the value of $h_d(n)$ at $n > N$ to 0 as

$$h_f(n) = h_d(n)w(n) \tag{8.6}$$

Let $W(e^{j\omega})$ be the discrete function Fourier transformation of a window $w(n)$. Because $h_f(n)$ is a product of $h_d(n)$ and $w(n)$ by Equation 8.6, the discrete function Fourier transformation $H_f(e^{j\omega})$ of $h_f(n)$ is the convolution of $H_d(e^{j\omega})$ and $W(e^{j\omega})$, that is,

$$H_f(e^{j\omega}) = \frac{1}{2\pi}\int_{-\pi}^{\pi} H_d(e^{j\delta})W(e^{j(\omega-\delta)})\,d\theta \tag{8.7}$$

From Equation 8.5, $W(e^{j\omega})$ is

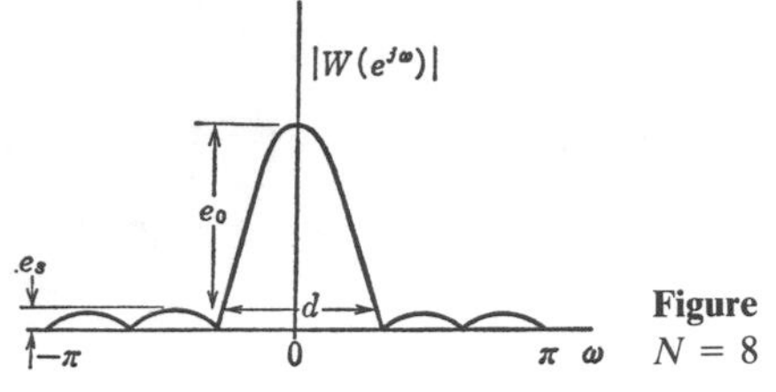

Figure 8.3 One period of $W(e^{j\omega})$ for $N = 8$

$$W(e^{j\omega}) = \sum_{n=0}^{N-1} w(n)e^{-j\omega n} = e^{-j\omega(N-1)/2} \frac{\sin\left(N\dfrac{\omega}{2}\right)}{\sin\left(\dfrac{\omega}{2}\right)} \tag{8.8}$$

whose absolute value in one period is shown in Fig. 8.3.

Because $H_d(e^{j\omega})$ is a desirable frequency response, convolution of $H_d(e^{j\omega})$ and $W(e^{j\omega})$ in Equation 8.7, which gives $H_f(e^{j\omega})$ must be almost identical with $H_d(e^{j\omega})$. This means that convolution of $H_d(e^{j\omega})$ and $W(e^{j\omega})$ should not change $H_d(e^{j\omega})$ much. It is clear that if one period of $W(e^{j\omega})$ is an impulse function $\delta(\omega)$, then the convolution of $H_d(e^{j\omega})$ and $W(e^{j\omega})$ will result in $H_d(e^{j\omega})$. Hence, it is desirable that one period of $W(e^{j\omega})$ is almost identical with an impulse function $\delta(\omega)$. In other words, the width of the main lobe of $W(e^{j\omega})$ in Fig. 8.3 should be as small as possible, and the height of the main lobe of $W(e^{j\omega})$ should be as large as possible. Furthermore, it is desirable that the height of each side lobe of $W(e^{j\omega})$ should be as small as possible. There are several windows that try to satisfy the preceding requirement as much as possible. One such window is known as a generalized Hamming window defined by

$$w(n) = \begin{cases} \alpha + (1-\alpha)\cos\left(\dfrac{2\pi n}{N-1}\right), & 0 \le n \le N-1 \\ 0, & \text{Otherwise} \end{cases}$$

where $0 \le \alpha \le 1$. When $\alpha = 0.54$ it is known as a Hamming window, and when $\alpha = 0.5$ it is known as a Hanning window. Another window known as a Kaiser window is defined as

$$w(n) = \begin{cases} \dfrac{I_0\left(w_a\sqrt{\left(\dfrac{N-1}{2}\right)^2 - \left[n - \left(\dfrac{N-1}{2}\right)\right]^2}\right)}{I_0\left(w_a\left(\dfrac{N-1}{2}\right)\right)}, & 0 \le n \le N-1 \\ 0, & \text{Otherwise} \end{cases}$$

where w_a is a constant, and $I_0(x)$ is the modified Bessel function of the first order.

8.2 FREQUENCY SAMPLING OF IIR FUNCTION

When a unit sample response $h_d(n)$ of a given frequency response $H_d(e^{j\omega})$ is an IIR function, using a window to approximate $h_d(n)$ to a FIR function $h_f(n)$ has been discussed in the previous section. In this section, we will obtain a frequency response $H_f(e^{j\omega})$ of a FIR function $h_f(n)$, which approximates IIR function $h_d(n)$ directly from a given frequency response $H_d(e^{j\omega})$. Because a FIR function $h_f(n)$ is a finite duration function, the DFT $H_f(k)$ of $h_f(n)$ is by Equation 6.22.

$$H_f(k) = \sum_{N=0}^{N-1} h_f(n) W_N^{nk}$$

and by Equation 6.23, its inverse DFT is

$$h_f(n) = \frac{1}{N} \sum_{n=0}^{N-1} H_f(k) W_N^{-nk}$$

Also we have studied in Section 6.1 that sampling the z-transformation $H_f(z)$ of $h_f(n)$ at points that are equally spaced on a circle of radius 1 gives $H_f(k)$, that is,

$$H_f(k) = H_f(z) \Big|_{z=e^{j(2\pi/N)k}} \tag{8.9}$$

Let $H_d(z)$ be a function obtained from a given frequency response $H_d(e^{j\omega})$ by substituting $e^{j\omega} = z$. Also let $H_d(k)$ be a function obtained from $H_d(z)$ by Equation 8.9. Because we can consider $H_d(k)$ as an approximation of $H_d(e^{j\omega})$, the inverse DFT $h(n)$ of $H_d(k)$ is an approximation of $h_d(n)$. The z-transformation of this $h(n)$ is

$$H(z) = \sum_{n=0}^{N-1} h(n) z^{-n} = \sum_{n=0}^{N-1} \left[\frac{1}{N} \sum_{k=0}^{N-1} H_d(k) e^{j(2\pi/N)nk}\right] z^{-n}$$

Interchanging summations in the right-hand side of this equation gives

$$H(z) = \sum_{k=0}^{N-1} \frac{H_d(k)}{N} \sum_{n=0}^{N-1} (e^{j(2\pi/N)k} z^{-1})^n = \sum_{k=0}^{N-1} \frac{H_d(k)}{N} \left(\frac{1 - e^{j2\pi k} z^{-N}}{1 - e^{j(2\pi/N)k} z^{-1}}\right) \tag{8.10}$$

Because $e^{j2\pi k} = 1$, this equation can simplify as

$$H(z) = \frac{1 - z^{-N}}{N} \sum_{k=0}^{N-1} \frac{H_d(k)}{1 - z^{-1} e^{j(2\pi/N)k}}$$

This equation shows the z-transformation of a function obtained from a given frequency response $H_d(e^{j\omega})$ by frequency sampling. Hence, by replacing z by $e^{j\omega}$, $H(z)$ becomes $H(e^{j\omega})$ as

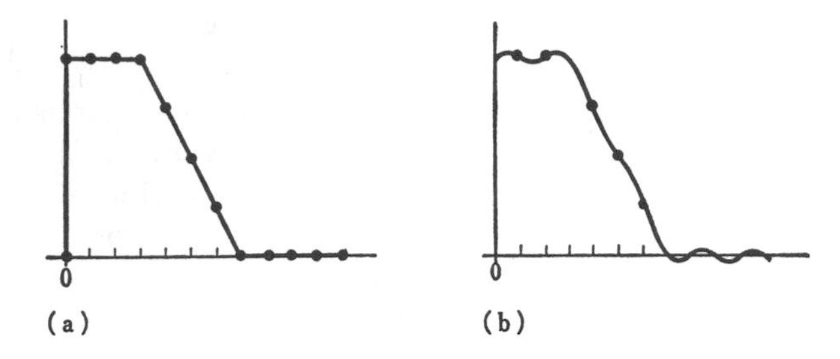

Figure 8.4 Approximation by Frequency Sampling. (a) Sampling Given Frequency Response; (b) Frequency Response of Approximated $h(n)$.

$$H(e^{j\omega}) = \frac{e^{-j\omega(N-1)/2}}{N} \sum_{k=0}^{N-1} \frac{H_d(k)e^{-j(\pi/N)k} \sin\left(\frac{N}{2}\omega\right)}{\sin\left(\frac{\omega}{2} - \frac{\pi}{N}k\right)} \tag{8.11}$$

which is the frequency response of $h(n)$ approximating $h_d(n)$ of a given frequency response $H_d(e^{j\omega})$. For example, suppose a given frequency response $H_d(e^{j\omega})$ is the one in Fig. 8.4(a). By the use of frequency sampling of this $H_d(e^{j\omega})$ and by Equation 8.11, we have $H(e^{j\omega})$ as shown in Fig. 8.4(b).

To obtain Equation 8.11, we sample the z-transformation $H_d(z)$ of a given unit sample response $H_d(n)$ at points that are equally spaced on a circle of radius 1 in the z-plane. Two examples for $N = 8$ and $N = 9$ of equally spaced points that we employ for sampling are shown in Fig. 8.5.

Obviously, there are other ways of obtaining equally spaced points on a circle. For simplicity, we pick particular ways of taking sampling points, satisfying a condition that a conjugate of a sample point is also a sample point. General cases of those shown in Fig. 8.5 are obtained by taking ω_k as

$$\omega_k = \frac{2\pi k}{N}$$

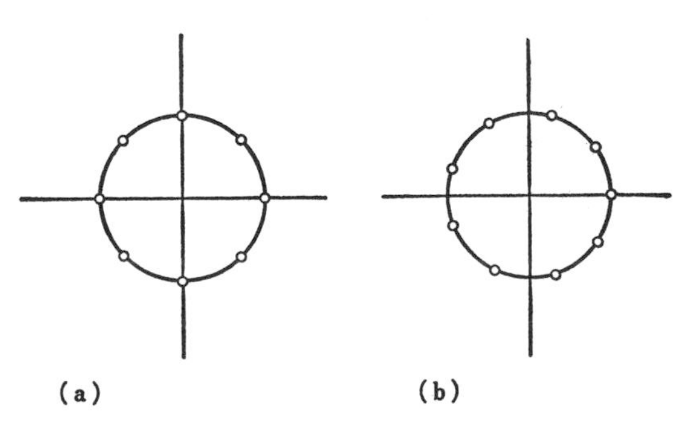

Figure 8.5 Equally Spaced Sample Point of First Type. (a) $N = 8$; (b) $N = 9$.

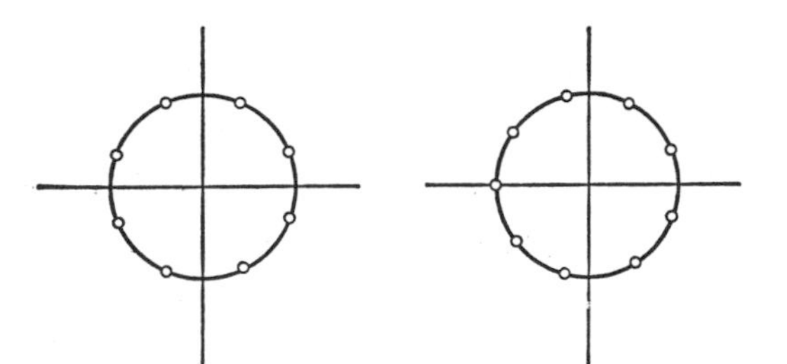

Figure 8.6 Equally Spaced Sample Point of Second Type. (a) $N = 8$; (b) $N = 9$.

There are other ways of taking sampling points that also satisfy this condition. Examples of these are shown in Fig. 8.6. These are obtained by taking ω_k as

$$\omega_k = \frac{2\pi\left(k + \frac{1}{2}\right)}{N}$$

and Equation 8.10 becomes

$$H(z) = \sum_{k=0}^{N-1} \frac{H_d(k)}{N} \frac{1 - e^{j2\pi(k+1/2)} z^{-N}}{1 - e^{j(2\pi/N)(k+1/2)} z^{-1}}$$

Because $e^{j2\pi(k+1/2)}$ is -1, a frequency response $H(e^{j\omega})$ of this $H(z)$ is

$$H(e^{j\omega}) = \frac{e^{-j\omega(N-1)/2}}{jN} \sum_{k=0}^{N-1} \frac{H_d(k)\, e^{-j(\pi/N)(k+1/2)} \cos\left(\frac{N}{2}\,\omega\right)}{\sin\left(\frac{\omega}{2} - \frac{\pi\left(k + \frac{1}{2}\right)}{N}\right)}$$

8.3 PROPERTIES OF FIR LINEAR PHASE FILTER

Often it is desirable to design a FIR filter having a linear phase. Let $H_R(e^{j\omega})$ be the real part of $H(e^{j\omega})$ and $H_I(e^{j\omega})$ be the imaginary part of $H(e^{j\omega})$. Then,

$$H(e^{j\omega}) = H_R(e^{j\omega}) + jH_I(e^{j\omega})$$

The absolute value of $H(e^{j\omega})$ is

$$|H(e^{j\omega})| = \sqrt{H_R(e^{j\omega})^2 + H_I(e^{j\omega})^2}$$

and the phase of $H(e^{j\omega})$ is

$$\underline{/H(e^{j\omega})} = \tan^{-1} \frac{H_I(e^{j\omega})}{H_R(e^{j\omega})}$$

When the phase $\underline{/H(e^{j\omega})}$ satisfies

$$\underline{/H(e^{j\omega})} = -(\alpha\omega + \beta) \tag{8.12}$$

where α and β are constants, it is called "linear phase," and a system whose frequency response is this $H(e^{j\omega})$ which is called a linear phase system. In other words, if $H(e^{j\omega})$ is linear phase, $H(e^{j\omega})$ can be expressed as

$$H(e^{j\omega}) = |H(e^{j\omega})|e^{-j(\alpha\omega+\beta)} \tag{8.13}$$

Because $e^{-j(\alpha\omega+\beta)}$ is

$$e^{-j(\alpha\omega+\beta)} = \cos(\alpha\omega + \beta) - j\sin(\alpha\omega + \beta)$$

Eq. 8.13 can be expressed as

$$H(e^{j\omega}) = |H(e^{j\omega})|[\cos(\alpha\omega + \beta) - j\sin(\alpha\omega + \beta)] \tag{8.14}$$

Because $H(e^{j\omega})$ is a frequency response of a unit sample response $h(n)$, $H(e^{j\omega})$ is $\sum_{n=0}^{\infty} h(n)e^{-j\omega n}$. Hence, Equation 8.14 can be written as

$$|H(e^{j\omega})|\{\cos(\alpha\omega + \beta) - j\sin(\alpha\omega + \beta)\} = \sum_{n=0}^{\infty} h(n)\{\cos\omega n - j\sin\omega n\} \tag{8.15}$$

In other words, if $h(n)$ satisfies Equation 8.15, $h(n)$ will be linear phase. To design a FIR linear phase filter, we can choose $h(n)$, which satisfies Equation 8.15. There is a much simpler way of obtaining $h(n)$, however, which we will discuss next.

Suppose $h(n)$ is real. Then the real part in the right-hand side of Equation 8.15 is $\sum_{n=0}^{\infty} h(n)\cos\omega n$. Because the real part in the left-hand side and that in the right-hand side of Equation 8.15 must be equal, we have

$$|H(e^{j\omega})|\cos(\alpha\omega + \beta) = \sum_{n=0}^{\infty} h(n)\cos\omega n \tag{8.16}$$

Also, the imaginary part in the left-hand side and that in the right-hand side of Equation 8.15 must be equal. Hence,

$$|H(e^{j\omega})|\sin(\alpha\omega + \beta) = \sum_{n=0}^{\infty} h(n)\sin\omega n \tag{8.17}$$

By using Equations 8.16 and 8.17, we can remove $|H(e^{j\omega})|$ as

$$\sum_{n=0}^{\infty} h(n)\cos\omega n\sin(\alpha\omega + \beta) - \sum_{n=0}^{\infty} h(n)\sin\omega n\cos(\alpha\omega + \beta) = 0$$

which can be rewritten as

$$\sum_{n=0}^{\infty} h(n) \sin(\alpha\omega + \beta - \omega n) = 0 \tag{8.18}$$

Because $\sin[(\alpha - n)\omega + \beta]$ is $\sin[(\alpha - n)\omega]\cos\beta + \cos[(\alpha - n)\omega]\sin\beta$, Equation 8.18 becomes

$$\sum_{n=0}^{\infty} h(n) \sin[(\alpha - n)\omega]\cos\beta + \sum_{n=0}^{\infty} h(n)\cos[(\alpha - n)\omega]\sin\beta = 0 \tag{8.19}$$

which must be satisfied if $h(n)$ is real and linear phase. A set of $h(n)$, β and α satisfying Equation 8.19 is

$$\beta = 0, \qquad \alpha = 0, \qquad h(n) = 0, \qquad \text{where } n = 0$$

which will not be useful for designing digital filters. There are two other sets satisfying Equation 8.19 which are

Type I:

$$\left.\begin{aligned} &\beta = 0 \\ &\alpha = \frac{N-1}{2} \\ &h(n) = h(N - 1 - n), \quad 0 \le n \le N - 1 \end{aligned}\right\} \tag{8.20}$$

and type II:

$$\left.\begin{aligned} &\beta = \pm\frac{\pi}{2} \\ &\alpha = \frac{N-1}{2} \\ &h(n) = -h(N - 1 - n), \quad 0 \le n \le N - 1 \end{aligned}\right\} \tag{8.21}$$

$h(n)$ of type I is symmetric around the center of the sequence, which is at $(N - 1)/2$ for an odd value of N and between $n/2$ and $(N/2) - 1$ for an even value of N as shown in Fig. 8.7(a) and (b).

Conversely, $h(n)$ of type II is antisymmetric around the center of the sequence, which is at $(N - 1)/2$ for an odd value of N, and between $N/2$ and $(N/2) - 1$ for an even value of N as shown in Fig. 8.8(a) and (b). This is the property of $h(n)$ when $h(n)$ is real and linear phase.

Let us find a frequency response of $h(n)$, which is real and linear phase. When N is even, $H(e^{j\omega})$ can be expressed as

$$H(e^{j\omega}) = \sum_{n=0}^{(N/2)-1} h(n)e^{-j\omega n} + \sum_{n=N/2}^{N-1} h(n)e^{-j\omega n} \tag{8.22}$$

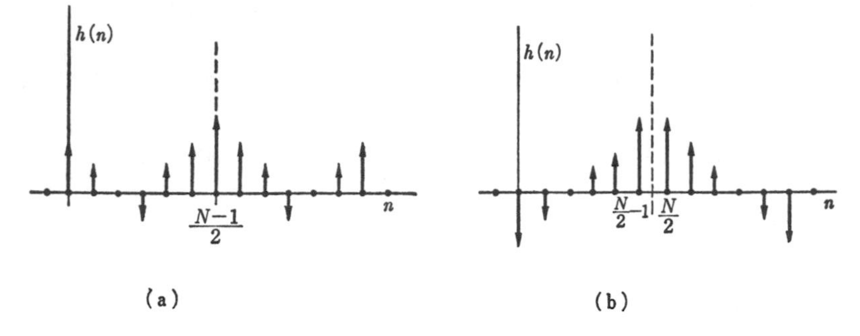

Figure 8.7 *h*(*n*) of Type I. (a) Odd Value of *N*; (b) Even Value of *N*.

If $h(n)$ is of type I, this equation can be changed by using Equation 8.20 as

$$H(e^{j\omega}) = \sum_{n=0}^{(N/2)-1} h(n)e^{-j\omega n} + \sum_{n=N/2}^{N-1} h(N-1-n)e^{-j\omega n}$$

$$= \sum_{n=0}^{(N/2)-1} h(n)[e^{-j\omega n} + e^{j\omega n}e^{-j\omega(N-1)}]$$

$$= e^{-j\omega((N-1)/2)} \sum_{n=0}^{(N/2)-1} 2h(n) \cos\left[\omega\left(\frac{N}{2} - n - \frac{1}{2}\right)\right]$$

A property of this equation is when $\omega = \pi$, $H(e^{j\omega}) = 0$.

When N is odd, $H(e^{j\omega})$ in Equation 8.22 becomes

$$H(e^{j\omega}) = \sum_{n=0}^{[(N-1)/2]-1} h(n)e^{-j\omega n} + h\left(\frac{N-1}{2}\right) e^{-j\omega(N-1)/2} + \sum_{n=(N-1)/2+1}^{N-1} h(n)e^{-j\omega n}$$

If $h(n)$ is of type I, this equation can be changed by using Equation 8.20 as

$$H(e^{j\omega}) = \sum_{n=0}^{(N-3)/2} h(n)e^{-j\omega n} + h\left(\frac{N-1}{2}\right) e^{-j(N-1)/2} + \sum_{n=0}^{(N-3)/2} h(n)e^{-j\omega(N-1-n)}$$

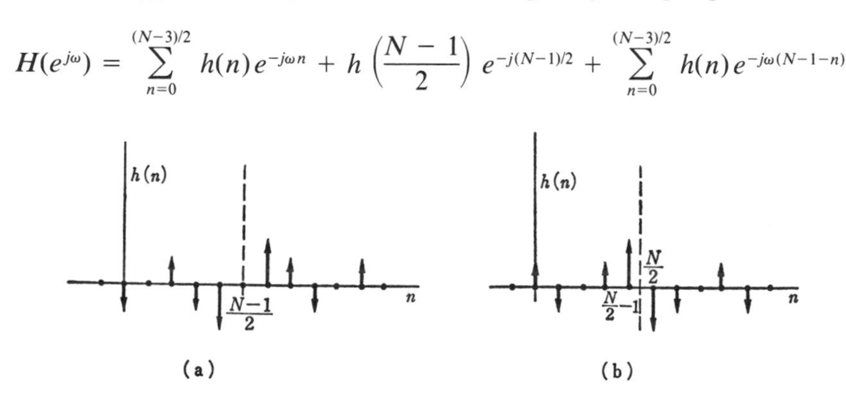

Figure 8.8 *h*(*n*) of Type II. (a) Odd Value of *N*; (b) Even Value of *N*.

$$= e^{-j\omega(N-1)/2}\left\{\sum_{n=0}^{(N-3)/2} 2h(n)\cos\left[\omega\left(\frac{N-1}{2}-n\right)\right] + h\left(\frac{N-1}{2}\right)\right\}$$

If $h(n)$ is of type II, $H(e^{j\omega})$ becomes

$$H(e^{j\omega}) = e^{-j\{[\omega(N-1)/2]-\pi/2\}}\sum_{n=0}^{(N/2)-1} 2h(n)\sin\left[\omega\left(\frac{N}{2}-n-\frac{1}{2}\right)\right]$$

for an even value of N and

$$H(e^{j\omega}) = e^{-j\{[\omega(N-1)/2]-\pi/2\}}\sum_{n=0}^{(N-3)/2} 2h(n)\sin\left[\omega\left(\frac{N-1}{2}-n\right)\right]$$

for an odd value of N.

8.4 IIR DIGITAL FILTER

A design of IIR filter can be accomplished often by changing suitable analog filter to a digital filter. In Chapter 4, we have studied that an output analog signal $y_a(t)$ is equal to the inverse Laplace transformation of product $H_a(s)X_a(s)$, where $H_a(s)$ is the Laplace transformation of a transfer function $h_a(t)$ of an analog system, and $X_a(s)$ is the Laplace transformation of an input signal $x(t)$ as

$$y_a(t) = \mathcal{L}^{-1}[H_a(s)X_a(s)]$$

Especially, when $x_a(t) = e^{j\omega t}$, the output $y_a(t)$ is

$$y_a(t) = \mathcal{L}^{-1}\left[H_a(s)\frac{1}{s-j\omega}\right]$$

If $j\omega$ is not a zero of $H_a(s)$, the partial fraction expansion of $H_a(s)/(s-j\omega)$ is

$$H_a(s)/(s-j\omega) = \frac{H_a(s)|_{s=j\omega}}{s-j\omega} + \cdots$$

Hence $y_a(t)$ is equal to

$$y_a(t) = \mathcal{L}^{-1}\left[\frac{H_a(s)|_{s=j\omega}}{s-j\omega} + \cdots\right] = H_a(s)|_{s=j\omega}e^{j\omega t} + \cdots \quad (8.23)$$

$H_a(s)|_{s=j\omega}e^{j\omega t}$ in the right-hand side of this equation is a steady-state response when $x_a(t) = e^{j\omega t}$.

With respect an analog system, let $H_a(s)$ be the Laplace transformation of a frequency response, $x_a(t)$ be an input signal, and $y_a(t)$ be the output signal. For a digital filter, let $H(z)$ be the z-transformation of a unit sample response $h(n)$, $x(nT)$ be the sampling of $x_a(t)$ with a sampling interval T, and

$y(nT)$ be the output signal for an input signal $x(nT)$. Then to convert an analog system corresponding to a given $H_a(s)$ into a digital filter means to obtain an $H(z)$ from the $H_a(s)$ such that the output signal $y(nT)$ is equal to the sampling of the signal $y_a(t)$ for an input signal $x_a(t)$.

Suppose an input signal $x_a(t)$ is

$$x_a(t) = e^{j\omega t}$$

Then sampling this $x_a(t)$ gives

$$x(nT) = e^{j\omega nT}$$

Hence the z-transformation of the $x(nT)$ becomes

$$X(z) = \sum_{n=0}^{\infty} e^{j\omega nT} z^{-n} = \frac{1}{1 - e^{j\omega T} z^{-1}}$$

With the z-transformation $H(z)$ of a unit sample response $h(n)$ of a digital filter and the z-transformation $X(z)$ of an input signal $x(nT)$, the output $y(nT)$ of the digital filter is the inverse z-transformation of $X(z)H(z)$, or

$$y(nT) = \mathfrak{Z}^{-1}[H(z)X(z)] \tag{8.24}$$

The right-hand side of this equation is

$$\mathfrak{Z}^{-1}[H(z)X(z)] = \mathfrak{Z}^{-1}\left[H(z)\frac{z}{z - e^{j\omega T}}\right] = \mathfrak{Z}^{-1}\left[z\left(\frac{H(z)}{z - e^{j\omega T}}\right)\right] \tag{8.25}$$

Because the partial fraction expansion of the term inside the parentheses in the right-hand side is

$$\frac{H(z)}{z - e^{j\omega T}} = \frac{H(z)|_{z=e^{j\omega T}}}{z - e^{j\omega T}} + \cdots$$

Equation 8.24 can be expressed as

$$y(nT) = H(z)|_{z=e^{j\omega T}} e^{j\omega nT} + \cdots \tag{8.26}$$

The term $H(z)|_{z=e^{j\omega T}} e^{j\omega nT}$ in the right-hand side is a steady-state output signal when an input signal is $e^{j\omega nT}$.

Because Equation 8.23 is the steady-state output signal of an analog system when an input signal is $e^{j\omega t}$, comparing $H_a(s)|_{s=j\omega}$ and $H(z)|_{z=e^{j\omega T}}$ in Equation 8.26 we can find whether an analog system having $H_a(s)$ and a digital filter having $H(z)$ have the same properties. Because comparing $H_a(s)|_{s=j\omega}$ and $H(z)|_{z=e^{j\omega T}}$, letting

$$e^{sT} = z$$

or

$$s = \frac{1}{T}\log_e z$$

changing $H(z)$ as

$$H(z) = H_a\left(\frac{1}{T}\log_e z\right)$$

we can see that $y_a(nT)$, which is the sampling of $y_a(t)$ and $y(nT)$, will be the same. Thus, it appears that we can obtain a desired digital filter from an analog system. $H_a\left(\frac{1}{T}\log_e z\right)$ is not a rational function of z, however. Hence, we cannot make a signal flow graph for a digital filter with a finite number of branches. This means that to use a computer to obtain an output signal $y(nT)$ is not effective. Hence, it is desirable to develop methods of transforming $H_a(s)$ to $H(z)$, which is a rational function of z. Among many methods, we will introduce three well-known transformations known as the "impulse invariance," the "finite difference," and the "bilinear" transformations.

8.4.1 Impulse Invariance Transformation

A transfer function $H_a(s)$ of an analog system can be expressed as

$$\frac{Y_a(s)}{X_a(s)} = H_a(s) = \frac{\sum_{m=0}^{M} d_m s^m}{\sum_{r=0}^{N} c_r s^r}$$

which is the Laplace transformation of the output signal $y(t)$ when the input signal is an impulse $\delta(t)$. The partial fraction expansion of the right-hand side of this equation gives

$$H_a(s) = \sum_{r=0}^{N} \frac{c_r}{s + p_r} \tag{8.27}$$

under the assumption that all poles are simple and $M < N$. By taking the inverse Laplace transformation of this, we have

$$h_a(t) = \sum_{r=0}^{N} c_r e^{-p_r t} u_s(t)$$

By sampling this $h_a(t)$ with a sampling interval T, we have an $h(nT)$ as

$$h(nT) = \sum_{r=0}^{N} c_r e^{-p_r nT} u_s(nT)$$

The z-transformation of this $h(nT)$ is

$$H(z) = \sum_{n=0}^{\infty} h(nT) z^{-n} = \sum_{r=0}^{N} \frac{c_r}{1 - e^{-p_r T} z^{-1}}$$

Comparing this equation and Equation 8.27, we can see that

$$\frac{1}{s + p_r} \leftrightarrow \frac{1}{1 - e^{-p_r T} z^{-1}}$$

This shows that by obtaining the partial fraction expansion of $H_a(s)$ and replace each $1/(s + p_r)$ by $1/(1 - e^{-p_r T} z^{-1})$, we will have $H(z)$ of a digital filter. This is known as the impulse invariant transformation.

As an example, let us obtain a digital filter from a transfer function $H_a(s)$ of an analog system where

$$H_a(s) = \frac{8s}{s^2 - 4}$$

First, we change $H_a(s)$ by using the partial fraction expansion as

$$H_a(s) = \frac{4}{s + 2} + \frac{4}{s - 2}$$

Then changing $1/(s + 2)$ by $1/(1 - e^{-2T} z^{-1})$ and $1/(s - 2)$ by $1/(1 - e^{2T} z^{-1})_1$ we have

$$H(z) = \frac{4}{1 - e^{-2T} z^{-1}} + \frac{4}{1 - e^{2T} z^{-1}}$$

which is the desired $H(z)$ of a digital filter.

Let us compare $H_a(s)$ and $H(z)$. From

$$H_a(s) = \frac{8s}{s^2 - 4}$$

a frequency response $H_a(j\omega)$ is

$$H_a(j\omega) = \frac{-8(j\omega)}{4 + \omega^2}$$

From this, the absolute value is

$$|H_a(j\omega)| = \frac{8\omega}{4 + \omega^2}$$

and the phase is

$$\underline{/H_a(j\omega)} = -90°, \qquad \omega \geq 0$$

Conversely, $H(z)$ of a digital filter can be expressed as

$$H(z) = \frac{4}{1 - e^{-2T} z^{-1}} + \frac{4}{1 - e^{2T} z^{-1}} = 4 \left(\frac{2 - (e^{2T} + e^{-2T}) z^{-1}}{1 - (e^{2T} + e^{-2T}) z^{-1} + z^{-2}} \right)$$

The frequency response $H(e^{j\omega T})$ can be obtained from $H(z)$ as

$$H(e^{j\omega T}) = 4\left(\frac{2 - Ke^{-j\omega T}}{1 - Ke^{-j\omega T} + e^{-2j\omega T}}\right) = 4\left(\frac{2\cos\omega T - K + j2\sin\omega T}{2\cos\omega T - K}\right)$$

where K is

$$K = e^{2T} + e^{-2T}$$

Hence, the absolute value of this $H(e^{j\omega T})$ is

$$|H(e^{j\omega T})| = \frac{4\sqrt{4 - 4K\cos\omega T + K^2}}{|2\cos\omega T - K|}$$

and its phase is

$$\underline{/H(e^{j\omega T})} = \tan^{-1}\frac{2\sin\omega T}{2\cos\omega T - K}$$

8.4.2 Finite Difference Transformation

This method converts $H_a(s)$ to $H(z)$ using a transformation, which is equivalent to changing a differential equation into a difference equation. Expressing $\left.\frac{dh_a(t)}{dt}\right|_{t=nT}$ as

$$\left.\frac{dh_a(t)}{dt}\right|_{t=nT} \rightarrow \nabla^{(1)}h(nT) = \frac{h(nT) - h[(n-1)T]}{T}$$

is known as a "backward difference interpolation." The z-transformation of this is

$$\mathscr{Z}[\nabla^{(1)}h(nT)] = \left(\frac{1 - z^{-1}}{T}\right) H(z)$$

$\left.\frac{d^2h_a(t)}{dt^2}\right|_{t=nT}$ is

$$\left.\frac{d^2h_a(t)}{dt^2}\right|_{t=nT} \rightarrow \nabla^{(2)}h(nT) = \nabla^{(1)}\left\{\frac{h(nT) - h[(n-1)T]}{T}\right\}$$

$$= \frac{\dfrac{h(nT) - h[(n-1)T]}{T} - \dfrac{h[(n-1)T] - h[(n-2)T]}{T}}{T}$$

and the z-transformation of this will be

$$\mathscr{Z}[\nabla^{(2)}h(nT)] = \mathscr{Z}[\nabla^{(1)}\nabla^{(1)}h(nT)] = \frac{1 - z^{-1}}{T}\left[\frac{1 - z^{-1}}{T}\right] H(z)$$

$$= \left(\frac{1 - z^{-1}}{T}\right)^2 H(z)$$

In general,

$$\left.\frac{d^r h_a(t)}{dt^r}\right|_{t=nT} \to \nabla^{(r)} h(nT)$$

and the z-transformation of this will be

$$\mathfrak{Z}[\nabla^{(r)} h(nT)] = \left(\frac{1 - z^{-1}}{T}\right)^r H(z) \tag{8.28}$$

Conversely, the Laplace transformation of $d^r h_a(t)/dt^r$ with all initial conditions being 0 will be

$$\mathcal{L}\left[\frac{d^r h_a(t)}{dt^r}\right] = s^r H_a(s) \tag{8.29}$$

By comparing Equations 8.28 and 8.29, we can see that changing s to $(1 - z^{-1})/T$ will produce a digital filter from an analog system. In other words, changing $H_a(s)$ into $H(z)$ by the finite difference transformation is accomplished just by replacing all s by

$$s \to \frac{1 - z^{-1}}{T} \tag{8.30}$$

Let us take the same example as the one in 8.4.1, which is

$$H_a(s) = \frac{8s}{s^2 - 4}$$

by Equation 8.30, $H(z)$ will be

$$H(z) = \frac{8\left(\dfrac{1 - z^{-1}}{T}\right)}{\left(\dfrac{1 - z^{-1}}{T}\right)^2 - 4} = -\frac{8T(1 - z)}{(1 - 4T^2)z - 2 + z^{-1}}$$

The frequency response of this $H(z)$ is

$$H(e^{j\omega T}) = -4T\frac{(1 - \cos\omega T)[2T^2 - (1 - \cos\omega T)] + j\sin\omega T[(1 - \cos\omega T) + 2T^2]}{(1 - \cos\omega T)^2 + 4T^2\cos\omega T(1 - \cos\omega T) + 4T^4}$$

Hence, the absolute value of $H(e^{j\omega T})$ is

$$|H(e^{j\omega T})| = 4T\frac{\sqrt{(1 - \cos\omega T)^2[2T^2 - (1 - \cos\omega T)]^2 + \sin^2\omega T[(1 - \cos\omega T) + 2T^2]^2}}{|(1 - \cos\omega T)^2 + 4T^2\cos\omega T(1 - \cos\omega T) + 4T^4|}$$

and its phase is

$$\underline{/H(e^{j\omega T})} = \tan^{-1} \frac{\sin \omega T[(1 - \cos \omega T) + 2T^2]}{(1 - \cos \omega T)[2T^2 - (1 - \cos \omega T)]}$$

8.4.3 Bilinear Transformation

Suppose we take the Laplace transformation of a differential equation

$$\alpha_1 y'_\alpha(t) + \alpha_0 y_\alpha(t) = \beta_0 x(t) \tag{8.31}$$

and obtain a transfer function $H_a(s)$, which will be

$$H_a(s) = \frac{\beta_0}{\alpha_1 s + \alpha_0} \tag{8.32}$$

Correspondingly, we will obtain the z-transformation $H(z)$ of a unit sample response $h(n)$ of a digital filter as follows: First, we express $y_a(t)$ by the use of integration as

$$y_a(t) = \int_{t_0}^{t} y'_a(\tau)\, d\tau + y_a(t_0)$$

Then sampling this with a sample interval T as

$$y_a(nT) = \int_{(n-1)T}^{nT} y'_a(\tau)\, d\tau + y_a[(n-1)T]$$

Now we change this expression by the use of the trapezoidal rule as

$$y_a(nT) = \frac{T}{2}\{y'_a(nT) + y'_a[(n-1)T]\} + y_a[(n-1)T] \tag{8.33}$$

Sampling Equation 8.31, we have

$$y'_a(nT) = -\frac{\alpha_0}{\alpha_1} y_a(nT) + \frac{\beta_0}{\alpha_1} x(nT)$$

Using this equation, we can change Equation 8.33 as

$$y_a(nT) = \frac{T}{2}\left\{-\frac{\alpha_0}{\alpha_1} y_a(nT) + \frac{\beta_0}{\alpha_1} x(nT) - \frac{\alpha_0}{\alpha_1} y_a[(n-1)T] + \frac{\beta_0}{\alpha_1} x[(n-1)T]\right\} + y_a[(n-1)T]$$

which can be rewritten as

$$y_a(nT) - y_a[(n-1)T] = \frac{T}{2}\left\{-\frac{\alpha_0}{\alpha_1}\left(y_a(nT) + y_a[(n-1)T]\right) + \frac{\beta_0}{\alpha_1}[x(nT) + x[(n-1)T]]\right\} \tag{8.34}$$

Applying the z-transformation on Equation 8.34 and solving $H(z) = Y_a(z)/X_a(z)$, we have

$$H(z) = \frac{\beta_0}{\alpha_1 \frac{2}{T}\left(\frac{1 - z^{-1}}{1 + z^{-1}}\right) + \alpha_0}$$

This $H(z)$ should correspond to $H_a(s)$ in Equation 8.32. Hence, the bilinear transformation is to replace s by $(2/T)(1 - z^{-1})/(1 + z^{-1})$ that is,

$$s \rightarrow \frac{2}{T}\frac{1 - z^{-1}}{1 + z^{-1}} \tag{8.35}$$

$H_a(s)$, which is used in 8.4.1 as an example, is

$$H_a(s) = \frac{8s}{s^2 - 4}$$

Changing this $H_a(s)$ by the bilinear transformation using Equation 8.35, we have

$$H(z) = \frac{8\left(\frac{2}{T}\frac{1 - z^{-1}}{1 + z^{-1}}\right)}{\left(\frac{2}{T}\frac{1 - z^{-1}}{1 + z^{-1}}\right)^2 - 4} = \frac{4T(1 - z^{-1})(1 + z^{-1})}{(1 - z^{-1})^2 - T^2(1 + z^{-1})^2}$$

A frequency response $H(e^{j\omega T})$ of this $H(z)$ is

$$H(e^{j\omega T}) = j4T\frac{\sin \omega T}{(1 - T^2)\cos \omega T - (1 + T^2)}$$

Hence, the absolute value and the phase for $T > 0$ of this $H(e^{j\omega T})$ are

$$|H(e^{j\omega T})| = 4T\left|\frac{\sin \omega T}{(1 - T^2)\cos \omega T - (1 + T^2)}\right|$$

and

$$\underline{/H(e^{j\omega T})} = \begin{cases} -90^\circ, & 0 \le \omega \le \pi \\ 90^\circ, & \pi < \omega \le 2\pi \end{cases}$$

8.5 SUMMARY

1. A unit sample response $h(n)$ is a FIR function if there exists a finite integer N such that $h(n)$ is 0 at all n for $n > N$.
2. A unit sample response $h(n)$ is an IIR function if for any integer N there is $n > N$ such that $h(n) \neq 0$.
3. A digital filter whose unit sample response is a FIR function is a FIR filter.

4. A digital filter whose unit sample response is an IIR function is an IIR filter.
5. When multiplying $w(n)$ by an IIR function $h(n)$ makes the result a FIR function, the function $w(n)$ is called a window. Typical examples of windows are a rectangular window, Hamming window, and Kaiser window.
6. Sampling a frequency response $H_a(e^{j\omega})$ of an IIR function $h_a(n)$ to obtain a FIR function is called an approximation by frequency sampling.
7. A function is of linear phase when its phase can be expressed as $\alpha\omega + \beta$ with real numbers α and β.
8. A unit sample response $h(n)$ of a FIR linear phase filter satisfies

$$|H(e^{j\omega})|\{\cos(\alpha\omega + \beta) - j\sin(\alpha\omega + \beta)\} = \sum_{n=0}^{\infty} h(n)\{\cos\omega n - j\sin\omega n\}$$

Especially when $h(n)$ is real, $h(n)$ satisfies one of the following two:

Type I:

$$\left.\begin{aligned} \beta &= 0 \\ \alpha &= \frac{N-1}{2} \\ h(n) &= h(N-1-n), \quad 0 \le n \le N-1 \end{aligned}\right\}$$

Type II:

$$\left.\begin{aligned} \beta &= \pm\frac{\pi}{2} \\ \alpha &= \frac{N-1}{2} \\ h(n) &= -h(N-1-n), \quad 0 \le n \le N-1 \end{aligned}\right\}$$

9. Converting an analog filter into an IIR digital filter is possible by changing s of $H(s)$ to a function of z to form $H(z)$. Typical transformations are
   a. The impulse invariant transformation, which uses

$$\frac{1}{s + p_r} \rightarrow \frac{1}{1 - e^{-p_r T} z^{-1}}$$

   b. The finite difference transformation, which uses

$$s \rightarrow \frac{1 - z^{-1}}{T}$$

and

c. The bilinear transformation, which uses

$$s \rightarrow \frac{2}{T}\frac{1 - z^{-1}}{1 + z^{-1}}$$

8.6 PROBLEMS

1. Calculate a frequency response $W(e^{j\omega})$ of a window $w(n)$ in Fig. 8.9.
2. Draw a frequency response $W(e^{j\omega})$ of a window $w(n)$ in Fig. 8.9.
3. For $H_d(e^{j\omega})$ in Fig. 8.10, obtain $H_f(e^{j\omega})$ when a rectangular window is employed. Note: If $h_d(n)$ is the inverse discrete function Fourier transformation of $H_d(e^{j\omega})$ and $h_f(n)$ is the inverse DFT of $H_f(e^{j\omega})$, then $h_f(n)$ is $h_d(n)w(n)$.
4. For $h(n)$ in Fig. 8.11, draw a frequency response $H(e^{j\omega})$ of $h(n)$.
5. From a frequency response $H(e^{j\omega})$ of $h(n)$ in Fig. 8.12, show that the phase of $H(e^{j\omega})$ is linear phase.
6. Under the second type of the frequency sampling, if we express $H(k)$ as $|H(k)|e^{j\theta(k)}$, prove that

$$|H(k)| = |H(N - k)|$$

$$\theta(k) = -\theta(N - k)$$

   are true.
7. Under the second type of the frequency sampling, if a unit sample response $h(n)$ is real, and N is even, then prove

$$H(e^{j\omega}) = e^{-j[(\omega(N-1)/2]-\pi/2]} \sum_{n=0}^{(N/2)-1} 2h(n) \sin\left[\omega\left(\frac{N}{2} - n - \frac{1}{2}\right)\right]$$

   is true.
8. Under the second type of the frequency sampling, if a unit sample response $h(n)$ is real, and N is odd, then prove

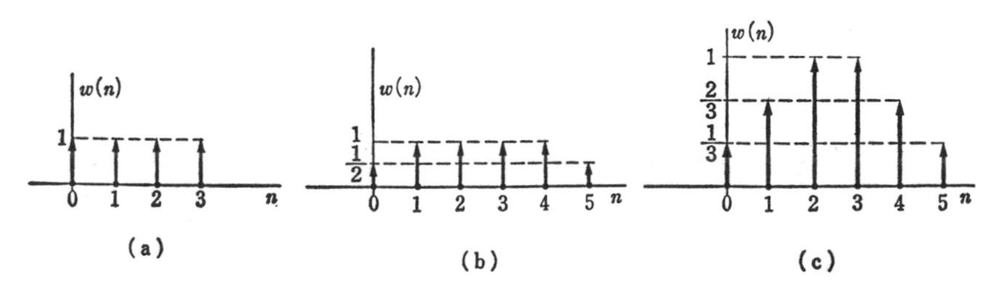

Figure 8.9 Window $w(n)$

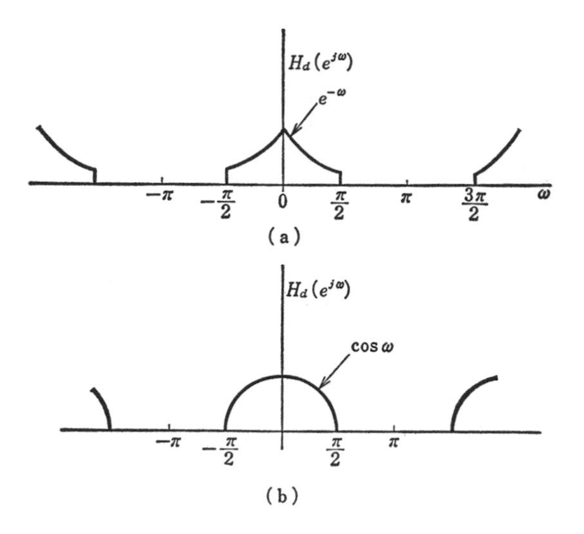

Figure 8.10 Frequency Response $H_d(e^{j\omega})$

$$H(e^{j\omega}) = e^{-j[(\omega(N-1)/2]-\pi/2]} \sum_{n=0}^{(N-3)/2} 2h(n) \sin\left[\omega\left(\frac{N-1}{2} - n\right)\right]$$

is correct.

9. From the following $H(s)$, obtain $H(z)$ by the impulse invariant transformation.

a. $H(s) = \dfrac{1}{(s+\alpha)(s+\beta)}$ $\quad ,T = T_0$

b. $H(s) = \dfrac{15}{s^3 + 6s^2 + 15s + 15}$ $\quad ,T = T_0$

c. $H(s) = \dfrac{1}{s^3 + 2s^2 + 2s + 1}$ $\quad ,T = T_0$

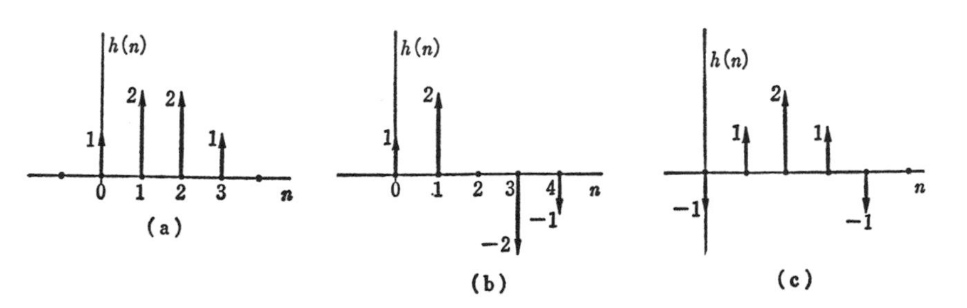

Figure 8.11 Unit Sample Response $h(n)$

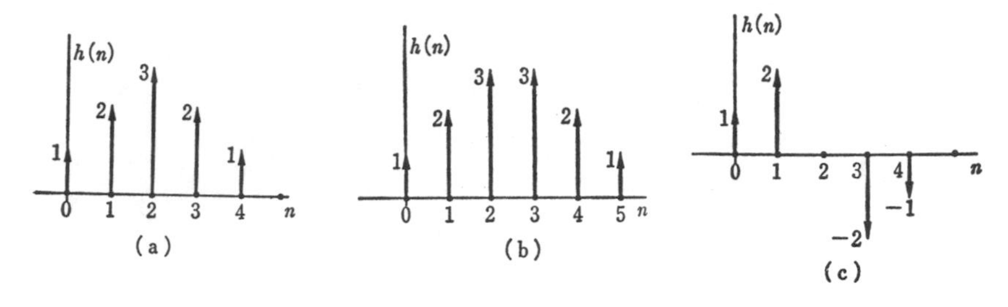

Figure 8.12 Function $h(n)$

d. $H(s) = \dfrac{0.668}{s^3 + 1.5s^2 + 1.453s + 0.668}$ $, T = T_0$

10. From $H(s)$ in problem 9, obtain $H(z)$ by the finite difference transformation.
11. From $H(s)$ in problem 9, obtain $H(z)$ by the bilinear transformation.
12. From $H(s)$ below, obtain $H(z)$ by the impulse invariant transformation.

a. $H(s) = \dfrac{1}{(s+1)(s+2)}$ b. $H(s) = \dfrac{1}{s^2 + s + 1}$

$T = 1/10$ $T = 1/10$

13. From $H(s)$ below, calculate $H(z)$ by the finite difference transformation.

a. $H(s) = \dfrac{1}{s+1}$ b. $H(s) = \dfrac{1}{s^2 + 1}$ c. $H(s) = \dfrac{1}{s^2 - 1}$

$T = 1/10$ $T = 1/10$ $T = 1/10$

14. From $H(s)$ in problem 13, obtain $H(z)$ by the bilinear transformation.

9
Hilbert Transformation

9.1 FOURIER TRANSFORMATION OF EVEN AND ODD FUNCTIONS, AND Sgn *t*

For analysis of electrical networks, it is often simpler to use $e^{j\omega t}$ rather than to use $\cos \omega t$ or $\sin \omega t$. Similarly, instead of using a given real function alone, using a complex function by adding a suitable imaginary function may make digital processing simpler. The Hilbert transformation is very effective to change a given real function or a given imaginary function to a complex function. The Hilbert transformation is developed by using the property that the real part of the Fourier transformation of a given function $h(t)$ is from an even function of $h(t)$, and the imaginary part of the Fourier transformation of $h(t)$ is from an odd function of $h(t)$. Hence, we will study the properties of even functions and odd functions first.

An even function $h_e(t)$ is one that satisfies

$$h_e(t) = h_e(-t) \tag{9.1}$$

and an odd function $h_o(t)$ is one satisfying

$$h_o(t) = -h_o(-t) \tag{9.2}$$

For example, $h_e(t) = t^2 + 2$ is an even function and $h_o(t) = t^3 + 3t$ is an odd function.

A real function $h(t)$ is always expressed as the sum of an even function and an odd function as

$$h(t) = h_e(t) + h_o(t) \tag{9.3}$$

For example $h(t) = t^3 + t^2 + 3t + 2$ is the sum of an even function $h_e(t) = t^2 + 2$ and an odd function $h_o(t) = t^3 + 3t$. By changing t to $-t$ in Equation 9.3, we have

$$h(-t) = h_e(-t) + h_o(-t)$$

We can change this equation by using Equation 9.1 and 9.2 as

$$h(-t) = h_e(t) - h_o(t)$$

From this equation and Equation 9.3, an even function $h_e(t)$ and an odd function $h_o(t)$ of a given function $h(t)$ are

$$\left.\begin{aligned} h_e(t) &= \frac{1}{2}[h(t) + h(-t)] \\ h_o(t) &= \frac{1}{2}[h(t) - h(-t)] \end{aligned}\right\} \tag{9.4}$$

This is a formula to obtain an even function and an odd function from a given function. Changing t to $-t$ of a function $h(t) = t^3 + t^2 + 3t + 2$ used in the previous example, we have

$$h(-t) = -t^3 + t^2 - 3t + 2$$

Then by Equation 9.4 an even function $h_e(t)$ and an odd function $h_o(t)$ of $h(t)$ are

$$h_e(t) = \frac{1}{2}[h(t) + h(-t)] = \frac{1}{2}[2t^2 + 4] = t^2 + 2$$

$$h_o(t) = \frac{1}{2}[h(t) - h(-t)] = \frac{1}{2}[2t^3 + 6t] = t^3 + 3t$$

To find a relationship between a real part and an imaginary part of the Fourier transformation $H(j\omega)$ of a function $h(t)$ and an even function $h_e(t)$ and an odd function $h_o(t)$ of $h(t)$, let us obtain the Fourier transformation $H_e(j\omega)$ of an even function $h_e(t)$ as

$$H_e(j\omega) = \int_{-\infty}^{\infty} h_e(t)e^{-j\omega t}\,dt \qquad \text{where } \omega = 2\pi f$$

Dividing an interval from $-\infty$ to ∞ of an integration in the right-hand side of the equation into two intervals, one of which is from $-\infty$ to 0 and the other from 0 to ∞, as

$$H_e(j\omega) = \int_{-\infty}^{0} h_e(t) e^{-j\omega t}\, dt + \int_{0}^{\infty} h_e(t) e^{-j\omega t}\, dt$$

Changing t to $-t$ in the first term in the right-hand side and using Equation 9.1 to change $h_e(-t)$ to $h_e(t)$, we can obtain

$$H_e(j\omega) = -\int_{0}^{\infty} h_e(-t) e^{j\omega t}\, dt + \int_{0}^{\infty} h_e(t) e^{-j\omega t}\, dt = 2\int_{0}^{\infty} h_e(t) \cos \omega t\, dt \tag{9.5}$$

This shows that $H_e(j\omega)$ is real.

Conversely, the Fourier transformation $H_o(j\omega)$ of an odd function $h_o(t)$ is

$$\begin{aligned} H_o(j\omega) = \int_{-\infty}^{\infty} h_o(t) e^{-j\omega t}\, dt &= -\int_{0}^{\infty} h_o(t) e^{j\omega t}\, dt \\ &+ \int_{0}^{\infty} h_o(t) e^{-j\omega t}\, dt = -2j\int_{0}^{\infty} h_o(t) \sin \omega t\, dt \end{aligned} \tag{9.6}$$

Hence, we can see clearly that $H_o(j\omega)$ is imaginary. Thus, when we obtain the Fourier transformation $H(j\omega)$ of a function $h(t)$, a real part of $H(j\omega)$ is the Fourier transformation of an even function $h_e(t)$ of $h(t)$, and the imaginary part of $H(j\omega)$ is the Fourier transformation of an odd function $h_o(t)$ of $h(t)$.

As an example, consider the Fourier transformation of the following $h(t)$, which is also shown in Fig. 9.1(a).

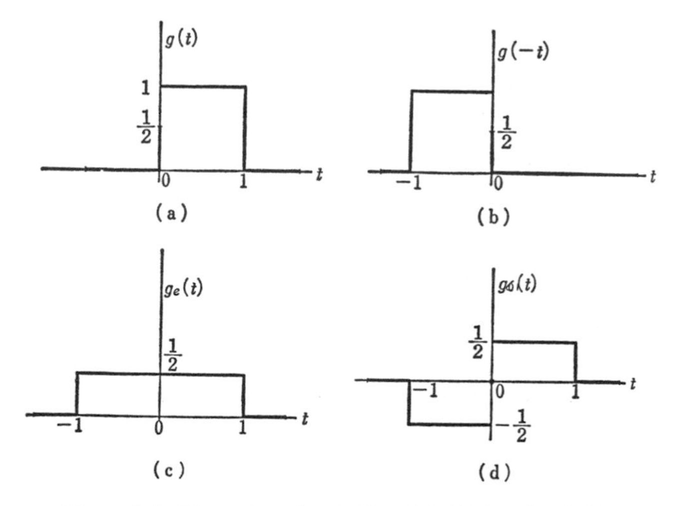

Figure 9.1 Even Function $h_e(t)$ and Odd Function $h_o(t)$

$$h(t) = \begin{cases} 1, & 0 < t \leq 1 \\ \dfrac{1}{2}, & t = 0 \\ 0, & \text{Otherwise} \end{cases}$$

From this $h(t)$, we can obtain $h(-t)$ as

$$h(-t) = \begin{cases} 1, & -1 \leq t < 0 \\ \dfrac{1}{2}, & t = 0 \\ 0, & \text{Otherwise} \end{cases}$$

which is also shown in Fig. 9.1(b). By Equation 9.4, an even function $h_e(t)$ of $h(t)$ is

$$h_e(t) = \begin{cases} \dfrac{1}{2}, & -1 \leq t \leq 1 \\ 0, & \text{Otherwise} \end{cases}$$

which is shown in Fig. 9.1(c) and an odd function $h_o(t)$ of $h(t)$ is

$$h_o(t) = \begin{cases} \dfrac{1}{2}, & 0 < t \leq 1 \\ -\dfrac{1}{2}, & -1 \leq t < 0 \\ 0, & \text{Otherwise} \end{cases}$$

which is shown in Fig. 9.1(d). The Fourier transformation $H_e(j\omega)$ or $h_e(t)$ is

$$H_e(j\omega) = \int_0^1 \cos \omega t \, dt = \frac{\sin \omega}{\omega}$$

and the Fourier transformation $H_o(j\omega)$ of $h_o(t)$ of $h(t)$ is

$$H_o(j\omega) = -j \int_0^1 \sin \omega t \, dt = j \frac{\cos \omega - 1}{\omega}$$

A function sign t is defined as

$$\text{sgn } t = \begin{cases} 1, & t \geq 0 \\ -1, & t < 0 \end{cases}$$

The Fourier transformation $\mathcal{F}[\text{sgn } t]$ of sgn t is an important function for developing the Hilbert transformation. To show that $\mathcal{F}[\text{sgn } t]$ is

$$\mathcal{F}[\text{sgn } t] = \frac{2}{j\omega} \tag{9.7}$$

we will prove that

$$\mathscr{F}^{-1}\left[\frac{2}{j\omega}\right] = \operatorname{sgn} t \tag{9.8}$$

The inverse Fourier transformation $\mathscr{F}^{-1}\left[\frac{2}{j\omega}\right]$ is by Equation 3.20

$$\mathscr{F}^{-1}\left[\frac{2}{j\omega}\right] = \frac{1}{2\pi}\int_{-\infty}^{\infty}\frac{2}{j\omega}\, e^{j\omega t}\, d\omega \tag{9.9}$$

The right-hand side of this equation can be expressed as

$$\frac{1}{2\pi}\int_{-\infty}^{\infty}\frac{2}{j\omega}(\cos\omega t + j\sin\omega t)\, d\omega = \frac{1}{j\pi}\int_{-\infty}^{\infty}\frac{\cos\omega t}{\omega}\, d\omega + \frac{1}{\pi}\int_{-\infty}^{\infty}\frac{\sin\omega t}{\omega}\, d\omega$$

The first term $\frac{\cos\omega t}{\omega}$ in the right-hand side is an odd function of ω, however. Hence, its integration $\int_{-\infty}^{\infty}\frac{\cos\omega t}{\omega}\, d\omega$ is 0. Hence, Equation 9.9 is

$$\mathscr{F}^{-1}\left[\frac{2}{j\omega}\right] = \frac{1}{\pi}\int_{-\infty}^{\infty}\frac{\sin\omega t}{\omega}\, d\omega$$

By making $y = \omega t$, $\mathscr{F}^{-1}\left[\frac{2}{j\omega}\right]$ for $t > 0$ is

$$\mathscr{F}^{-1}\left[\frac{2}{j\omega}\right] = \frac{1}{\pi}\int_{-\infty}^{\infty}\frac{\sin y}{y}\, dy = 1$$

and for $t < 0$ it is

$$\mathscr{F}^{-1}\left[\frac{2}{j\omega}\right] = \frac{-1}{\pi}\int_{-\infty}^{\infty}\frac{\sin y}{y}\, dy = -1$$

Also from Equation 9.9, $\mathscr{F}^{-1}\left[\frac{2}{j\omega}\right]$ at $t = 0$ is

$$\mathscr{F}^{-1}\left[\frac{2}{j\omega}\right] = \lim_{a\to\infty}\frac{1}{2\pi}\int_{-a}^{a}\frac{d\omega}{j\omega} = 1$$

which proves that $\mathscr{F}^{-1}\left[\frac{2}{j\omega}\right]$ is sgn t.

9.2 HILBERT TRANSFORMATION OF ANALOG SIGNAL

A function $h(t)$ is said to be causal or a causal function if $h(t) = 0$ for $t < 0$. For example, a function

$$h(t) = \begin{cases} 2t^3 + 3t^2 + t + 4, & t \geq 0 \\ 0 \quad , & t < 0 \end{cases}$$

is causal. Because an even function and an odd function of a causal function lead to the Hilbert transformation of an analog function, we will investigate properties of such functions. Consider a function $h(t)$ shown in the preceding example. Changing t to $-t$, $h(t)$ becomes

$$h(-t) = \begin{cases} -2t^3 + 3t^2 - t + 4, & t \leq 0 \\ 0 \quad , & t > 0 \end{cases}$$

By using Equation 9.4, an even function $h_e(t)$ and an odd function $h_o(t)$ of this function are

$$h_e(t) = \begin{cases} \frac{1}{2}[2t^3 + 3t^2 + t + 4] \quad , & t > 0 \\ 4 \quad , & t = 0 \\ \frac{1}{2}[-2t^3 + 3t^2 - t + 4], & t < 0 \end{cases}$$

and

$$h_o(t) = \begin{cases} \frac{1}{2}[2t^3 + 3t^2 + t + 4] \quad , & t > 0 \\ 0 \quad , & t = 0 \\ -\frac{1}{2}[-2t^3 + 3t^2 - t + 4], & t < 0 \end{cases}$$

It can be seen from this example that an even function $h_e(t)$ and an odd function $h_o(t)$ of a causal function satisfy the following:

$$\left.\begin{aligned} &h_e(t) = h_o(t) \quad , & t > 0 \\ &h_e(t) = h(0), \quad h_o(t) = 0, & t = 0 \\ &h_e(t) = -h_o(t) \quad , & t < 0 \end{aligned}\right\}$$

Hence, if a causal function $h(t)$ does not contain an impulse function $\delta(t)$, using sgn t and a unit sample $u_I(t)$ where

$$u_I(t) = \begin{cases} 1, & t = 0 \\ 0, & t \neq 0 \end{cases}$$

$h(t)$ can be expressed as

$$h_o(t) = h_e(t) \operatorname{sgn} t - h_e(0)u_I(t) \tag{9.10}$$

This equation shows that if an even function of a causal function $h(t)$ is known, an odd function $h_o(t)$ of $h(t)$ can be uniquely obtained. Also, an even function $h_e(t)$ of $h(t)$ can be expressed as

$$h_e(t) = h_o(t) \operatorname{sgn} t + h(0)u_I(t) \tag{9.11}$$

Hence, if an odd function $h_o(t)$ of a causal function $h(t)$ is known, Equation 9.11 indicates that an even function of $h(t)$ can be uniquely obtained.

Let the Fourier transformation $H(j\omega)$ of a causal function $h(t)$ be

$$H(j\omega) = P(\omega) + jQ(\omega) \tag{9.12}$$

where $P(\omega)$ and $Q(\omega)$ are real. Then by Equations 9.5 and 9.6, the Fourier transformation of an even function $h_e(t)$ and an odd function $h_o(t)$ are

$$\mathscr{F}[h_e(t)] = P(\omega) \tag{9.13}$$

and

$$\mathscr{F}[h_o(t)] = jQ(\omega) \tag{9.14}$$

If we take the Fourier transformation of Equation 9.10, we have

$$\mathscr{F}[h_o(t)] = \mathscr{F}[h_e(t) \text{ sgn } t] - \mathscr{F}[h_e(0)u_I(t)] \tag{9.15}$$

Because $\mathscr{F}[h_e(0)u_I(t)]$ is 0, Equation 9.15 is

$$\mathscr{F}[h_o(t)] = \frac{1}{2\pi} \mathscr{F}[h_e(t)] * \mathscr{F}[\text{sgn } t] \tag{9.16}$$

By Equation 9.7, the Fourier transformation of sgn t is $2/j\omega$. Hence, using Equations 9.13 and 9.14, Equation 9.16 becomes

$$\mathscr{F}[h_o(t)] = jQ(\omega) = \frac{1}{2\pi} \int_{-\infty}^{\infty} P(v) \frac{2}{j(\omega - v)}\, dv$$

Thus, $Q(\omega)$ is

$$Q(\omega) = \frac{-1}{\pi} \int_{-\infty}^{\infty} \frac{P(v)}{\omega - v}\, dv \tag{9.17}$$

This is the Hilbert transformation of an analog function to obtain a $Q(\omega)$ from $P(\omega)$.

The Hilbert transformation to obtain $P(\omega)$ from $Q(\omega)$ can be developed from the Fourier transformation of Equation 9.11, which is

$$\mathscr{F}[h_e(t)] = \mathscr{F}[h_o(t) \text{ sgn } t] + \mathscr{F}[h(0)u_I(t)] = \frac{1}{2\pi} \mathscr{F}[h_o(t)] * \mathscr{F}[\text{sgn } t]$$

Hence,

$$P(\omega) = \frac{1}{\pi} \int_{-\infty}^{\infty} \frac{Q(v)}{\omega - v}\, dv$$

As an example, suppose $P(\omega)$ is

$$P(\omega) = \begin{cases} 1, & |\omega| \le 1 \\ 0, & \text{Otherwise} \end{cases}$$

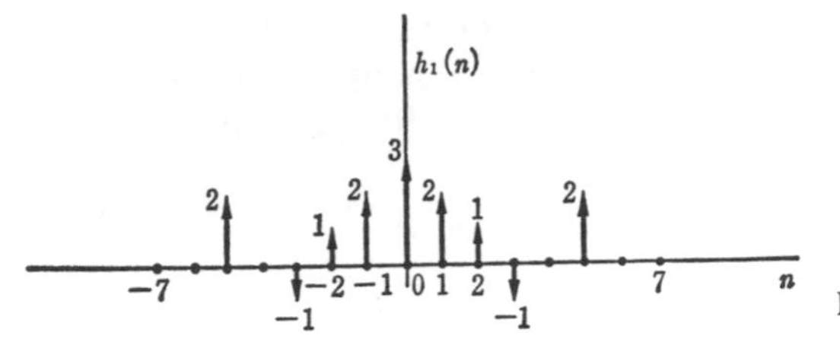

Figure 9.2 Even Function

Then $Q(\omega)$ can be obtained by Equation 9.17 as

$$Q(\omega) = \frac{-1}{\pi} \int_{-1}^{1} \frac{1}{\omega - v}\, dv = \frac{1}{\pi} \ln \left| \frac{\omega - 1}{\omega + 1} \right|$$

9.3 EVEN AND ODD FUNCTIONS OF DISCRETE FUNCTION

Just as an analog function can be expressed by the sum of an even and odd function, a discrete function can also be expressed as the sum of an even function and an odd function. An even function $h_e(n)$ of a discrete function $h(n)$ will satisfy

$$h_e(n) = h_e(-n) \tag{9.18}$$

and an odd function $h_o(n)$ will satisfy

$$h_o(n) = -h_o(-n) \tag{9.19}$$

For example, we can see that changing n to $-n$ of a discrete function $h_1(n)$ in Fig. 9.2 will result in the same function $h_1(n)$. Hence, $h_1(n)$ satisfies Equation 9.18. Thus, $h_1(n)$ is an even function. Changing n to $-n$ of a

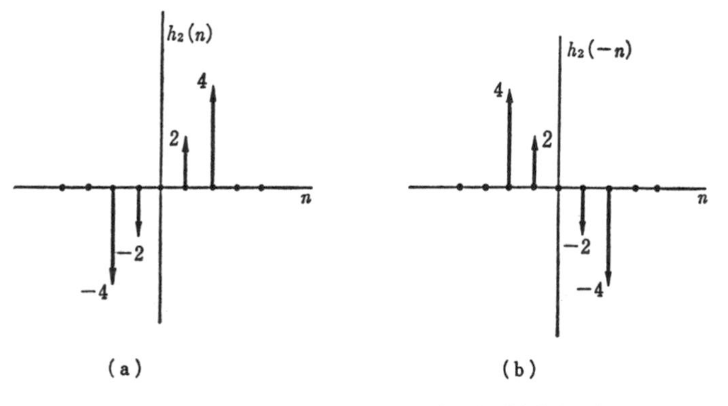

Figure 9.3 Odd Function. (a) $h_2(n)$; (b) $h_2(-n)$.

discrete function $h_2(n)$ in Fig. 9.3(a) will produce the result shown in Fig. 9.3(b). It can be seen that multiplying this result by -1 makes it the same as $h_2(n)$. Hence, $h_2(n)$ satisfies Equation 9.19. Thus, $h_2(n)$ is an odd function.

We can express a discrete function $h(n)$ as

$$h(n) = h_e(n) + h_o(n) \tag{9.20}$$

which is the sum of an even function $h_e(n)$ and an odd function $h_o(n)$. Changing n to $-n$, we have

$$h(-n) = h_e(-n) + h_o(-n)$$

Modifying this equation by the use of Equations 9.18 and 9.19, we have

$$h(-n) = h_e(n) - h_o(n)$$

By this equation and Equation 9.20, an even function $h_e(n)$ and an odd function $h_o(n)$ can be written as

$$h_e(n) = \frac{1}{2}[h(n) + h(-n)] \tag{9.21}$$

and

$$h_o(n) = \frac{1}{2}[h(n) - h(-n)] \tag{9.22}$$

From a discrete function $h(n)$ in Fig. 9.4(a), $h(-n)$ shown in Fig. 9.4(b) can

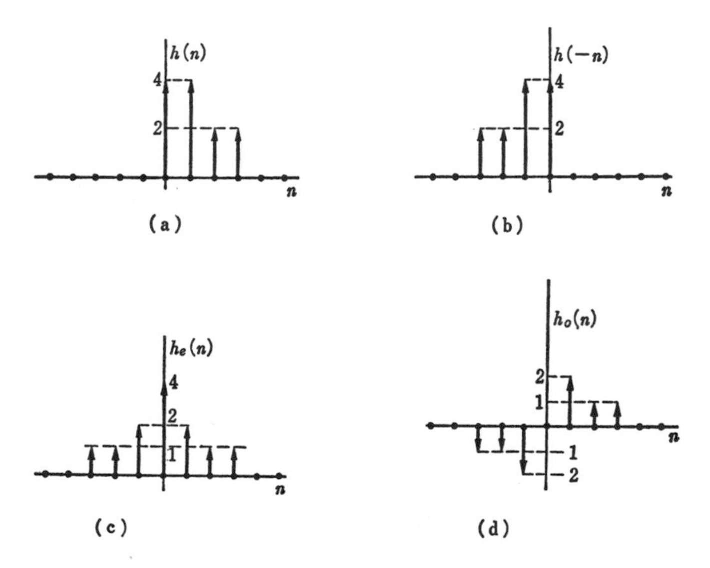

Figure 9.4 Even and Odd Functions of $h(n)$. (a) $h(n)$; (b) $h(-n)$; (c) Even Function $h_e(n)$; (d) Odd Function $h_o(n)$.

be obtained. From these $h(n)$ and $h(-n)$, using Equations 9.21 and 9.22, an even function $h_e(n)$ shown in Fig. 9.3(c), and an odd function $h_o(n)$ shown in Fig. 9.3(d) can be obtained.

Consider a discrete function $h(n)$ that will be 0 for $n < 0$. If a system whose unit sample response is $h(n)$ is a shift invariant and satisfies initial rest conditions, then $h(n) = 0$ for $n < 0$ as discussed in Chapter 2. Here, $h(n)$ need not be a unit sample response. A function $h(n)$ in Fig. 9.4(a) is an example. By defining a function $u_t(n)$ as

$$u_t(n) = \begin{cases} 2, & n > 0 \\ 1, & n = 0 \\ 0, & n < 0 \end{cases} \tag{9.23}$$

then $h(n)$ can be obtained from an even function $h_e(n)$ of $h(n)$ as

$$h(n) = h_e(n)u_t(n) \tag{9.24}$$

Also from an odd function $h_o(n)$, a function $h(n)$ can be obtained by

$$h(n) = h_o(n)u_t(n) + h(0)u_I(n) \tag{9.25}$$

Consider the Fourier transformation of this $h(n)$. From Equation 9.20,

$$H(e^{j\omega}) = H_e(e^{j\omega}) + H_o(e^{j\omega}) \tag{9.26}$$

A frequency response $H_e(e^{j\omega})$ is

$$H_e(e^{j\omega}) = \sum_{n=-\infty}^{\infty} h_e(n)e^{-j\omega n} = \sum_{n=-\infty}^{-1} h_e(n)e^{-j\omega n} + h_e(0) + \sum_{n=1}^{\infty} h_e(n)e^{-j\omega n} \tag{9.27}$$

The first term of this equation is

$$\sum_{n=-\infty}^{-1} h_e(n)e^{-j\omega n} = \sum_{n=1}^{\infty} h_e(-n)e^{j\omega n}$$

By Equation 9.18, however, this can be expressed as

$$\sum_{n=-\infty}^{-1} h_e(n)e^{-j\omega n} = \sum_{n=1}^{\infty} h_e(n)e^{j\omega n}$$

Using this equation, Equation 9.27 becomes

$$\begin{aligned} H_e(e^{j\omega}) &= \sum_{n=1}^{\infty} h_e(n)e^{j\omega n} + h_e(0) + \sum_{n=1}^{\infty} h_e(n)e^{-j\omega n} \\ &= 2\sum_{n=1}^{\infty} h_e(n)\cos \omega n + h_e(0) \end{aligned} \tag{9.28}$$

which shows that $H_e(e^{j\omega})$ is real. Indicating this by $P(e^{j\omega})$ or

$$H_e(e^{j\omega}) = P(e^{j\omega}) \tag{9.29}$$

A frequency response $H_o(e^{j\omega})$ in Equation 9.26 is

$$H_o(e^{j\omega}) = \sum_{n=-\infty}^{\infty} h_o(n)e^{-j\omega n} = \sum_{n=-\infty}^{-1} h_o(n)e^{-j\omega n} + h_o(0) + \sum_{n=1}^{\infty} h_o(n)e^{-j\omega n} \tag{9.30}$$

Because $h_o(0) = 0$, and the first term can be expressed by using Equation 9.19 as

$$\sum_{n=-\infty}^{-1} h_o(n)e^{-j\omega n} = -\sum_{n=1}^{\infty} h_o(n)e^{j\omega n}$$

Equation 9.30 becomes

$$\begin{aligned} H_o(e^{j\omega}) &= -\sum_{n=1}^{\infty} h_o(n)e^{j\omega n} + \sum_{n=1}^{\infty} h_o(n)e^{-j\omega n} \\ &= -2j\sum_{n=1}^{\infty} h_o(n) \sin \omega n \end{aligned} \tag{9.31}$$

Indicating this by the symbol $jQ(e^{j\omega})$ or

$$H_o(e^{j\omega}) = jQ(e^{j\omega}) \tag{9.32}$$

Notice that $Q(e^{j\omega})$ is real by Equation 9.31. By Equations 9.28 and 9.32, Equation 9.26 can be expressed as

$$H(e^{j\omega}) = P(e^{j\omega}) + jQ(e^{j\omega}) \tag{9.33}$$

Let us obtain $P(e^{j\omega})$ and $Q(e^{j\omega})$ of a function $h(n)$ in Fig. 9.4(a). A frequency response $H(e^{j\omega})$ of $h(n)$ is

$$H(e^{j\omega}) = \sum_{n=-\infty}^{\infty} h(n)e^{-j\omega n} = 4 + 4e^{-j\omega} + 2e^{-j2\omega} + 2e^{-j3\omega}$$

By rewriting this, we have

$$\begin{aligned} H(e^{j\omega}) &= 4 + 4\cos\omega - j4\sin\omega + 2\cos 2\omega - j2\sin 2\omega + 2\cos 3\omega \\ &\quad - j2\sin 3\omega \\ &= 4 + 4\cos\omega + 2\cos 2\omega + 2\cos 3\omega \\ &\quad + j(-4\sin\omega - 2\sin 2\omega - 2\sin 3\omega) \end{aligned}$$

Hence, $P(e^{j\omega})$ and $Q(e^{j\omega})$ are

$$P(e^{j\omega}) = 4 + 4\cos\omega + 2\cos 2\omega + 2\cos 3\omega$$

and

$$Q(e^{j\omega}) = -4\sin\omega - 2\sin 2\omega - 2\, sin\, 3\omega$$

From $h_e(n)$ in Fig. 9.4(c), we can obtain $P(e^{j\omega})$ as

$$P(e^{j\omega}) = \sum_{n=-\infty}^{\infty} h_e(n)e^{-j\omega n} = e^{j3\omega} + e^{j2\omega} + 2e^{j\omega} + 4 + 2e^{-j\omega} + e^{-j2\omega} + e^{-j3\omega}$$

which is the same as the one we have obtained previously. Similarly, $jQ(e^{j\omega})$ obtained from $h_o(n)$ in Fig. 9.4(d), is the same as that we have calculated previously.

9.4 HILBERT TRANSFORMATION OF DISCRETE PERIODIC FUNCTION

A discrete periodic function $\mathring{h}(n)$ of a period N can be expressed as the sum of an even function $\mathring{h}_e(n)$ and an odd function $\mathring{h}_o(n)$ as

$$\mathring{h}(n) = \mathring{h}_e(n) + \mathring{h}_o(n), \qquad n = 0, 1, 2, \cdots, N-1 \tag{9.34}$$

An even function $\mathring{h}_e(n)$ and an odd function $\mathring{h}_o(n)$ can be expressed as

$$\left.\begin{aligned} \mathring{h}_e(n) &= \frac{1}{2}[\mathring{h}(n) + \mathring{h}(-n)] \\ \mathring{h}_o(n) &= \frac{1}{2}[\mathring{h}(n) - \mathring{h}(-n)] \end{aligned}\right\}, \quad n = 0, 1, 2, \cdots, N-1 \tag{9.35}$$

Because $\mathring{h}(n)$ is not causal, we will define a *periodic causality* as follows: When a period N is even, a periodic function $\mathring{h}(n)$ is periodic causal if $\mathring{h}(n)$ satisfies

$$\mathring{h}(n) = 0, \qquad -\frac{N}{2} < n < 0$$

For example, a function $\mathring{h}(n)$ in Fig. 9.5 is periodic causal.

When a function $\mathring{h}(n)$ is periodic causal, both $\mathring{h}(n)$ and $\mathring{h}(-n)$ can be 0 only at $n = 0$ and $n = N/2$. Hence,

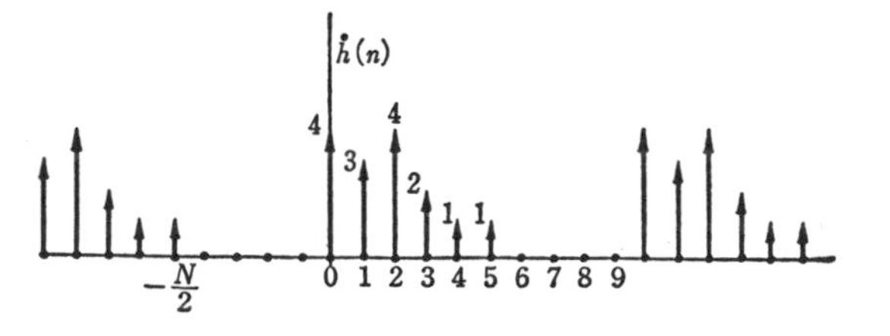

Figure 9.5 Periodic Function $\mathring{h}(n)$ with $N = 10$

$$\mathring{h}(n) = \begin{cases} \mathring{h}_e(n), & n = 0, \dfrac{N}{2} \\ 2\mathring{h}_e(n), & n = 1, 2, \cdots, \dfrac{N}{2} - 1 \\ 0 \quad , & n = \dfrac{N}{2} + 1, \cdots, N - 1 \end{cases} \tag{9.36}$$

and

$$\mathring{h}(n) = \begin{cases} \mathring{h}(0), & n = 0 \\ 2\mathring{h}_o(n), & n = 1, 2, \cdots, \dfrac{N}{2} - 1 \\ \mathring{h}\left(\dfrac{N}{2}\right), & n = \dfrac{N}{2} \\ 0, & n = \dfrac{N}{2} + 1, \cdots, N - 1 \end{cases} \tag{9.37}$$

Be defining $\mathring{u}_N(n)$ as

$$\mathring{u}_N(n) = \begin{cases} 1, & n = 0, \dfrac{N}{2} \\ 2, & n = 1, 2, \cdots, \dfrac{N}{2} - 1 \\ 0, & n = \dfrac{N}{2} + 1, \cdots, N - 1 \end{cases} \tag{9.38}$$

$\mathring{h}(n)$ can be expressed by Equation 9.36 as

$$\mathring{h}(n) = \mathring{h}_e(n)\mathring{u}_N(n)$$

Also by Equation 9.37, $\mathring{h}(n)$ can be expressed as

$$\mathring{h}(n) = \mathring{h}_o(n)\mathring{u}_N(n) + \mathring{h}(0)\mathring{u}_I(n) + \mathring{h}\left(\frac{N}{2}\right)\mathring{u}_I\left(n - \frac{N}{2}\right) \tag{9.39}$$

This $\mathring{u}_I(n)$ is a periodic unit sample, that is, $\mathring{u}_I(n) = 1$ at $n = 0$ and $\mathring{u}_I(n) = 0$ at $n = 1, 2, \cdots, N - 1$.

Consider $h(n)$ in Fig. 9.6a. Because a period of this $\mathring{h}(n)$ is 8, $N/2$ is 4. $\mathring{h}(n) = 0$ at $n = -3, -2, -1$. Makes $\mathring{h}(n)$ periodic causal. Figure 9.6(b) shows $\mathring{h}(-n)$; from these $\mathring{h}(n)$ and $\mathring{h}(-n)$, an even function $\mathring{h}_e(n)$ and an odd function $\mathring{h}_o(n)$, shown in Fig. 9.6(c) and (d) can be obtained.

By Equation 6.10, the DFS $\mathring{H}(k)$ of a periodic function $\mathring{h}(n)$ is

$$\mathring{H}(k) = \sum_{n=0}^{N-1} \mathring{h}(n) W_N^{nk}$$

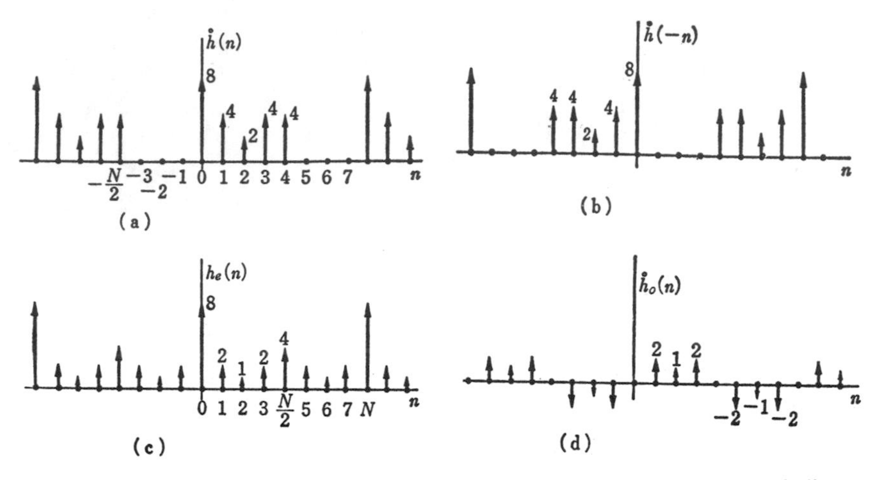

Figure 9.6 Even and Odd Functions of Periodic Function $\mathring{h}(n)$. (a) Periodic Function $h\mathring{h}(n)$ with $N = 8$; (b) $h\mathring{h}(-n)$; (c) Even Function $h\mathring{h}_e(n)$; (d) Odd Function $h\mathring{h}_o(n)$.

where W_N is $e^{-j(2\pi/N)}$. The DFS $\mathring{H}_e(k)$ of an even function $\mathring{h}_e(n)$ of $\mathring{h}(n)$ is

$$\mathring{H}_e(k) = \sum_{n=0}^{N-1} \mathring{h}_e(n) W_N^{nk} \tag{9.40}$$

Using Equation 9.35, replacing $\mathring{h}_e(n)$ by $\mathring{h}(n)$ and $\mathring{h}(-n)$ in Equation 9.40 produces

$$\mathring{H}_e(k) = \sum_{n=0}^{N-1} \frac{\mathring{h}(n)}{2} W_N^{nk} + \sum_{n=0}^{N-1} \frac{\mathring{h}(-n)}{2} W_N^{nk} \tag{9.41}$$

Because $\mathring{h}(-n)$ is a periodic function, it is equal to $\mathring{h}(N - n)$. Hence, Equation 9.41 can be written as

$$\begin{aligned} \mathring{H}_e(k) &= \sum_{n=0}^{N-1} \frac{\mathring{h}(n)}{2} W_N^{nk} + \sum_{n=0}^{N-1} \frac{\mathring{h}(N-n)}{2} W_N^{nk} \\ &= \sum_{n=0}^{N-1} \frac{\mathring{h}(n)}{2} W_N^{nk} + \sum_{n'=1}^{N} \frac{\mathring{h}(n')}{2} W_N^{(N-n')k} \end{aligned} \tag{9.42}$$

Because W_N^{Nk} is $e^{-j(2\pi/N)Nk}$, W_N^{Nk} is 1. Hence, Equation 9.42 becomes

$$\begin{aligned} \mathring{H}_e(k) &= \sum_{n=0}^{N-1} \frac{\mathring{h}(n)}{2} W_N^{nk} + \sum_{n=0}^{N-1} \frac{\mathring{h}(n')}{2} W_N{}^{-n'k} = \sum_{n=0}^{N-1} \mathring{h}(n) \left(\frac{W_N^{nk} + W_N^{-nk}}{2} \right) \\ &= \sum_{n=0}^{N-1} \mathring{h}(n) \cos\left(\frac{2\pi}{N} nk \right) \end{aligned} \tag{9.43}$$

Because the right-hand side of this equation is real, $\mathring{H}_e(k)$ is real.

Similarly, we can obtain the DFS $\mathring{H}_o(k)$ of an odd function $\mathring{h}_o(n)$ as

$$\mathring{H}_o(k) = -j \sum_{n=0}^{N-1} \mathring{h}_o(n) \sin\left(\frac{2\pi}{N} nk\right) \tag{9.44}$$

This shows that if $\mathring{h}_o(n)$ is real, $\mathring{H}_o(k)$ is real.

From Equations 9.43 and 9.44, if $\mathring{h}(n)$ is real, the DFS $\mathring{H}(k)$ of $\mathring{h}(n)$ can be expressed as

$$\mathring{H}(k) = \mathring{P}(k) + j\mathring{Q}(k) \tag{9.45}$$

where $\mathring{P}(k)$ and $\mathring{Q}(k)$ are real. These $\mathring{P}(k)$ and $\mathring{Q}(k)$ are

$$\mathring{P}(k) = \mathring{H}_e(k) \tag{9.46}$$

and

$$\mathring{Q}(k) = \frac{\mathring{H}_o(k)}{j} \tag{9.47}$$

This $\mathring{P}(k)$ is called a real part of $\mathring{H}(k)$, and the $\mathring{Q}(k)$ is called an imaginary part of $\mathring{H}(k)$. The DFS $\mathring{U}_N(k)$ of $\mathring{u}_N(n)$ defined by Equation 9.38 is

$$\mathring{U}_N(k) = \sum_{n=0}^{N-1} \mathring{u}_N(n) W_N^{nk} = W_N^0 + 2W_N^k + 2W_N^{2k} + \cdots + 2W_N^{(N/2-1)k} + W_N^{(N/2)k} \tag{9.48}$$

When $k = 0$, it is clear that

$$\mathring{U}_N(0) = N \tag{9.49}$$

Because $W_N^{(N/2)k}$ is $e^{-j(2\pi/N)(N/2)k} = e^{-j\pi k} = (-1)^k$ $(W_N^0 = 1)$, Equation 9.48 becomes

$$\mathring{U}_N(k) = 1 + (-1)^k + 2\,[W_N^k + W_N^{2k} + \cdots + W_N^{[(N/2)-1]k}] \tag{9.50}$$

When k is even, this equation becomes

$$\mathring{U}_N(k) = 2\,[1 + W_N^k + W_N^{2k} + \cdots + W_N^{[(N/2-1)k]}] = 2\,\frac{1 - W_N^{(N/2)k}}{1 - W_N^k}$$

$$= 2\,\frac{W_N^{(N/4)k}(W_N^{-(N/4)k} - W_N^{(N/4)k})}{W_N^{k/2}(W_N^{-k/2} - W_N^{k/2})} = 2\,\frac{\left(\cos\frac{\pi}{2}k - j\sin\frac{\pi}{2}k\right)\sin\frac{\pi}{2}k}{\left(\cos\frac{\pi}{N}k - j\sin\frac{\pi}{N}k\right)\sin\frac{\pi}{N}k}$$

$\sin(\pi/2)k$ is 0 when k is even and $\sin(\pi/N)k$ will not be 0 for $0 < k \leq N/2 - 1$,

$$\mathring{U}_N(k) = 0 \tag{9.51}$$

where k is even but not 0. When k is odd, Equation 9.50 becomes

$$\mathring{U}_N(k) = 2\,\{1 + W_N^k + W_N^{2k} + \cdots + W_N^{(N/2-1)k} - 1\}$$
$$= 2\left\{\frac{\left(\cos\frac{\pi}{2}k - j\sin\frac{\pi}{2}k\right)\sin\frac{\pi}{2}k}{\left(\cos\frac{\pi}{N}k - j\sin\frac{\pi}{N}k\right)\sin\frac{\pi}{N}k} - 1\right\} \tag{9.52}$$

Because k is odd, $\cos(\pi/2)k$ is 0 and $\sin^2[(\pi/2)k]$ is 1, Equation 9.52 becomes

$$\mathring{U}_N(k) = 2\left\{\frac{-j}{\left(\cos\frac{\pi}{N}k - j\sin\frac{\pi}{N}k\right)\sin\frac{\pi}{N}k} - 1\right\} \tag{9.53}$$
$$= -j2\cot\frac{\pi}{N}k, \text{ for } k \text{ even.}$$

By Equations 9.49, 9.51, and 9.53, $\mathring{U}_N(k)$ is

$$\mathring{U}_N(k) = \begin{cases} N, & k = 0 \\ -j2\cot\frac{\pi}{N}k, & k \text{ odd} \\ 0, & k \text{ even} \end{cases} \tag{9.54}$$

Using this equation and Equation 9.46, the DFS of Equation 9.35 can be expressed as

$$\mathring{H}(k) = \sum_{n=0}^{N-1} [\mathring{h}_e(n)\mathring{u}_N(n)]\,W_N^{nk}$$

By Equation 6.19, this is equal to circular convolution of the DFS of $\mathring{h}_e(n)$ and the DFS of $\mathring{u}_N(n)$, that is,

$$\mathring{H}(k) = \frac{1}{N}\sum_{m=0}^{N-1} \mathring{P}(m)\mathring{U}_N(k-m) \tag{9.55}$$

This shows that if $\mathring{h}(n)$ is periodic causal, $\mathring{H}(k)$ can be obtained from a real part $\mathring{P}(k)$ of $\mathring{H}(k)$. Similarly, by using Equation 9.39, we have

$$\mathring{H}(k) = \frac{1}{N}\sum_{m=0}^{N-1} j\mathring{Q}(m)\mathring{U}_N(k-m) + \mathring{h}(0) + (-1)^k\mathring{h}\left(\frac{N}{2}\right) \tag{9.56}$$

This equation shows that $H(k)$ can be obtained from an imaginary part $Q(k)$ of $H(k)$. These two Equations 9.55 and 9.56 are the Hilbert transformations of periodic functions.

We have studied that a finite duration function $h(n)$ of a duration N is considered as one period of a periodic function $\mathring{h}(n)$ of a period N. That is,

$$h(n) = \mathring{h}(n)R_N(n)$$

where $R_N(n)$ is

$$R_N(n) = \begin{cases} 1, & n = 0, 1, 2, \cdots N - 1 \\ 0, & \text{Otherwise} \end{cases} \tag{9.57}$$

By Equation 9.55, the DFT $H(k)$ of $h(n)$ is

$$H(k) = \begin{cases} \dfrac{1}{N} \displaystyle\sum_{m=0}^{N-1} \mathring{P}(m)\mathring{U}_N(k - m), & 0 \le k \le N - 1 \\ 0, & \text{Otherwise} \end{cases} \tag{9.58}$$

where $\mathring{P}(m)$ is the DFS of a periodic function $\mathring{h}_e(n)$ satisfying a periodic causality whose one period is $h_e(n)$. Similarly by Equation 9.56, $H(k)$ can be expressed as

$$H(k) = \begin{cases} \dfrac{1}{N} \displaystyle\sum_{m=0}^{N-1} j\mathring{Q}(m)\mathring{U}_N(k - m) + h(0) + (-1)^k h\left(\dfrac{N}{2}\right), & 0 \le k \le N - 1 \\ 0, & \text{Otherwise} \end{cases} \tag{9.59}$$

where $j\mathring{Q}(n)$ is the DFS of a periodic function $\mathring{h}_e(n)$ satisfying a periodic causality whose one period is $h_e(n)$. Equations 9.58 and 9.59 are the Hilbert transformation of a finite duration function.

As an example, consider a finite duration function $h(n)$ in Fig. 9.7(a).

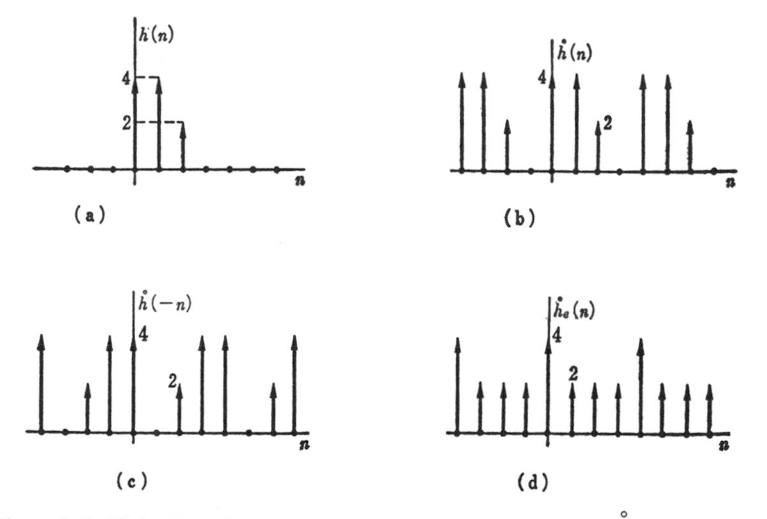

Figure 9.7 Finite Duration Function $h(n)$ and Periodic Function $\mathring{h}(n)$. (a) Finite Duration Function $h(n)$; (b) Periodic Function $\mathring{h}(n)$; (c) $\mathring{h}(-n)$; (d) Even Function $\mathring{h}_e(n)$.

Choosing $n = 4$, a periodic function $\mathring{h}(n)$ corresponding to $h(n)$ and $\mathring{h}(-n)$ are shown in Fig. 9.7(b) and (c). From these with Equation 9.35, an even function $\mathring{h}_e(n)$ is shown in Fig. 9.7(d). Computing $\mathring{P}(k)$ from this $\mathring{h}_e(n)$ as

$$\mathring{P}(0) = 10,\ \mathring{P}(1) = 4 + 2(-j) + 2(-1) + 2(j) = 2,\ \mathring{P}(2) = 2 \text{ and } \mathring{P}(3) = 2$$

Also we can obtain the DFS $\mathring{U}_4(k)$ of $\mathring{u}_4(n)$ by Equation 9.54 as

$$\mathring{U}_4(0) = 4,\ \mathring{U}_4(1) = -j2,\ \mathring{U}_4(2) = 0, \text{ and } \mathring{U}_4(3) = j2$$

By Equation 9.58, the circular convolution of this $\mathring{P}(k)$ and $\mathring{U}_4(k)$ gives

$$H(0) = \frac{1}{4}\{10(4) + 2(j2) + 2(0) + 2(-j2)\} = 10$$

$$H(1) = \frac{1}{4}\{10(-j2) + 2(4) + 2(j2) + 2(0)\} = 2 - j4$$

$$H(2) = \frac{1}{4}\{10(0) + 2(-j2) + 2(4) + 2(j2)\} = 2$$

and

$$H(3) = \frac{1}{4}\{10(j2) + 2(0) + 2(-j2) + 2(4)\} = 2 + j4$$

These are the DFT $H(n)$ of $h(n)$ obtained by the use of the Hilbert transformation using an even function $h_e(n)$. Obtaining $H(k)$ directly from $h(n)$ by $H(k) = \sum_{n-0}^{N-1} \mathring{h}(n) W_N^{nk}$, we have

$$H(0) = 10,\ H(1) = 4 + 4(-j) + 2(-1) = 2 - j4,\ H(2) = 4 + 4(-1) + 2(1) = 2, \text{ and } H(3) = 4 + 4(j) + 2(-1) = 2 + j4$$

which is the same as those obtained previously.

9.5 SUMMARY

1. A function $h(t)$ is an even function if $h(t) = h(-t)$.
2. A function $h(t)$ is an odd function if $h(t) = -h(-t)$.
3. A function $h(t)$ can be expressed as the sum of an even function $h_e(t)$ and an odd function $h_o(t)$ as

$$h(t) = h_e(t) + h_o(t)$$

An even function $h_e(t)$ and an odd function $h_o(t)$ are

$$h_e(t) = \frac{1}{2}[h(t) + h(-t)]$$

$$h_o(t) = \frac{1}{2}[h(t) - h(-t)]$$

4. A function $h(t)$ is causal if $h(t) = 0$ for all $t < 0$.
5. When the Fourier transformation $H(j\omega)$ of a function $h(t)$ is expressed as

$$H(j\omega) = P(\omega) + jQ(\omega)$$

where $P(\omega)$ and $Q(\omega)$ are real, $P(\omega)$ is the real part, and $Q(\omega)$ is the imaginary part of $G(j\omega)$. An even function $h_e(t)$, and an odd function $h_o(t)$ of $h(t)$ and $P(\omega)$ and $Q(\omega)$ are related by

$$P(\omega) = \mathcal{F}[h_e(t)]$$

and

$$Q(\omega) = \frac{\mathcal{F}[h_o(t)]}{j}$$

6. A function sgn t is defined as

$$\text{sgn } t = \begin{cases} 1, & t \geq 0 \\ -1, & t < 0 \end{cases}$$

The Fourier transformation of sgn t is

$$\mathcal{F}[\text{sgn } t] = \frac{2}{j\omega}$$

7. When a function $h(t)$ is causal, the real part $P(\omega)$ and the imaginary part $Q(\omega)$ of the Fourier transformation $H(j\omega) = P(\omega) + jQ(\omega)$ satisfy

$$Q(\omega) = \frac{-1}{\pi}\int_{-\infty}^{\infty} \frac{P(v)}{\omega - v}\,dv$$

and

$$P(\omega) = \frac{1}{\pi}\int_{-\infty}^{\infty} \frac{Q(v)}{\omega - v}\,dv$$

where $h(t)$ contains no impulses at $t = 0$. These two equations are called the Hilbert transformation.

8. A function $u_t(n)$ is defined as

$$u_t(n) = \begin{cases} 2, & n > 0 \\ 1, & n = 0 \\ 0, & n < 0 \end{cases}$$

9. A discrete function $h(n)$ can be expressed as

$$h(n) = h_e(n)u_t(n)$$

$$h(n) = h_o(n)u_t(n) + h(0)u_I(n)$$

where $h_e(n)$ is an even function and $h_o(n)$ is an odd function of $h(n)$ and $u_I(n)$ is a unit sample.

10. A periodic function $\mathring{h}(n)$ with the value of N being even is circular causal if

$$\mathring{h}(n) = 0, \qquad -\frac{N}{2} < n < 0$$

11. A periodic function $\mathring{u}_N(n)$ is defined as

$$\mathring{u}_N(n) = \begin{cases} 1, & n = 0, \dfrac{N}{2} \\ 2, & n = 1, 2, \cdots, \dfrac{N}{2} - 1 \\ 0, & n = \dfrac{N}{2} + 1, \cdots, N - 1 \end{cases}$$

12. A periodic function $\mathring{h}(n)$, an even function $\mathring{h}_e(n)$ and an odd function $\mathring{h}_o(n)$ of $\mathring{h}(n)$ satisfy

$$\mathring{h}(n) = \mathring{h}_e(n)\mathring{u}_N(n)$$

$$\mathring{h}(n) = \mathring{h}_o(n)\mathring{u}_N(n) + \mathring{h}(0)\mathring{u}_I(n) + \mathring{h}\left(\frac{N}{2}\right)\mathring{u}_I\left(n - \frac{N}{2}\right)$$

where $\mathring{u}_I(n)$ is a periodic unit sample defined as

$$\mathring{u}_I(n) = \begin{cases} 1, & n = 0 \\ 0, & n = 1, 2, \cdots, N - 1 \end{cases}$$

13. The DFS $\mathring{U}_N(k)$ of $\mathring{u}_N(n)$ is

$$\mathring{U}_N(k) = \begin{cases} N, & k = 0 \\ -j2 \cot \dfrac{\pi}{N} k, & k \text{ odd} \\ 0, & k \text{ even} \end{cases}$$

14. $\mathring{H}(k) = \mathring{P}(\omega) + j\mathring{Q}(\omega)$ where $\mathring{P}(k)$ is the real part, and $\mathring{Q}(k)$ is the imaginary part of $\mathring{H}(k)$ can be expressed as

$$\mathring{H}(k) = \frac{1}{N} \sum_{m=1}^{N-1} \mathring{P}(m)\mathring{U}_N(k - m)$$

and

$$\mathring{H}(k) = \frac{1}{N} \sum_{m=0}^{N-1} j\mathring{Q}(m)\mathring{U}_N(k - m) + \mathring{h}(0) + (-1)^k\mathring{h}\left(\frac{N}{2}\right)$$

which is the Hilbert transformation of a periodic discrete function.

15. The Hilbert transformation of a finite duration function $h(n)$ whose DFT is $H(k)$ is

$$H(k) = \begin{cases} \dfrac{1}{N} \sum_{m=0}^{N-1} \mathring{P}(m)\mathring{U}_N(k-m), & 0 \le k \le N-1 \\ 0, & \text{Otherwise} \end{cases}$$

and

$$H(k) = \begin{cases} \dfrac{1}{N} \sum_{m=0}^{N-1} j\mathring{Q}(m)\mathring{U}_N(k-m) + \mathring{h}(0) + \mathring{h}(-1)^k \left(\dfrac{N}{2}\right), & 0 \le k \le N-1 \\ 0, & \text{Otherwise} \end{cases}$$

9.6 PROBLEMS

1. Obtain an even function and an odd function of the following functions.

a. $t^4 + 2t^3 + 3t + 5$ **b.** $2t^5 + 3t^4 + t^2 + 6$

c. $\sum_{t=0}^{N} a_t t^t$ where N is an even number **d.** $\sin^2 t + 2 \sin t$

e. e^t **f.** $\cosh t$

2. Obtain an even function and an odd function of the following:

a. $h(t) = \begin{cases} t^4 + 2t^3 + 3t + 5, & t \ge 0 \\ 0, & t < 0 \end{cases}$

b. $h(t) = \begin{cases} \sin^2 t, & t \ge 0 \\ 0, & t < 0 \end{cases}$

c. $h(t) = \begin{cases} e^t, & t \ge 0 \\ 0, & t < 0 \end{cases}$

d. $h(t) = \begin{cases} \cosh t, & t \ge 0 \\ 0, & t < 0 \end{cases}$

3. Find a complex function whose real part is $\sin^2 \omega - \cos^2 \omega$ by using the Hilbert transformation.

4. Using the Hilbert transformation, obtain a complex function whose real part is the following function.

a. $P(\omega) = \sin \omega$ **b.** $P(\omega) = \begin{cases} 1, & |\omega| \le \omega_0 \\ 0, & \text{Otherwise} \end{cases}$

5. Obtain an even function and an odd function of the following discrete function:

a. $h(n) = 5n^2 + 2n + 3$ **b.** $h(n) = \cos \dfrac{2\pi}{N} n$

c. $h(n) = e^{j(2\pi/N)n}$ **d.** $h(n) = e^{(2\pi/N)n}$

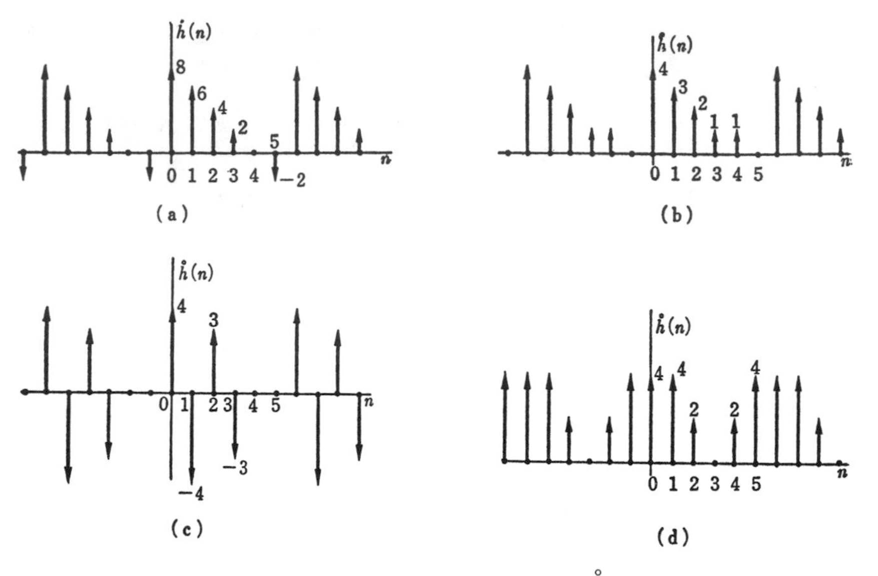

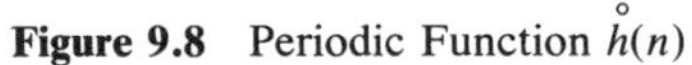
Figure 9.8 Periodic Function $\overset{\circ}{h}(n)$

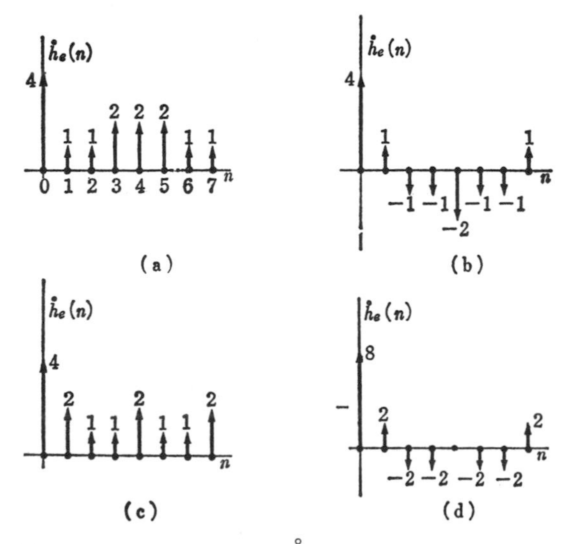

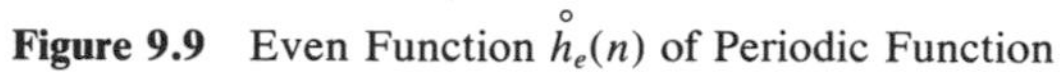
Figure 9.9 Even Function $\overset{\circ}{h}_e(n)$ of Periodic Function

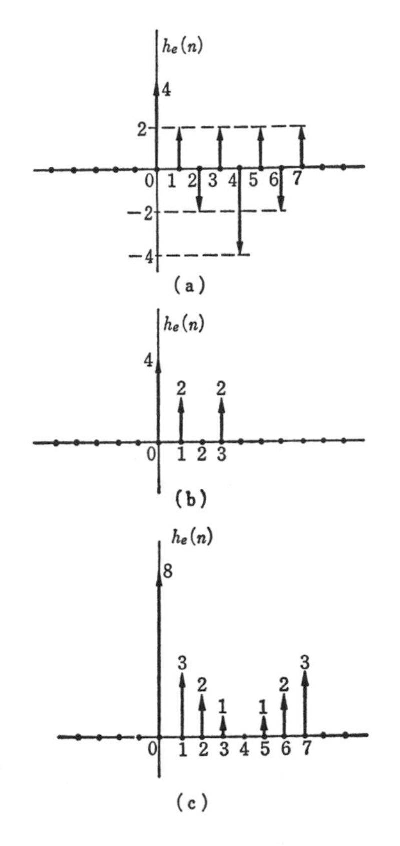

Figure 9.10 Even Function $h_e(n)$ of Finite Duration Function

6. Obtain an even function and an odd function of each of the periodic functions of $N = 8$ in Fig. 9.8.
7. Obtain the DFS $\mathring{H}(k)$ of a periodic function $\mathring{h}(n)$ whose real part $\mathring{P}(k)$ of $\mathring{H}(k)$ is the DFS of $\mathring{h}_e(n)$ in Fig. 9.9 by using the Hilbert transformation.
8. From an even function $h_e(n)$ of a finite duration function $h(n)$ in Fig. 9.10, obtain the DFT $H(k)$ of $h(n)$ by using the Hilbert transformation.

10
Two-Dimensional Discrete Signal

10.1 TWO-DIMENSIONAL DISCRETE SIGNAL

Research in two-dimensional discrete signals is progressing rapidly because it is widely used in many fields such as recognition of patterns, design of patterns, correction of pictures, and so on. We will expand the concept of one-dimensional discrete signals into two-dimensional discrete signals, and study some important properties and techniques associated with two-dimensional signals.

As we have found, a unit sample and a unit step are important signals associated with one-dimensional signals. We will see that a two-dimensional unit sample and a two-dimensional unit step are very useful in study and application of two-dimensional signal processing.

A unit sample $u_I(m, n)$ is

$$u_I(m, n) = \begin{cases} 1, & m = n = 0 \\ 0, & m, n \neq 0 \end{cases} \tag{10.1}$$

which is shown in Fig. 10.1, and a unit step $u_s(m, n)$ is

$$u_s(m, n) = \begin{cases} 1, & m \geq 0, n \geq 0 \\ 0, & m < 0, n < 0 \end{cases} \tag{10.2}$$

which is shown in Fig. 10.2.

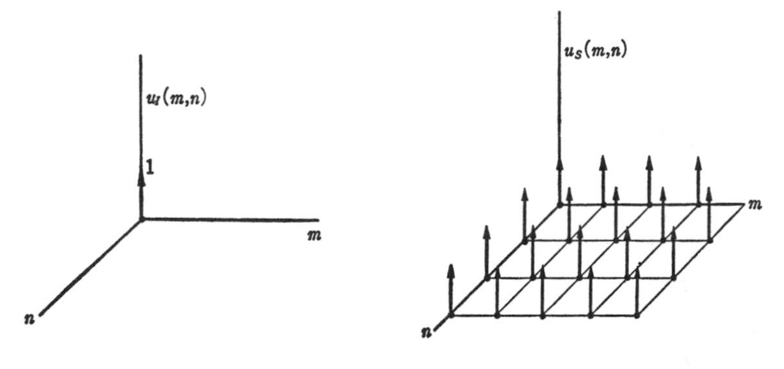

Figure 10.1 Unit Sample $u_I(m, n)$

Figure 10.2 Unit Step $u_S(m, n)$

By using the unit sample, a two-dimensional discrete signal $x(n, m)$ can be expressed as

$$x(m, n) = \sum_{k=-\infty}^{\infty} \sum_{r=-\infty}^{\infty} x(k, r)u_I(m - k, n - r) \tag{10.3}$$

Let $h(m, n)$ be an output signal of a system when an input signal is $u_I(m, n)$. Notice that $h(m, n)$ is a unit sample response of a system. When a system is linear and shift invariant, an output signal is $h(m - k, n - r)$ for an input signal $u_I(m - k, n - r)$. Furthermore when an input signal is $\sum_{k=-\infty}^{\infty} \sum_{r=-\infty}^{\infty} u_I(m - k, n - r)$, an output signal $y(m, n)$ will be

$$y(m, n) = \sum_{k=-\infty}^{\infty} \sum_{r=-\infty}^{\infty} h(m - k, n - r) \tag{10.4}$$

This means that when an input signal is $x(m, n)$, an output signal can be expressed by the use of Equation 10.3 as

$$y(m, n) = \sum_{k=-\infty}^{\infty} \sum_{r=-\infty}^{\infty} x(k, r)h(m - k, n - r) \tag{10.5}$$

By changing variables, we have

$$y(m, n) = \sum_{k=-\infty}^{\infty} \sum_{r=-\infty}^{\infty} h(k, r)x(m - k, n - r) \tag{10.6}$$

which shows that an output signal $y(m, n)$ is a two-dimensional convolution of an input signal $x(m, n)$ and a unit sample response $h(m, n)$ of a linear shift invariant system.

To obtain a frequency response of a two-dimensional unit sample response $h(m, n)$ of a system, let us use an input signal $x(m, n) = e^{j(\omega_1 m + \omega_2 n)}$ and calculate an output signal $y(m, n)$. By Equation 10.6, $y(m, n)$ is

$$
\begin{aligned}
y(m, n) &= \sum_{k=-\infty}^{\infty} \sum_{r=-\infty}^{\infty} h(k, r) e^{j[\omega_1(m-k)+\omega_2(n-r)]} \\
&= \left[\sum_{k=-\infty}^{\infty} \sum_{r=-\infty}^{\infty} h(k, r) e^{-j\omega_1 k} e^{-j\omega_2 r} \right] e^{j\omega_1 m} e^{j\omega_2 n} \qquad (10.7) \\
&= H(e^{j\omega_1}, e^{j\omega_2}) e^{j(\omega_1 m + \omega_2 n)}
\end{aligned}
$$

This $H(e^{j\omega_1}, e^{j\omega_2})$ is

$$
H(e^{j\omega_1}, e^{j\omega_2}) = \sum_{k=-\infty}^{\infty} \sum_{r=-\infty}^{\infty} h(k, r) e^{-j\omega_1 k} e^{-j\omega_2 r} \qquad (10.8)
$$

which is a frequency response of $h(m, n)$. Changing ω_1 by $\omega_1 + 2\pi p$, we have

$$
H(e^{j\omega_1 + 2\pi p}, e^{j\omega_2}) = \sum_{k=-\infty}^{\infty} \sum_{r=-\infty}^{\infty} h(k, r) e^{-jk(\omega_1 + 2\pi p)} e^{-j\omega_2 r}
$$

Because p is an integer, kp is also an integer. Hence, $e^{-j2\pi kp} = 1$. Thus $H(e^{j\omega_1 + 2\pi p}, e^{j\omega_2})$ is

$$
H(e^{j\omega_1 + 2\pi p}, e^{j\omega_2}) = \sum_{k=-\infty}^{\infty} \sum_{r=-\infty}^{\infty} h(k, r) e^{-j\omega_1 k} e^{-j\omega_2 r} = H(e^{j\omega_1}, e^{j\omega_2})
$$

This shows that $H(e^{j\omega_1}, e^{j\omega_2})$ is a periodic function of a period 2π on the ω_1 axis. Instead of ω_1, if we change ω_2 by $\omega_2 + 2\pi p$, we can see that $H(e^{j\omega_1}, e^{j\omega_2})$ is also a periodic function of a period 2π on the ω_2 axis.

Now, we will discuss a property known as a "separability," which can only be considered in a two-dimensional or a higher-dimensional signal processing. A two-dimensional signal $x(m, n)$ is said to be separable if it can be expressed as product of two one-dimensional signals as

$$
x(m, n) = x_1(m) x_2(n)
$$

If a unit sample response $h(m, n)$ is separable, a frequency response $H(e^{j\omega_1}, e^{j\omega_2})$ in Equation 10.8 becomes

$$
\begin{aligned}
H(e^{j\omega_1}, e^{j\omega_2}) &= \sum_{k=-\infty}^{\infty} \sum_{r=-\infty}^{\infty} h(k, r) e^{-j\omega_1 k} e^{-j\omega_2 r} \\
&= \sum_{k=-\infty}^{\infty} h_1(k) e^{-j\omega_1 k} \sum_{r=-\infty}^{\infty} h_2(r) e^{-j\omega_2 r} = H_1(e^{j\omega_1}) H_2(e^{j\omega_2})
\end{aligned} \qquad (10.9)
$$

which means that a frequency response $H(e^{j\omega_1}, e^{-j\omega_2})$ is also separable.

Now we will discuss the fourier transformation of a two-dimensional digital signal. To obtain a unit sample response $h(m, n)$ from a frequency response, consider an integration of product of $H(e^{j\omega_1}, e^{j\omega_2})$ and $e^{j(\omega_1 m + \omega_2 n)}$. Because $H(e^{j\omega_1}, e^{j\omega_2})$ is a periodic function of a period 2π on both the ω_1 axis and the ω_2 axis, the integral of $H(e^{j\omega_1}, e^{j\omega_2})$ for one period on the ω_1 axis and on the ω_2 axis using Equation 10.8 becomes

$$\int_{-\pi}^{\pi}\int_{-\pi}^{\pi} H(e^{j\omega_1}, e^{j\omega_2}) e^{j\omega_1 m} e^{j\omega_2 n}\, d\omega_1 d\omega_2$$

$$= \int_{-\pi}^{\pi}\int_{-\pi}^{\pi} \sum_{k=-\infty}^{\infty}\sum_{r=-\infty}^{\infty} h(k, r) e^{-j\omega_1 k - j\omega_2 r} e^{j\omega_1 m} e^{j\omega_2 n}\, d\omega_1 d\omega_2 \quad (10.10)$$

$$= \sum_{k=-\infty}^{\infty}\sum_{r=-\infty}^{\infty} h(k, r) \int_{-\pi}^{\pi}\int_{-\pi}^{\pi} e^{j\omega_1(m-k)} e^{j\omega_2(n-r)}\, d\omega_1 d\omega_2$$

It is clear that the integration in the right-hand side is 0 for $m \neq k$ or $n \neq r$, and $4\pi^2$ when both $m = k$ and $n = r$. Hence, this equation becomes

$$\int_{-\pi}^{\pi}\int_{-\pi}^{\pi} H(e^{j\omega_1}, e^{j\omega_2}) e^{j\omega_1 m} e^{j\omega_2 n}\, d\omega_1 d\omega_2 = 4\pi^2 h(m, n)$$

Thus, a unit sample response $h(m, n)$ can be expressed as

$$h(m, n) = \frac{1}{4\pi^2}\int_{-\pi}^{\pi}\int_{-\pi}^{\pi} H(e^{j\omega_1}, e^{j\omega_2}) e^{j\omega_1 m} e^{j\omega_2 n}\, d\omega_1 d\omega_2 \quad (10.11)$$

Because $h(m, n)$ in Equation 10.8 and 10.11 need not be a unit sample response, using the symbol $x(m, n)$ for a discrete signal and $X(e^{j\omega_1}, e^{j\omega_2})$ for a frequency response of $x(m, n)$, we have

$$X(e^{j\omega_1}, e^{j\omega_2}) = \sum_{k=-\infty}^{\infty}\sum_{r=-\infty}^{\infty} x(k, r) e^{-j\omega_1 k} e^{-j\omega_2 r} \quad (10.12)$$

$$x(m, n) = \frac{1}{4\pi^2}\int_{-\pi}^{\pi}\int_{-\pi}^{\pi} X(e^{j\omega_1}, e^{j\omega_2}) e^{j\omega_1 m} e^{j\omega_2 n}\, d\omega_1 d\omega_2 \quad (10.13)$$

These are known as the Fourier transformation and the inverse Fourier transformation of a two-dimensional digital signal.

Let $H(e^{j\omega_1}, e^{j\omega_2})$, $X(e^{j\omega_1}, e^{j\omega_2})$, and $Y(e^{j\omega_1}, e^{j\omega_2})$ are the Fourier transformations of unit sample response $h(m, n)$, an input signal $x(m, n)$, and an output signal $y(m, n)$, respectively. Then $Y(e^{j\omega_1}, e^{j\omega_2})$ is

$$Y(e^{j\omega_1}, e^{j\omega_2}) = \sum_{k=-\infty}^{\infty}\sum_{r=-\infty}^{\infty} y(k, r) e^{-j\omega_1 k} e^{-j\omega_2 r}$$

Using Equation 10.5 for $y(k, r)$, this equation can be expressed as

$$Y(e^{j\omega_1}, e^{j\omega_2}) = \sum_{k=-\infty}^{\infty} \sum_{r=-\infty}^{\infty} \left[\sum_{p=-\infty}^{\infty} \sum_{q=-\infty}^{\infty} x(p, q)h(k - p, r - q) \right] e^{-j\omega_1 k} e^{-j\omega_2 r}$$

By changing $k - p = p'$ and $r - q = q'$, $Y(e^{j\omega_1}, e^{j\omega_2})$ becomes

$$\begin{aligned} Y(e^{j\omega_1}, e^{j\omega_2}) &= \sum_{p=-\infty}^{\infty} \sum_{q=-\infty}^{\infty} \sum_{p'=-\infty}^{\infty} \sum_{q'=-\infty}^{\infty} x(p, q)\, h(p', q') e^{-j\omega_1(p'+p)} e^{-j\omega_2(q'+q)} \\ &= \sum_{p=-\infty}^{\infty} \sum_{q=-\infty}^{\infty} x(p, q) e^{-j\omega_1 p} e^{j\omega_2 q} \sum_{p'=-\infty}^{\infty} \sum_{q'=-\infty}^{\infty} h(p', q') e^{-j\omega_1 p'} e^{-j\omega_2 q'} \\ &= X(e^{j\omega_1}, e^{j\omega_2}) H(e^{j\omega_1}, e^{j\omega_2}) \end{aligned} \tag{10.14}$$

which shows that the relationship between $Y(e^{j\omega_1}, e^{j\omega_2})$, $H(e^{j\omega_1}, e^{j\omega_2})$ and $X(e^{j\omega_1}, e^{j\omega_2})$ are the same as those of the one-dimensional case.

10.2 TWO-DIMENSIONAL Z-TRANSFORMATION

Similar to the one-dimensional z-transformation, the two-dimensional z-transformation is useful in design of two-dimensional digital filters. So that the Fourier transformation $X(e^{j\omega_1}, e^{j\omega_2})$ of a discrete function $x(m, n)$ exists, the right-hand side of the equation

$$X(e^{j\omega_1}, e^{j\omega_2}) = \sum_{k=-\infty}^{\infty} \sum_{r=-\infty}^{\infty} x(k, r) e^{-j\omega_1 k} e^{-j\omega_2 r}$$

must converge. If we multiply $x(m, n)$ by $e^{-\alpha_1 k} e^{-\alpha_2 r}$ and consider the result $x(m, n) e^{-\alpha_1 k} e^{-\alpha_2 r}$ as a given function, then it is possible that there exist finite α_1 and α_2 by which the right-hand side of an equation

$$\sum_{k=-\infty}^{\infty} \sum_{r=-\infty}^{\infty} x(k, r) e^{-(\alpha_1 + j\omega_1)k} e^{-(\alpha_2 + j\omega_2)r}$$

becomes finite even if the Fourier transformation of $x(m, n)$ does not exist. This means that the result of the preceding equation can be used as a transformation of $x(m, n)$. Expressing these $e^{-(\alpha_1 + j\omega_1)}$ and $e^{-(\alpha_2 + j\omega_2)}$ by z_1 and z_2, the preceding equation becomes

$$X(z_1, z_2) = \sum_{k=-\infty}^{\infty} \sum_{r=-\infty}^{\infty} x(k, r) z_1^{-k} z_2^{-r} \tag{10.15}$$

which is called the "two-dimensional z-transformation," or simply the "z-transformation" of $x(m, n)$. The two-dimensional inverse z-transformation is also the generalization of the one-dimensional z-transformation, which is

$$x(m, n) = \left(\frac{1}{2\pi j}\right)^2 \oint_{c_1} \oint_{c_2} X(z_1, z_2) z_1^{m-1} z_2^{n-1}\, dz_1 dz_2 \tag{10.16}$$

where c_1 and c_2 are Jordan curves, which contain an origin and all singularities.

Let us investigate some properties of the two-dimensional z-transformation. First, to see whether this z-transformation is linear, let $X_1(z_1, z_2)$ and $X_2(z_1, z_2)$ be the z-transformations of $x_1(m, n)$ and $x_2(m, n)$, respectively. Then the z-transformation of $ax_1(m, n) + bx_2(m, n)$ is by Equation 10.15

$$\mathscr{Z}[ax_1(m, n) + bx_2(m, n)] = \sum_{m=-\infty}^{\infty} \sum_{n=-\infty}^{\infty} [ax_1(m, n) + bx_2(m, n)]z_1^{-m}z_2^{-n}$$

The right-hand side of this equation is

$$\sum_{m=-\infty}^{\infty} \sum_{n=-\infty}^{\infty} [ax_1(m, n)z_1^{-m}z_2^{-n} + bx_2(m, n)z_1^{-m}z_2^{-n}]$$

$$= a \sum_{m=-\infty}^{\infty} \sum_{n=-\infty}^{\infty} x_1(m, n)z_1^{-m}z_2^{-n} + b \sum_{m=-\infty}^{\infty} \sum_{n=-\infty}^{\infty} x_2(m, n)z_1^{-m}z_2^{-n}$$

Hence, $\mathscr{Z}[ax_1(m, n) + bx_2(m, n)]$ is

$$\mathscr{Z}[ax_1(m, n) + bx_2(m, n)] = a\mathscr{Z}[x_1(m, n)] + b\mathscr{Z}[x_2(m, n)] \tag{10.17}$$

which shows that the two-dimensional z-transformation is linear.

We have found that the one-dimensional z-transformation of $x(m + m_0)$ is $z^{m_0}X(z)$. What will be the two-dimensional z-transformation of $x(m + m_0, n + n_0)$? By Equation 10.15,

$$\mathscr{Z}[x(m + m_0, n + n_0)] = \sum_{m=-\infty}^{\infty} \sum_{n=-\infty}^{\infty} x(m + m_0, n + n_0)z_1^{-m}z_2^{-n}$$

$$= \sum_{m'=-\infty}^{\infty} \sum_{n'=-\infty}^{\infty} x(m', n')z_1^{-(m'-m_0)}z_2^{-(n'-n_0)}$$

$$= z_1^{m_0}z_2^{n_0} \sum_{m=-\infty}^{\infty} \sum_{n=-\infty}^{\infty} x(m, n)z_1^{-m}z_2^{-n}$$

which can be expressed as

$$\mathscr{Z}[x(m + m_0, n + n_0)] = z_1^{m_0}z_2^{n_0}X(z_1, z_2) \tag{10.18}$$

Hence, we have the same form as that of the one-dimensional z-transformation.

The two-dimensional z-transformation of $a^m b^n (m, n \geq 0)$ is

$$\mathscr{Z}[a^m b^n] = \sum_{m=-\infty}^{\infty} \sum_{n=-\infty}^{\infty} a^m b^n z_1^{-m}z_2^{-n} = \sum_{m=-\infty}^{\infty} (az_1^{-1})^m \sum_{n=-\infty}^{\infty} (bz_2^{-1})^n$$

$$= \left(\frac{1}{1 - az_1^{-1}}\right)\left(\frac{1}{1 - bz_2^{-1}}\right)$$

We can generalize this result as follows: Suppose a two-dimensional discrete function $x(m, n)$ is separable as $x_1(m)x_2(n)$. Then the z-transformation of $x_1(m)x_2(n)$ is

$$\begin{aligned}\mathscr{Z}[x_1(m)x_2(n)] &= \sum_{m=-\infty}^{\infty}\sum_{n=-\infty}^{\infty} x_1(m)x_2(n)z_1^{-m}z_2^{-n} \\ &= \sum_{m=-\infty}^{\infty} x_1(m)z_1^{-m}\sum_{n=-\infty}^{\infty} x_2(n)z_2^{-n} = \mathscr{Z}[x_1(m)]\mathscr{Z}[x_2(n)]\end{aligned} \tag{10.19}$$

This means the z-transformation of a separable discrete function is also separable.

Finally, consider a discrete function $x(m, n)$, which is

$$x(m, n) = \begin{cases} K^m u_I(m - n), & m, n \geq 0 \\ 0, & \text{Otherwise} \end{cases} \tag{10.20}$$

The z-transformation $X(z_1, z_2)$ of this $x(m, n)$ is

$$\begin{aligned}X(z_1, z_2) &= \sum_{m=0}^{\infty}\sum_{n=0}^{\infty} K_{u_I}^{m}(m - n)z_1^{-m}z_2^{-n} \\ &= \sum_{m=0}^{\infty} K^m(z_1z_2)^{-m} = \frac{1}{1 - Kz_1^{-1}z_2^{-1}}\end{aligned} \tag{10.21}$$

Using the two-dimensional z-transformation, let us solve the following difference equation:

$$\sum_{k=0}^{M_1}\sum_{r=0}^{N_1} a_{kr}y(m - k, n - r) = \sum_{p=0}^{M_2}\sum_{q=0}^{N_2} b_{pq}x(m - p, n - q)$$

Taking the z-transformation of this equation, we have

$$\sum_{k=0}^{M_1}\sum_{r=0}^{N_1} a_{kr}\mathscr{L}[y(m - k, n - r)] = \sum_{p=0}^{M_2}\sum_{q=0}^{N_2} b_{pq}z[x(m - p, n - q)]$$

By Equation 10.18, this can be expressed as

$$Y(z_1, z_2)\left[\sum_{k=0}^{M_1}\sum_{r=0}^{N_1} a_{kr}z_1^{-k}z_2^{-r}\right] = X(z_1, z_2)\left(\sum_{p=0}^{M_2}\sum_{q=0}^{N_2} b_{pq}z_1^{-p}z_2^{-q}\right]$$

Then the ratio $H(z_1, z_2) = Y(z_1, z_2)/X(z_1, z_2)$ is

$$H(z_1, z_2) = \frac{\sum_{p=0}^{M_2}\sum_{q=0}^{N_2} b_{pq}z_1^{-p}z_2^{-q}}{\sum_{k=0}^{M_1}\sum_{r=0}^{N_1} a_{kr}z_1^{-k}z_2^{-r}}$$

By taking the inverse z-transformation, we can obtain $h(m, n) = y(m, n)/x(m, n)$. The inverse z-transformation, however, is not as easy as the one-dimensional inverse z-transformation.

As an example, consider a linear shift invariant system. Let us obtain a unit sample response of an output signal, $y(m, n) = 2^{m+1}u_I(m - n) - u_I(m - n)$, when an input signal is $x(m, n) = 2u_I(m - n)$. By Equation 10.21, the z-transformation of $y(m, n)$ is

$$Y(z_1, z_2) = \frac{2}{1 - 2z_1^{-1}z_2^{-1}} - \frac{1}{1 - z_1^{-1}z_2^{-1}}$$

and the z-transformation of $x(m, n)$ is

$$X(z_1, z_2) = \frac{2}{1 - z_1^{-1}z_2^{-1}}$$

Hence, the ratio $H(z_1, z_2) = Y(z_1, z_2)/X(z_1, z_2)$ is

$$H(z_1, z_2) = Y(z_1, z_2)/X(z_1, z_2) = \frac{\frac{1}{2}}{1 - 2z_1^{-1}z_2^{-1}}$$

The inverse z-transformation of this $H(z_1, z_2)$, which is $h(m, n)$, is

$$h(m, n) = \frac{1}{2}\, 2^m u_I(m - n)$$

10.3 TWO-DIMENSIONAL DFT AND FFT

To develop the DFT of a one-dimensional finite duration function, we have to introduce a one-dimensional periodic function. To extend the DFT to two-dimensional signals, we will use a two-dimensional periodic function $x(m, n)$, which must be periodic in the m-axis and periodic in the n-axis. In other words, for any integers p and q, a function $\mathring{x}(m, n)$ satisfying

$$\mathring{x}(m, n) = \mathring{x}(m - pM, n - qN) \tag{10.22}$$

is called periodic. The DFS $\mathring{X}(k, r)$ of a periodic function $\mathring{x}(m, n)$ will be

$$\mathring{X}(k, r) = \sum_{n=0}^{M-1} \sum_{n=0}^{N-1} \mathring{x}(m, n) W_M^{mk} W_N^{nr} \tag{10.23}$$

and the inverse DFS will be

$$\mathring{x}(m, n) = \frac{1}{MN} \sum_{k=0}^{M-1} \sum_{r=0}^{N-1} \mathring{X}(k, r) W_M^{-mk} W_N^{-nr} \tag{10.24}$$

It can be seen from Equation 10.23 that $\mathring{X}(k, r)$ will have the same period as $\mathring{x}(m, n)$.

As we have defined a one-dimensional finite duration function, we need to define a two-dimensional finite duration function $x(m, n)$ so that we can develop the DFT of a two-dimensional finite duration function. For simplicity, we will define a function $R(m, n)$, which is

$$R_{MN}(m, n) = \begin{cases} 1, & 0 \le m \le M-1, 0 \le n \le N-1 \\ 0, & \text{Otherwise} \end{cases} \tag{10.25}$$

By this $R(m, n)$ we can define two-dimensional finite duration function $x(m, n)$, which takes one period of $\mathring{x}(m, n)$ as

$$x(m, n) = \mathring{x}(m, n) R_{MN}(m, n)$$

Now the Fourier transformation of a two-dimensional finite duration function $x(m, n)$ can be expressed as

$$X(k, r) = \left[\sum_{m=0}^{M-1} \sum_{n=0}^{N-1} \mathring{x}(m, n) W_M^{mk} W_N^{nr} \right] R_{MN}(k, r) \tag{10.26}$$

and the inverse Fourier transformation is

$$x(m, n) = \frac{1}{MN} \left[\sum_{k=0}^{M-1} \sum_{r=0}^{N-1} \mathring{X}(k, r) W_M^{-mk} W_N^{-nr} \right] R_{MN}(m, n) \tag{10.27}$$

How should we obtain FFT from this DFT in Equation 10.26? Because $R_{MN}(m, n)$ is separable, $R_{MN}(m, n)$ can be expressed as

$$R_{MN}(m, n) = R_M(m) R_N(n)$$

Using this property, Equation 10.26 can be changed as

$$X(k, r) = \sum_{m=0}^{M-1} \left[\left(\sum_{n=0}^{N-1} \mathring{x}(m, n) W_N^{nr} \right) R_N(r) \right] W_M^{mk} R_M(k) \tag{10.28}$$

Let the symbol $V(m, r)$ be used for indicating the inside of the parentheses in the right-hand side of the preceding equation, that is,

$$V(m, r) = \sum_{n=0}^{N-1} \mathring{x}(m, n) W_N^{nr} R_N(r) \tag{10.29}$$

Then $X(k, r)$ of Equation 10.26 becomes

$$X(k, r) = \sum_{m=0}^{M-1} V(m, r) W_M^{mk} R_M(k) \tag{10.30}$$

Equations 10.30 and 10.29 are the same as the one-dimensional DFT.

Hence, we can obtain $X(k, r)$ by first obtaining $V(m, r)$ using the one-dimensional FFT and then by obtaining $X(k, r)$ using one-dimensional FFT.

Similarly, Equation 10.27 can be expressed as

$$
\begin{aligned}
x(m, n) &= \frac{1}{MN}\left(\sum_{k=0}^{M-1}\sum_{r=0}^{N-1} \mathring{X}(k, r) W_M^{-mk} W_N^{-nr}\right) R_M(m) R_N(n) \\
&= \frac{1}{M}\sum_{k=0}^{M-1}\left[\left(\frac{1}{N}\sum_{r=0}^{N-1} \mathring{X}(k, r) W_N^{-nr}\right) R_N(n)\right] W_M^{-mk} R_M(m)
\end{aligned}
\tag{10.31}
$$

By the symbol $J(k, n)$ to indicate

$$
J(k, n) = \frac{1}{N}\sum_{r=0}^{N-1} \mathring{X}(k, r) W_N^{-nr} R_N(n) \tag{10.32}
$$

Equation 10.31 becomes

$$
x(m, n) = \frac{1}{M}\sum_{k=0}^{M-1} J(k, n) W_M^{-mk} R_M(m) \tag{10.33}
$$

Equations 10.32 and 10.33 show that we can obtain $x(m, n)$ from $X(k, r)$ by first obtaining $J(k, n)$ using the one-dimensional inverse FFT and then by calculating $x(m, n)$ from $J(k, n)$ by using one-dimensional inverse FFT.

As an example, consider $x(m, n)$

$$
x(m, n) = \begin{cases} 1, & 0 \le m, n \le 2 \\ 0, & \text{Otherwise} \end{cases}
$$

We will obtain the DFT $X(k, r)$ of this $x(m, n)$ by using Equations 10.29 and 10.30. Choosing $N = 4$ as a period on both the m-axis and the n-axis, the FFT $V(m, r)$ of $x(m, n)$ will be the one shown in Fig. 10.3(a). From this FFT, we have

$$
V(m, r) = p = \begin{cases} 3, & r = 0,\ m = 0, 1, 2 \\ -j, & r = 1,\ m = 0, 1, 2 \\ 1, & r = 2,\ m = 0, 1, 2 \\ j, & r = 3,\ m = 0, 1, 2 \end{cases}
$$

When $m = 3$, $x(3, 0) = x(3, 1) = x(3, 2) = x(3, 3) = 0$. Hence, $V(3, r)$ are all 0. Fig. 10.3(b) shows the FFT $X(k, r)$ of $V(m, r)$. Notice that $V(m, r)$ for $m = 0, 1, 2, 3$ are input signals for this FFT. From this, we can obtain all $X(k, r)$, which are shown in Fig. 10.4.

We can obtain $x(m, n)$ from these $X(k, r)$ by the use of Equations 10.32 and 10.33. Fig. 10.5(a) shows the inverse FFT $J(0, n)$ of $X(0, r)$, and Fig. 10.5(b) is the FFT $J(k, n)$ of $X(k, r)$, which are from Equation 10.32. Fig.

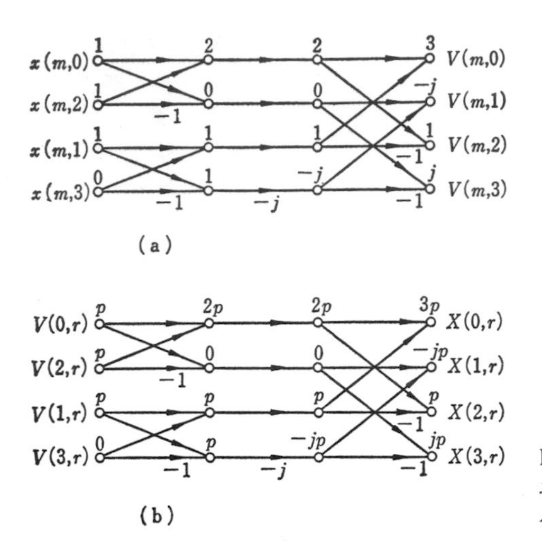

Figure 10.3 FFT for $X(k, r)$ from $x(m, n)$. (a) FFT for $V(m, r)$; (b) FFT for $X(k, r)$.

10.5(c) is the inverse FFT $x(m, n)$ of $J(k, n)$ by Equation 10.33. The reader should use these to calculate $x(m, n)$.

To investigate properties of the DFT of a finite duration function $x(m, n)$, we assume that a period of $x_1(m, n)$ and $x_2(m, n)$ on an m-axis are M and that on an n-axis are N. Also we assume that $X_1(k, r)$ is the DFT of $x_1(m, n)$ and $X_2(k, r)$ is the DFT of $x_2(m, n)$. Then the DFT $X_3(k, r)$ of $x_3(m, n) = x_1(m, n) + ax_2(m, n)$ is

$$\begin{aligned} X_3(k, r) &= \left[\sum_{m=0}^{M-1} \sum_{n=0}^{N-1} (\mathring{x}_1(m, n) + a\mathring{x}_2(m, n)) W_M^{mk} W_N^{nr}\right] R_{MN}(k, r) \\ &= \sum_{m=0}^{M-1} \sum_{n=0}^{N-1} \mathring{x}_1(m, n) W_M^{mk} W_N^{nr} R_{MN}(k, r) \\ &\quad + a \left[\sum_{m=0}^{M-1} \sum_{n=0}^{N-1} \mathring{x}_2(m, n) W_M^{mk} W_N^{nr}\right] R_{MN}(k, r) \end{aligned}$$

k \ r	0	1	2	3
0	9	$-j3$	3	$j3$
1	$-j3$	-1	$-j$	1
2	3	$-j$	1	j
3	$j3$	1	j	-1

Figure 10.4 $X(k, r)$

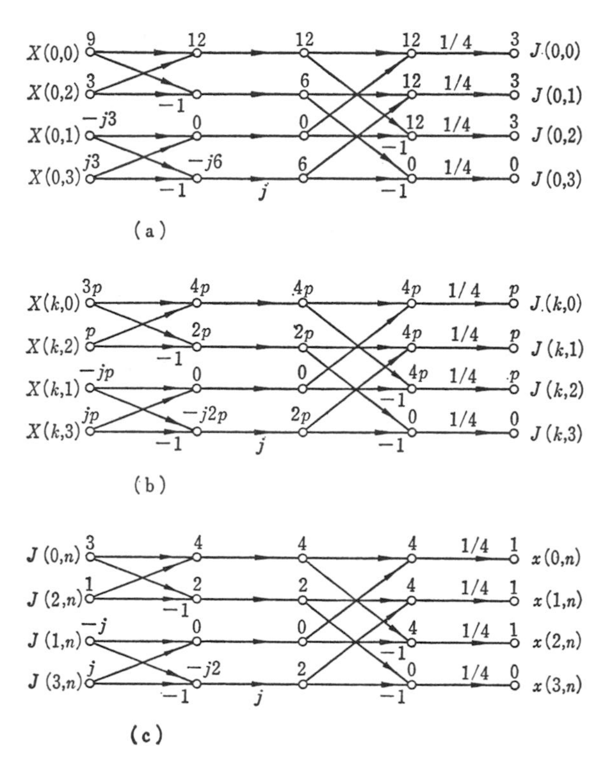

Figure 10.5 Inverse FFT for $x(m, n)$ from $X(k, r)$. (a) Inverse FFT for $J(0, n)$ from $X(0, r)$; (b) Inverse FFT for $J(k, n)$ from $X(k, r)$; (c) Inverse FFT for $x(m, n)$ from $J(k, n)$.

$$= X_1(k, r) + aX_2(k, r)$$

which shows that the two-dimensional DFT is linear.

Suppose $x(m, n)$ is separable, that is, $x(m, n)$ can be expressed as $x_1(m)x_2(n)$. Then the DFT $X(k, r)$ of $x(m, n)$ is

$$X(k, r) = \left[\sum_{m=0}^{M-1}\sum_{n=0}^{N-1} \mathring{x}(m, n)\, W_M^{mk} W_N^{nr}\right] R_{MN}(k, r)$$

$$= \left[\sum_{m=0}^{M-1} \mathring{x}_1(m)\, W_M^{mk}\right] R_M(k) \left[\sum_{n=0}^{N-1} \mathring{x}_2(n)\, W_N^{nr}\right] R_N(r)$$

$$= X_1(k)X_2(r)$$

This shows that if $x(m, n)$ is separable, then its DFT is also separable. A function $x(m, n)$ used in the previous example, which is

$$x(m, n) = \begin{cases} 1, & 0 \le m, n \le 2 \\ 0, & \text{Otherwise} \end{cases}$$

is separable. In other words, this $x(m, n)$ can be expressed as

$$x(m, n) = x_1(m)x_2(n)$$

where

$$x_1(m) = \begin{cases} 1, & 0 \le m \le 2 \\ 0, & \text{Otherwise} \end{cases}$$

$$x_2(n) = \begin{cases} 1, & 0 \le n \le 2 \\ 0, & \text{Otherwise} \end{cases}$$

Hence, the DFT $X(k, r)$ can be obtained by calculating the DFT $X_1(k)$ of $x_1(m)$ by

$$X_1(k) = \sum_{m=0}^{3} \mathring{x}_1(m) W_4^{mk} R_4(k)$$

which gives

$$X_1(k) = \begin{cases} 3, & k = 0 \\ -j, & k = 1 \\ 1, & k = 2 \\ j, & k = 3 \end{cases}$$

It is clear that the DFT $X_2(r)$ of $x_2(n)$ is the same as $X_1(k)$ of $x_1(m)$. Hence, the DFT $X(k, r)$ of $x(m, n)$ is

$$X(k, r) = X_1(k)X_2(r)$$

which gives the same result as in Fig. 10.4.

Finally, let us investigate circular convolution of $x_1(m, n)$ and $x_2(m, n)$. Suppose the DFT of $x_1(m, n)$ is $X_1(k, r)$, and the DFT of $x_2(m, n)$ is $X_2(k, r)$. Also suppose the period of $\mathring{x}_1(m, n)$ and that of $\mathring{x}_2(m, n)$ are the same and are M on the m-axis and N on the n-axis. Let $X_3(k, r)$ be

$$X_3(k, r) = X_1(k, r)X_2(k, r) \tag{10.34}$$

Then the inverse DFT $x_3(m, n)$ will be

$$\begin{aligned} x_3(m, n) &= \frac{1}{MN} \sum_{k=0}^{M-1} \sum_{r=0}^{N-1} \mathring{X}_3(k, r) W_M^{-mk} W_N^{-nr} R_{MN}(m, n) \\ &= \frac{1}{MN} \sum_{k=0}^{M-1} \sum_{r=0}^{N-1} \mathring{X}_1(k, r)\mathring{X}_2(k, r) W_M^{-mk} W_N^{-nr} R_{MN}(m, n) \end{aligned}$$

By substituting $X_1(k, r)$ and $X_2(k, r)$ by the DFT of $x_1(m, n)$ and that of $x_2(m, n)$, we have

$$x_3(m, n) = \frac{1}{MN}\sum_{k=0}^{M-1}\sum_{r=0}^{N-1}\left(\sum_{p=0}^{M-1}\sum_{q=0}^{N-1}\mathring{x}_1(p, q)\,W_M^{pk}W_N^{qr}\right)$$
$$\times\left(\sum_{s=0}^{M-1}\sum_{t=0}^{N-1}\mathring{x}_2(s, t)\,W_M^{sk}W_N^{tr}\right)W_M^{-mk}W_N^{-nr}R_{MN}(m, n)$$

or

$$x_3(m, n) = \sum_{p=0}^{M-1}\sum_{q=0}^{N-1}\mathring{x}_1(p, q)\sum_{s=0}^{M-1}\sum_{t=0}^{N-1}\mathring{x}_2(s, t)$$
$$\times\left(\frac{1}{MN}\sum_{k=0}^{M-1}\sum_{r=0}^{N-1}W_M^{(p+s-m)k}W_N^{(q+t-n)r}\right)R_{MN}(m, n) \tag{10.35}$$

The inside of the parentheses in the right-hand side of this equation is

$$\frac{1}{MN}\sum_{k=0}^{M-1}\sum_{r=0}^{N-1}W_M^{(p+s-m)k}W_N^{(q+t-n)r} \tag{10.36}$$
$$= \begin{cases} 0, & p + s - m \neq 0 \quad \text{or} \quad q + t - m \neq 0 \\ 1, & p + s - m = 0 \quad \text{and} \quad q + t - m = 0 \end{cases}$$

Hence, Equation 10.35 becomes

$$x_3(m, n) = \sum_{p=0}^{M-1}\sum_{q=0}^{N-1}\mathring{x}_1(p, q)\mathring{x}_2(m - p, n - q)R_{MN}(m, n) \tag{10.37}$$

This is the two-dimensional circular convolution of $x_1(m, n)$ and $x_2(m, n)$. It is clear from this development that the DFT of the two-dimensional circular convolution of $x_1(m, n)$ and $x_2(m, n)$ is equal to the product of the DFT of $x_1(m, n)$ and the DFT of $x_2(m, n)$.

As an example, let us obtain the DFT of circular convolution of $x_1(m, n)$ and $x_2(m, n)$ shown in Fig. 10.6. These $x_1(m, n)$ and $x_2(m, n)$ are separable and can be expressed as

$$x_1(m, n) = x_a(m)x_b(n)$$
$$x_2(m, n) = x_b(m)x_a(n)$$

where $x_a(m)$ and $x_b(n)$ are

$$x_a(m) = \begin{cases} 2, & m = 0 \\ 0, & m = 1 \\ 1, & m = 2 \\ 0, & m = 3 \end{cases}$$

$$x_b(n) = \begin{cases} 3, & n = 0 \\ 2, & n = 1 \\ 3, & n = 2 \\ 2, & n = 3 \end{cases}$$

m \ n	0	1	2	3
0	6	4	6	4
1	0	0	0	0
2	3	2	3	2
3	0	0	0	0

(a)

m \ n	0	1	2	3
0	6	0	3	0
1	4	0	2	0
2	6	0	3	0
3	4	0	2	0

(b)

Figure 10.6 $x_1(m, n)$ and $x_2(m, n)$. (a) $x_1(m, n)$; (b) $x_2(m, n)$.

Hence, the DFT $X_1(k, r)$ of $x_1(m, n)$ is the product of the DFT $X_a(k)$ of $x_a(m)$ and the DFT $X_b(r)$ of $x_b(n)$. Thus, the product of $X_a(k)$ and $X_b(r)$

$$X_1(k, r) = X_a(k)X_b(r)$$

obtained by FFTs in Fig. 10.7 will give $X_1(k, r)$, which is shown in Fig. 10.8(a). Similarly, the DFT $X_2(k, r)$ of $x_2(m, n)$ is

$$X_2(k, r) = X_b(k)X_a(r)$$

which is shown in Fig. 10.8(b). $X_3(k, r)$, which is the product of $X_1(k, r)$ and $X_2(k, r)$, that is,

$$X_3(k, r) = X_1(k, r)X_2(k, r)$$

is shown in Fig. 10.8(c).

To see that $X_3(k, r)$ is equal to the DFT of circular convolution of $x_1(m, n)$ and $x_2(m, n)$, we will calculate $x_3(m, n)$

$$x_3(m, n) = \sum_{p=0}^{3} \sum_{q=0}^{3} \mathring{x}_1(p, q)\mathring{x}_2(m - p, n - q)$$

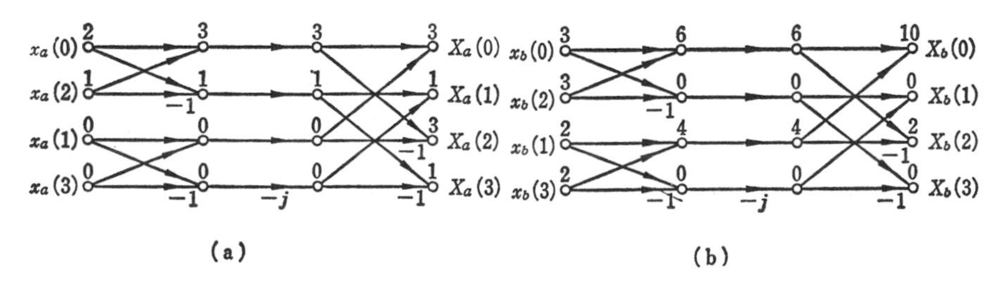

Figure 10.7 FFT for $X_a(k)$ and $X_b(r)$. (a) $X_a(k)$ from $x_a(m)$; (b) $X_b(r)$ from $x_b(n)$.

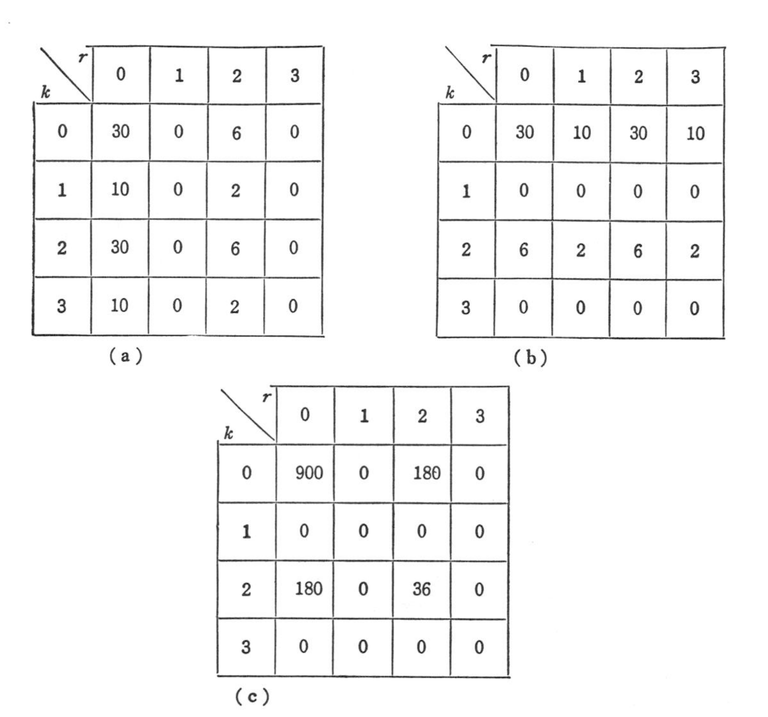

(a)

k \ r	0	1	2	3
0	30	0	6	0
1	10	0	2	0
2	30	0	6	0
3	10	0	2	0

(b)

k \ r	0	1	2	3
0	30	10	30	10
1	0	0	0	0
2	6	2	6	2
3	0	0	0	0

(c)

k \ r	0	1	2	3
0	900	0	180	0
1	0	0	0	0
2	180	0	36	0
3	0	0	0	0

Figure 10.8 $X_1(k, r)$, $X_2(k, r)$, and $X_3(k, r)$. (a) $X_1(k, r)$; (b) $X_2(k, r)$; (c) $X_3(k, r) = X_1(k, r)X_2(k, r)$.

The result is shown in Fig. 10.9, which indicates that $x_3(m, n)$ is separable as

$$x_3(m, n) = x_c(m)x_c(m)$$

where

$$x_c(m) = \begin{cases} 9, & m = 0 \\ 6, & m = 1 \\ 9, & m = 2 \\ 6, & m = 3 \end{cases}$$

$x_3(m, n)$

m \ n	0	1	2	3
0	81	54	81	54
1	54	36	54	36
2	81	54	81	54
3	54	36	54	36

Figure 10.9 $x_3(m, n)$

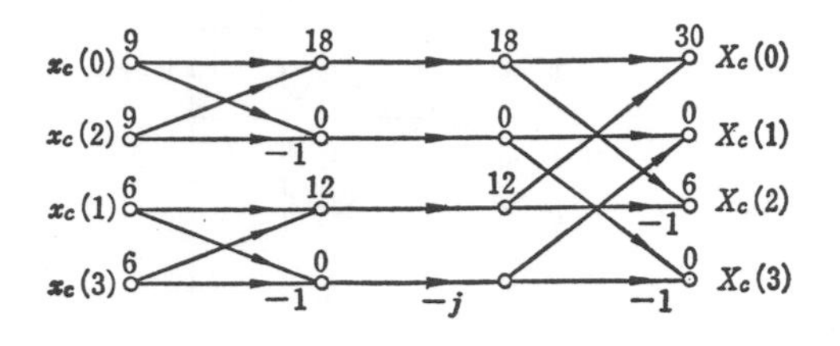

Figure 10.10 FFT for $X_c(k)$ from $x_c(m)$

The DFT of $x_c(m)$ is from the FFT in Fig. 10.10 as

$$X_c(k) = \begin{cases} 30, & k = 0 \\ 0, & k = 1 \\ 6, & k = 2 \\ 0, & k = 3 \end{cases}$$

Hence,

$$X_3(k, r) = X_c(k)X_c(k)$$

which is the same as that in Fig. 10.8.

Comparing the form of Equation 10.26, which is the DFT of $x(m, n)$, and that of Equation 10.27, which is the inverse DFT of $X(k, r)$, we can see that the difference is multiplication of $1/MN$. Hence, the DFT of the product of two functions $x_1(m, n)$ and $x_2(m, n)$ is equal to $1/MN$ times the circular convolution of $X_1(k, r)$ and $X_2(k, r)$. That is,

$$\sum_{m=0}^{M-1}\sum_{n=0}^{N-1} \mathring{x}_1(m, n)\mathring{x}_2(m, n) W_M^{mk} W_N^{nr} R_{MN}(k, r)$$
$$= \frac{1}{MN}\sum_{p=0}^{M-1}\sum_{q=0}^{N-1} \mathring{X}_1(p, q)\mathring{X}_2(k-p, r-q)R_{MN}(k, r) \tag{10.38}$$

10.4 SUMMARY

1. Two-dimensional unit sample $u_I(m, n)$ is defined as

$$u_I(m, n) = \begin{cases} 1, & m = n = 0 \\ 0, & \text{Otherwise} \end{cases}$$

2. Two-dimensional unit step $u_S(m, n)$ is defined as

$$u_S(m, n) = \begin{cases} 1, & m, n \geq 0 \\ 0, & \text{Otherwise} \end{cases}$$

3. The Fourier transformation $X(e^{j\omega_1}, e^{j\omega_2})$ of $x(m, n)$ and its inverse Fourier transformation are

$$X(e^{j\omega_1}, e^{j\omega_2}) = \sum_{k=-\infty}^{\infty} \sum_{r=-\infty}^{\infty} x(k, r) e^{-j\omega_1 k} e^{-j\omega_2 r}$$

and $$x(m, n) = \frac{1}{4\pi^2} \int_{-\pi}^{\pi} \int_{-\pi}^{\pi} X(e^{j\omega_1}, e^{j\omega_2}) e^{j\omega_1 m} e^{j\omega_2 n}\, d\omega_1\, d\omega_2$$

4. The Fourier transformation $H(e^{j\omega_1}, e^{j\omega_2})$ of a unit sample response $h(m, n)$ of a system is called a frequency response of the system.
5. If a function $x(m, n)$ is separable, it can be expressed as

$$x(m, n) = x_1(m)x_2(n)$$

6. If $x(m, n)$ is separable, then the Fourier transformation of $x(m, n)$ is also separable, and the Fourier transformation of $x(m, n) = x_1(m)x_2(n)$ is

$$\mathcal{F}[x(m, n)] = \mathcal{F}[x_1(m)]\mathcal{F}[x_2(n)]$$

7. The Fourier transformation of the convolution of functions $x(m, n)$ and $h(m, n)$ is equal to

$$\mathcal{F}\left[\sum_{p=-\infty}^{\infty} \sum_{q=-\infty}^{\infty} x(p, q)h(k - p, r - q)\right] = \mathcal{F}[x(m, n)]\mathcal{F}[h(m, n)]$$

8. The z-transformation of a function $x(m, n)$ and its inverse z-transformation are

$$X(z_1, z_2) = \sum_{k=-\infty}^{\infty} \sum_{r=-\infty}^{\infty} x(k, r) z_1^{-k} z_2^{-r}$$

and $$x(m, n) = \left(\frac{1}{2\pi j}\right)^2 \oint_{c1} \oint_{c2} X(z_1, z_2) z_1^{m-1} z_2^{n-1}\, dz_1 dz_2$$

9. Two-dimensional z-transformations are shown in Table 10.1.

TABLE 10.1 Two-Dimensional z-Transformations

$f(m, n)$	$\mathcal{L}[f(m, n)]$
$ax_1(m, n) + bx_2(m, n)$	$a\mathcal{L}[x_1(m, n)] + b\mathcal{L}[x_2(m, n)]$
$x(m + m_0, n + n_0)$	$z_1^{m0} z_2^{n0} \mathcal{L}[x(m, n)]$
$a^m b^n u_S(m, n)$	$\left(\frac{1}{1 - az_1^{-1}}\right)\left(\frac{1}{1 - bz_2^{-1}}\right)$
$x_1(m)x_2(n)$	$\mathcal{L}[x_1(m)]\mathcal{L}[x_2(n)]$
$x(m, n) = \begin{cases} K^m u_I(m - n), & m, n \geq 0 \\ 0, & \text{otherwise} \end{cases}$	$\frac{1}{1 - Kz_1^{-1}z_2^{-1}}$

10. A function $\mathring{x}(m, n)$ with period M on an m-axis and period N on an n-axis means

$$\mathring{x}(m, n) = \mathring{x}(m - pM, n - qN)$$

where p and q are integers.

11. The DFS $\mathring{X}(k, r)$ of a periodic function $\mathring{x}(m, n)$ and its inverse DFS are

$$\mathring{X}(k, r) = \sum_{m=0}^{M-1} \sum_{n=0}^{N-1} \mathring{x}(m, n) W_M^{mk} W_N^{nr}$$

and

$$\mathring{x}(m, n) = \frac{1}{MN} \sum_{k=0}^{M-1} \sum_{r=0}^{N-1} \mathring{X}(k, r) W_M^{-mk} W_N^{-nr}$$

where

$$W_M = e^{-j(2\pi/M)} \quad \text{and} \quad W_N = e^{-j(2\pi/N)}$$

12. A function $R_{MN}(m, n)$ is defined as

$$R_{MN}(m, n) = \begin{cases} 1, & 0 \le m \le M - 1, 0 \le n \le N - 1 \\ 0, & \text{Otherwise} \end{cases}$$

13. A finite duration function $x(m, n)$ can be expressed as

$$x(m, n) = \mathring{x}(m, n) R_{MN}(m, n)$$

14. The DFT $X(k, r)$ of a finite duration function $x(m, n)$ and its inverse DFT are

$$X(k, r) = \left[\sum_{m=0}^{M-1} \sum_{n=0}^{N-1} \mathring{x}(m, n) W_M^{mk} W_N^{nr} \right] R_{MN}(k, r)$$

and

$$x(m, n) = \frac{1}{MN} \left[\sum_{k=0}^{M-1} \sum_{r=0}^{N-1} \mathring{X}(k, r) W_M^{-mk} W_N^{-nr} \right] R_{MN}(m, n)$$

15. The FFT for a finite duration function $x(m, n)$ is

$$V(m, r) = \sum_{n=0}^{N-1} x(m, n) W_N^{nr} R_N(r)$$

$$X(k, r) = \sum_{m=0}^{M-1} V(m, r) W_M^{mk} R_M(k)$$

and its inverse FFT is

$$J(k, n) = \frac{1}{N} \sum_{r=0}^{N-1} X(k, r) W_N^{-nr} R_N(m)$$

$$x(m, n) = \frac{1}{M} \sum_{k=0}^{M-1} J(k, n) W_N^{-mk} R_M(m)$$

10.5 PROBLEMS

1. Calculate the Fourier transformation of the following $x(m, n)$:

a. $x(m, n) = \begin{cases} 1, & m = 0, 1, 2 \quad n = 0, 1, 2 \\ 0, & \text{Otherwise} \end{cases}$

b. $x(m, n) = \begin{cases} 1, & m = 0, 1, 2 \quad n = 0 \\ 2, & m = 0, 1, 2 \quad n = 1 \\ 3, & m = 0, 1, 2 \quad n = 3 \\ 0, & \text{Otherwise} \end{cases}$

c. $x(m, n) = \begin{cases} 2, & 0 \le m \le 3 \quad 0 \le n \le 3 \\ 0, & \text{Otherwise} \end{cases}$

d. $x(m, n) = \begin{cases} m + n, & 0 \le m \le 3 \quad 0 \le n \le 3 \\ 0, & \text{Otherwise} \end{cases}$

2. Which functions in problem 1 are separable?

3. Calculate the Fourier transformation of $x(m, n) = x_1(m)x_2(n)$ where

$$x_1(m) = \begin{cases} 1, & 0 \le m \le 3 \\ 0, & \text{Otherwise} \end{cases}$$

a. $x_2(n) = \begin{cases} 1, & 0 \le n \le 3 \\ 0, & \text{Otherwise} \end{cases}$

b. $x_2(n) = \begin{cases} n, & 0 \le n \le 3 \\ 0, & \text{Otherwise} \end{cases}$

c. $x_2(n) = \begin{cases} n^2, & 1 \le n \le 3 \\ 0, & \text{Otherwise} \end{cases}$

d. $x_2(n) = \begin{cases} 2 - n, & 0 \le n \le 3 \\ 0, & \text{Otherwise} \end{cases}$

4. Obtain the Fourier transformation of convolution of $x(m, n)$ and $h(m, n)$ where

$$x(m, n) = \begin{cases} 1, & 0 \le m \le 3 \quad 0 \le n \le 4 \\ 0, & \text{Otherwise} \end{cases}$$

and

$$h(m, n) = \begin{cases} m + 1, & 0 \le m \le 3 \quad 0 \le n \le 4 \\ 0, & \text{Otherwise} \end{cases}$$

5. Obtain the Fourier transformation of convolution of $x(m, n)$ and $h(m, n)$ where

$$x(m, n) = \begin{cases} 1, & 0 \le m \le 3 \quad 0 \le n \le 3 \\ 0, & \text{Otherwise} \end{cases}$$

and

$$h(m, n) = \begin{cases} mn, & 0 \le m \le 3 \quad 0 \le n \le 3 \\ 0, & \text{Otherwise} \end{cases}$$

6. Obtain the z-transformation of $x(m, n)$ in problem 1.

7. Obtain the z-transformation of $x(m, n)$ in problem 2.

8. Obtain the FFT of $x(m, n)$ in problem 1.

9. Calculate the FFT of $x(m, n)$ in problem 3.

10. Obtain a unit sample response $h(m, n)$ of a frequency response $H(e^{j\omega_1}, e^{j\omega_2})$ where

$$H(e^{j\omega_1}, e^{j\omega_2}) = \begin{cases} 2, & \omega_1^2 + \omega_2^2 \leq 5 \\ 0, & \text{Otherwise} \end{cases}$$

Appendix

A.1 ARITHMETIC OF COMPLEX NUMBERS

A complex number a consists of two parts a_R and a_I, which is normally expressed by $a_R + ja_I$. Graphical representation of a complex number is shown in Fig. A-1, where the horizontal axis is used for a_R, and the vertical axis is for a_I.

The angle θ between a vector a and the horizontal axis, and the length $|a|$ of the vector as shown in Fig. A-2 can be employed to express a_R and a_I as

$$\left.\begin{aligned} a_R &= |a| \cos\theta \\ a_I &= |a| \sin\theta \end{aligned}\right\} \tag{A-1}$$

where a_R is called a ''real part'' and a_I is an ''imaginary part'' of a complex number a. Also the angle θ is called the ''argument'' or the ''phase,'' and the length $|a|$ is the ''absolute value'' or the ''magnitude'' of the complex number a. The horizontal axis indicating a real part is called the ''real axis,'' and the vertical axis indicating an imaginary part is the ''imaginary axis.'' A plane formed by the real axis and the imaginary axis is called the ''complex plane.'' From

$$a = a_R + ja_I \tag{A-2}$$

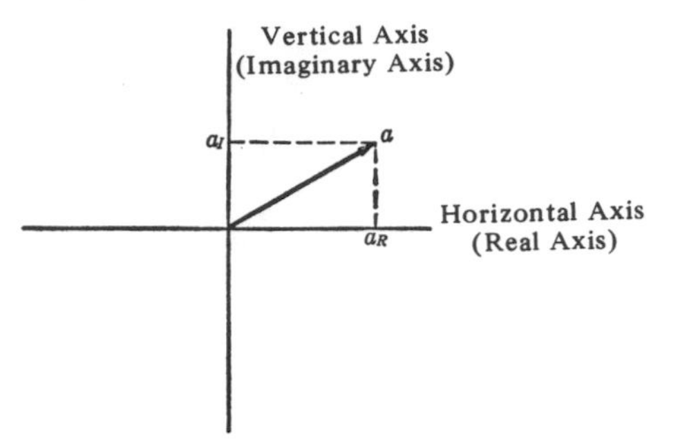

Figure A-1 Plane

and by Equation A-1, a complex number can be expressed as

$$a = |a| \cos \theta + j|a| \sin \theta \tag{A-3}$$

From Fig. A-3, this

$$|a| = \sqrt{a_R^2 + a_I^2} \tag{A-4}$$

Also, the argument θ is

$$\theta = \tan^{-1} \frac{a_I}{a_R} \tag{A-5}$$

A.2 MULTIPLICATION OF TWO COMPLEX NUMBERS

A trigonometric function $\cos \theta + j \sin \theta$ can be expressed as $e^{j\theta}$, that is,

$$e^{j\theta} = \cos \theta + j \sin \theta \tag{A-6}$$

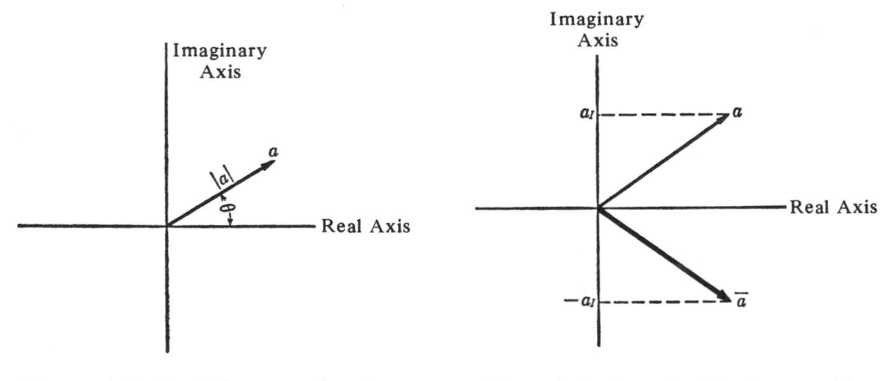

Figure A-2 Real Part a_R, Imaginary Part a_I, and Absolute Value $|a|$

Figure A-3 Complex Number a and Its Conjugate

which is known as the Euler's formula. By using this, Equation A-3 can be written as

$$a = |a|e^{j\theta} \tag{A-7}$$

which is known as the polar form where e is a constant equal to

$$e = \lim_{k=\infty}\left(1 + \frac{1}{k}\right)^k = \sum_{k=0}^{\infty} \frac{1}{k!} = 2.71828 \cdots$$

Consider two complex numbers a and b where

$$\left.\begin{aligned} a &= |a|e^{j\theta_a} = a_R + ja_I \\ b &= |b|e^{j\theta_b} = b_R + jb_I \end{aligned}\right\} \tag{A-8}$$

multiplying these two numbers, we have

$$ab = (|a|e^{j\theta_a})(|b|e^{j\theta_b}) = |a||b|e^{j(\theta_a + \theta_b)} \tag{A-9}$$

By using Equation A-6, Equation A-9 becomes

$$ab = |a||b[[\cos(\theta_a + \theta_b) + j\sin(\theta_a + \theta_b)] \tag{A-10}$$

Expanding Cos $(\theta_a + \theta_b)$ and Sin $(\theta_a + \theta_b)$, the preceding equation becomes

$$ab = |a||b||[(\cos\theta_a \cos\theta_b - \sin\theta_a \sin\theta_b) + j(\sin\theta_a \cos\theta_b + \cos\theta_a \sin\theta_b)]$$

From Equations A-6 and A-8, we have

$$\left.\begin{aligned} a_R &= |a| \cos\theta_a,\ a_I = |a| \sin\theta_a \\ b_R &= |b| \cos\theta_b,\ b_I = |b| \sin\theta_b \end{aligned}\right\} \tag{A-11}$$

Hence, Equation A-10 becomes

$$ab = (a_R b_R - a_I b_I) + j(a_I b_R + a_R b_I) \tag{A-12}$$

We can obtain Equation A-12 differently as follows: From Equation A-8, the product of complex numbers a and b is

$$ab = (a_R + ja_I)(b_R + jb_I) \tag{A-13}$$

Expanding this equation, we have

$$\begin{aligned} ab &= a_R b_R + (ja_I)(jb_I) + (ja_I)b_R + a_R(jb_I) \\ &= a_R b_R + (j)^2 a_I b_I + j(a_I b_R + a_R b_I) \end{aligned} \tag{A-14}$$

A question is how to take care of $(j)^2$ in the equation. Because j is $j1$, it is $0 + j1$. The absolute value of $0 + j1$ is 1, and its argument is 90°. Hence,

$$0 + j1 = e^{j90°} \tag{A-15}$$

Thus, $(j)^2$ is

$$(j)^2 = (j1)^2 = e^{j180°}$$

By Equation A-3, this equation becomes

$$(j)^2 = \cos 180° + j \sin 180° = -1 \quad \text{(A-16)}$$

which means that $(j)^2$ is -1. Hence, Equation A-14 is the same expression as Equation A-12.

A.3 QUOTIENT OF COMPLEX NUMBER

For a complex number a

$$a = a_R + ja_I \quad \text{(A-17)}$$

the conjugate $\bar{a}$ is defined to be

$$\bar{a} = a_R - ja_I \quad \text{(A-18)}$$

In other words, the difference between a complex number a and its conjugate $\bar{a}$ is the sign of the imaginary part. Because the conjugate $\bar{a}$ can be obtained from a complex number a just by changing the sign of the imaginary part of a, the conjugate $\bar{a}$ is the mirror image of a complex number with respect to the real axis as shown in Fig. A-3.

Because multiplying a complex number a and its conjugate $\bar{a}$ is

$$a\bar{a} = (a_R + ja_I)(a_R - ja_I) = a_R^2 + a_I^2 \quad \text{(A-19)}$$

the absolute value of a can be expressed as

$$|a| = \sqrt{a\bar{a}} \quad \text{(A-20)}$$

Also, it is clear from Equation A-20 that $a\bar{a}$ is equal to

$$a\bar{a} = |a|^2 \quad \text{(A-21)}$$

To obtain the real and imaginary parts of an inverse of a complex number that is

$$\frac{1}{a} = \frac{1}{a_R + ja_I} \quad \text{(A-22)}$$

we can multiply the conjugate by both the numerator and the denominator of the right-hand side of Equation A-22 as

$$\frac{1}{a} = \frac{1}{(a_R + ja_I)} \frac{(a_R - ja_I)}{(a_R - ja_I)} = \frac{a_R - ja_I}{|a|^2} = \frac{a_R}{|a|^2} - j\frac{a_I}{|a|^2} \quad \text{(A-23)}$$

Another way of obtaining the real and imaginary parts of $1/a$ is to express it in a polar form as

$$\frac{1}{a} = \frac{1}{|a|e^{j\theta}} = \frac{e^{-j\theta}}{|a|} \tag{A-24}$$

which can be expressed by the use of Equation A-3 as

$$\frac{1}{a} = \frac{1}{|a|}[\cos(-\theta) + j\sin(-\theta)] = \frac{|a|}{|a|^2}[\cos\theta - j\sin\theta] \tag{A-25}$$

Because $|a|$ Cos θ is a_R and $|a|$ sin θ is a_1, the preceding result is the same as that in Equation A-23.

A.4 TRIGONOMETRIC FUNCTION AND COMPLEX NUMBER

From a complex number $e^{j\theta}$ that is

$$e^{j\theta} = \cos\theta + j\sin\theta \tag{A-26}$$

we can obtain an equation for $e^{-j\theta}$ as

$$e^{-j\theta} = \cos(-\theta) + j\sin(-\theta) = \cos\theta - j\sin\theta \tag{A-27}$$

Adding these two equations, we can obtain

$$e^{j\theta} + e^{-j\theta} = 2\cos\theta \tag{A-28}$$

Hence,

$$\cos\theta = \frac{e^{j\theta} + e^{-j\theta}}{2} \tag{A-29}$$

which is a very useful formula for cos θ. Similarly, $e^{j\theta} - e^{-j\theta}$ can be expressed as

$$e^{j\theta} - e^{-j\theta} = 2j\sin\theta \tag{A-30}$$

Hence, sin θ is

$$\sin\theta = \frac{e^{j\theta} - e^{-j\theta}}{2j} \tag{A-31}$$

This is also a useful formula. For example, transformation of sin θ cos θ can be accomplished by the use of Equations A-29 and A-31 as

$$\sin\theta\cos\theta = \frac{e^{j\theta} + e^{-j\theta}}{2}\frac{e^{j\theta} - e^{-j\theta}}{2j} = \frac{e^{j2\theta} - e^{-j2\theta}}{4j} = \frac{1}{2}\sin 2\theta$$

A.5 SIGNAL FLOW GRAPH OF SIMULTANEOUS DIFFERENCE EQUATIONS

We will employ examples to show how to obtain a solution of simultaneous difference equations by the use of a signal flow graph. Suppose we would like to obtain $y_0(n)/x(n)$ of simultaneous difference equations

$$y_0(n) = y_0(n-1) + 2y_1(n) + x(n)$$
$$y_1(n) = 2y_0(n-1) + y_1(n-1)$$

To employ a signal flow graph to show the process of obtaining a solution, we use the relationships

$$y_0(n-1) = z^{-1}y_0(n)$$
$$y_1(n-1) = z^{-1}y_1(n)$$

with the given simultaneous equations to obtain a signal flow graph as shown in Fig. A-4. Now employing Mason's formula, we will obtain

$$\frac{y_0(n)}{x(n)} = \frac{1 - z^{-1}}{1 - z^{-1} - z^{-1} - 4z^{-1} + z^{-2}} = \frac{1 - z^{-1}}{1 - 6z^{-1} + z^{-2}}$$

This is equal to $y_0(z)/x(z)$ when all initial conditions are 0.

As another example, suppose we calculate $y_0(n)/x(n)$ of simultaneous equations

$$y_0(n) = y_0(n-1) + y_1(n) + 2y_1(n-1) + 2y_1(n-2)$$
$$y_1(n) = 2y_0(n) + 2y_1(n-2) + y_2(n)$$
$$y_2(n) = y_2(n-1) + x(n)$$

The relationships

$$y_0(n-1) = z^{-1}y_0(n)$$
$$y_1(n-1) = z^{-1}y_1(n)$$
$$y_1(n-2) = z^{-1}y_1(n-1)$$
$$y_2(n-1) = z^{-1}y_2(n)$$

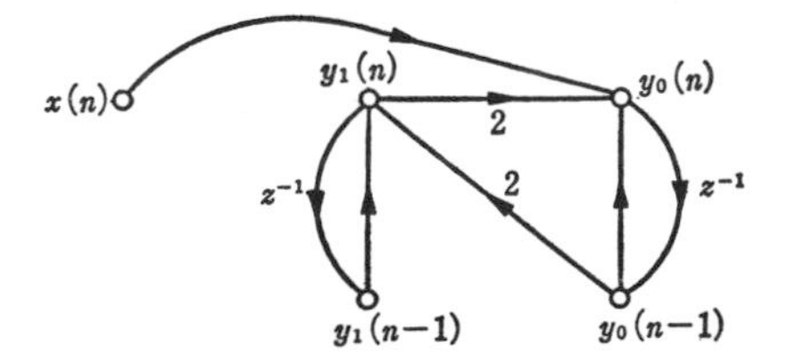

Figure A-4 Signal Flow Graph for Simultaneous Difference Equation

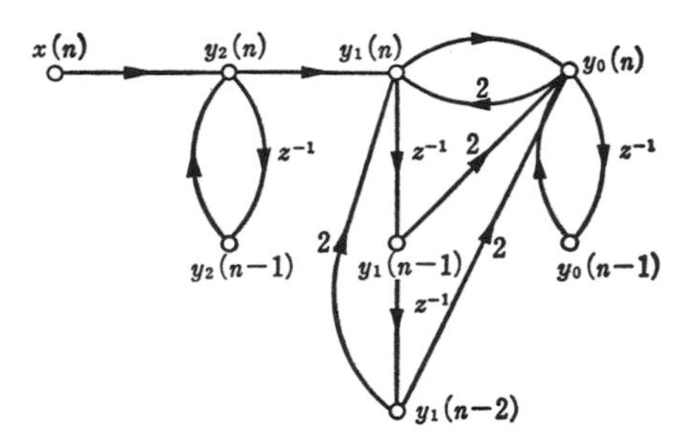

Figure A-5 Signal Flow Graph for Simultaneous Difference Equation

with the given simultaneous equations give a signal flow graph shown in Fig. A-5. By the use of Mason's formula, we can obtain

$$\frac{y_0(n)}{x(n)} = \frac{1 + 2z^{-1} + 2z^{-2}}{\begin{array}{l}1 - z^{-1} - z^{-1} - 2z^{-2} - 2 - 4z^{-1} - 4z^{-2} + z^{-2} \\ \quad + 2z^{-1} + 4z^{-2} + 4z^{-3} + 2z^{-3} + 2z^{-3} - 2z^{-4}\end{array}}$$

$$= \frac{1 + 2z^{-1} + 2z^{-2}}{-1 - 4z^{-1} - z^{-2} + 8z^{-3} - 2z^{-4}}$$

References

1. L. R. Rabiner & B. Gold: Theory and Application of Digital Signal Processing, Prentice Hall, 1975
2. A. V. Oppenheim & R. W. Schafer: Digital Signal Processing, Prentice Hall, 1975
3. E. O. Brigham: The Fast Fourier Transform, Prentice Hall, 1974
4. D. Childers & A. Durling: Digital Filtering and Signal Processing, West Publ., 1975
5. K. Steiglitz: An Introduction to Discrete Systems, John Wiley, 1974
6. W. D. Stanley: Digital Signal Processing, Reston Publ., 1975
7. H. G. Alles, J. H. Condon, W. C. Fisher & H. S. McDonald: Digital Signal Processing in Telephone Switching Proc. 1974 Intern, Conf., on Comm., Minneapolis, Minn, pp. 18, E. 1–18, E. 2, 1974
8. H. S. McDonald: Impact of Large-Scale Integrated Circuits on Communications Equipment, Proc. Natl. Electron. Conf., Vol. 24, pp. 569–572, 1968
9. S. L. Freeny, R. B. Kieburtz, K. V. Mina & S. K. Tewksbury: Systems Analysis of a TDM-FDM Transistor/Digital A-Type Channel Bank, IEEE Trans Comm. Tech., Vol. 19, No. 6, pp. 1050–1051, Dec 1971
10. A. Croisier, D. J. Esteban, M. E. Levilon & V. Riso: Digital Filter for PCM Encoded Signals, U.S. Patent No. 3777130, Dec. 4, 1973
11. J. Kriz: A 16Bit A-D-A Conversion System for High-Fidelity Audio Research,

IEEE Trans. Acoust. Speech Signal Processing, Vol. ASSP-23, No. 1, pp. 146–149, 1975

12. B. Blesser & F. Lee: An Audio Delay System Using Digital Technology, J. Audio Eng. Soc., Vol. 19, No. 5, pp. 393–397, 1971
13. D. Osborne & M. Croll: Digital Sound Signals: Bit-Rate Reduction Using an Experimental Digital Compander, BBC Monograph, Great Britain, 1973/41, Research Department, Engr Division, 1973
14. J. Myers & A. Feinberg: High-Quality Professional Recording Using Digital Techniques, J. Audio Engr Soc., Vol. 20, No. 8, pp. 622–628, 1972
15. N. Sato: PCM Recorder—A New Type of Audio Magnetic Tape Recorder, J. Audio Eng. Soc., Vol. 21, No. 7, pp. 542–558, 1973
16. A. Jones & J. Chambers: Digital Magnetic Recording: Conventional Saturation Techniques, BBC Monograph, Great Britain, 1972/9, Research Dept., Engr Division, 1972
17. F. Bellis & M. Smith: A Stereo Digital Sound Recorder, BBC Monograph, Great Britain, 1974/39, Res. Dept., Engr Division, 1974
18. H. Hessenmueller: The Transmission of Broadcast Programs in a Digital Integrated Network, IEEE Trans. Audio Electroacoustics, Vol. AU-21, No. 1, pp. 17–20, 1973
19. R. Steele: Delta Modulation Systems, John Wiley, 1975
20. R. W. Schafer: A Survey of Digital Speech Processing Techniques, IEEE Trans. Audio Electroacoustics, Vol. AU-20, No. 4, pp. 28–35, Mar. 1972
21. R. W. Schafer & L. R. Rabiner: Digital Representations of Speech Signals, Proc. IEEE, Vol. 63, No. 4, pp. 662–677, Apr. 1975
22. H. R. R. Dudley & S. S. A. Watkins: A Synthetic Speaker, J. Franklin Inst., Vol. 227, pp. 739–764, 1939
23. J. L. Kelly Jr., & C. Lochbaum: Speech Synthesis, Proc. 4th Intern. Congr. Acoust., Vol. G42, pp. 1–4, 1962
24. J. L. Flanagan, C. H. Coker & C. M. Bird: Digital Computer Simulation of a Formant-Vocoder Speech Synthesizer, 15th Ann, Meeting Audio Engr. Soc., Preprint 307, 1963
25. L. R. Rabiner, L. B. Jackson, R. W. Schafer & C. H. Coker: Digital Hardware for Speech Synthesis, IEEE Trans. Commun, Tech., Vol. Com-19, pp. 1016–1020, 1971
26. L. R. Rabiner, R. W. Schafer & C. M. Rader: The Chirp Z-Transform Algorithm and Its Applications, Bell System Tech, J., Vol. 48, pp. 1249–1292, May 1969
27. A. V. Oppenheim: Speech Spectrograms Using the Fast Fourier Transform, IEEE Spectrum, Vol. 7, pp. 57–62, Aug. 1970
28. M. R. Portnoff: Implementation of the Digital Phase Vocoder Using the Fast Fourier Transform, IEEE Trans. Acoust. Speech Signal Processing, Vol. Assp-24, No. 3, pp. 243–248, June 1976
29. J. Makhoul: Spectral Analysis of Speech by Linear Prediction, IEEE Trans. Audio Electroacoustics, Vol. AU-21, pp. 140–148, June 1973
30. H. C. Andrews & B. R. Hunt: Digital Image Restoration, Prentice Hall, 1977

31. A. Habibi & G. Robinson: A Survey of Digital Picture Coding, Computer, Vol. 7, pp. 22–35, May 1974
32. P. A. Wintz: Transform Picture Coding, Proc. IEEE, Vol. 60, pp. 809–820, 1972
33. H. C. Andrews, A. G. Tescher & R. P. Kruger: Image Processing by Digital Computer, IEEE Spectrum, Vol. 9, pp. 20–32, July 1972
34. R. S. Berkowitz, ed.: Modern Radar: Analysis, Evaluation and System Design, 1965
35. B. Gold, L. L. Lebow, P. G. McHugh & C. M. Rader: The FDP, A Fast Programmable Signal Processor, IEEE Trans. Comput., Vol. C-20, No. 1, pp. 33–38, Jan. 1971
36. J. R. Klauder, A. C. Price, S. Darlington & W. J. Albersheim: The Theory and Design of Chirp Radar, Bell System Tech. J., Vol. 39, No. 4, pp. 745–808, July 1960
37. R. J. Rurdy et al.: Digital Signal Processor Designs for Radar Applications, M.I.T. Lincoln Lab., Mass., Tech. Note 1974–58, Vols. 1 & 2, Dec. 1974
38. O. F. Hastrup: Digital Analysis of Acoustic Reflectivity in the Tyrrhenian Abyssal Plain, J. Acous. Soc. Amer., Vol. 37, pp. 1037–1051, 1965
39. W. B. Allen & E. C. Westerfield: Digital Compressed-Time Correlator and Matched Filters for Active Sonar, J. Acount. Soc. Amer., Vol. 36, pp. 121–139, 1964
40. L. C. Wood & S. Treitel: Seismic Signal Processing, Proc. IEEE, Vol. 63, pp. 649–661, 1975
41. R. R. Read, J. L. Shanks & S. Treitel: Two-Dimensional Recursive Filtering, Topics in Applied Physics, Vol. 6, pp. 131–176, 1975

Answers

CHAPTER 2

1. a. $y(n) = (-5)^{n+1}$ c. $y(n) = -\frac{3}{2} + \frac{3}{5}(-2)^n + \frac{9}{10}3^n$

   e. $y(n) = \left(\frac{4}{5} + j\frac{18}{5}\right)(j2)^n + \left(\frac{4}{5} - j\frac{18}{5}\right)(-j2)^n + \frac{2}{5}$

2. a. $h(n) = (-5)^n$ c. $h(n) = \frac{1}{3} + \frac{2}{3}(-2)^n$

   e. $h(n) = \frac{1}{2}((j2)^n + (-j2)^n)$

3. a. $y(n) = u_I(n) + 4u_I(n-1) + 10u_I(n-2) + 12u_I(n-3) + 8u_I(n-4)$

   c. $y(n) = u_I(n-1) + 2u_I(n-2) + 5u_I(n-3) + 2u_I(n-4) + 4u_I(n-5)$

5. a. $y(n) = u_I(n-2) + 3u_I(n-3) + 6u_I(n-4) + 9u_I(n-5) + 12u_I(n-6) + 15u_I(n-7) + 11u_I(n-8) + 6u_I(n-9)$

CHAPTER 3

1. a. $\frac{1}{2} + \sum_{\substack{m=-\infty \\ m\neq 0}}^{\infty} \frac{j}{2\pi m} e^{j2\pi mt}$ c. $\frac{2}{\pi}\left[1 + \sum_{\substack{m=-\infty \\ m\neq 0}}^{\infty} \frac{(-1)^{m+1}}{4m^2 - 1} e^{-j2mt}\right]$

3. a. $X(j\omega) = \begin{cases} 1, & |\omega| \leq 2\pi \\ 0, & \text{Otherwise} \end{cases}$

c. $X(j\omega) = \begin{cases} -jT_0, & \omega = \omega_0 \\ jT_0, & \omega = -\omega_0 \\ j\dfrac{2\omega \sin \omega T_0}{\omega_0^2 - \omega^2}, & \text{Otherwise} \end{cases}$

5. b. $2 \cos 2\pi f_0 t$

9. a. $H(e^{jw}) = \dfrac{1}{1 - ae^{jw}}$

c. $H(e^{jw}) = \dfrac{1}{1 - 5e^{-jw} + 4e^{-j2w}}$

CHAPTER 4

1. a. $2/s^3$ c. $10/(s^2 + 4)$ e. $\frac{1}{2}\left(\frac{s}{s^2 + 4a^2} + \frac{1}{s}\right)$ g. $\frac{1}{s^2} - \frac{2}{s-1}$

2. a.* $e^{-at} - 2ate^{-at} + \frac{a^2}{2} t^2 e^{-at}$ c. $\frac{1}{2}(1 + e^{-2t})$

e. $e^{-t}(\cos t + \sin t)$ g. $\frac{1}{2}(1 + e^{-2t})$ i. $\cos t$

3. a.* $\frac{1}{20} e^{3t} + \frac{7}{4} e^{-t} - \frac{4}{5} e^{-2t}$ c. $\frac{1}{20} e^{3t} + \frac{19}{4} e^{-t} - \frac{14}{5} e^{-2t}$

e. $\frac{5}{4} - \frac{1}{3} e^{-t} + \frac{1}{12} e^{-4t}$

4. a. $\dfrac{z(z+1)}{(z-1)^3}$ c. $\dfrac{z \sin a}{z^2 - 2z \cos a + 1}$ e. $\dfrac{5z(z^2 - 1) \sin 2}{(z^2 - 2z \cos 2 + 1)^2}$

5. a. $5^n u_s(n)$ c. $n2^n u_s(n)$ e. $\left(\frac{4}{3} n3^n + 3^n\right) u_s(n)$

* for $t > 0$.

g. $\frac{1}{5}(3^n - (-2)^n)u_s(n)$

6. a. $\frac{9}{28}3^n + \frac{9}{4}(-1)^n - \frac{8}{14}\left(\frac{-1}{2}\right)^n$

c. $\frac{9}{28}3^n + \frac{1}{4}(-1)^n - \frac{1}{14}\left(\frac{-1}{2}\right)^n$

e. $\frac{25}{126}5^n - \frac{23}{18}(-1)^n + \frac{5}{63}\left(-\frac{1}{4}\right)^n$

g. $-\frac{1}{2}n - \frac{3}{4} - \frac{1}{4}(-1)^n$

CHAPTER 5

1. a. Fig. B.1

b. Fig. B.2

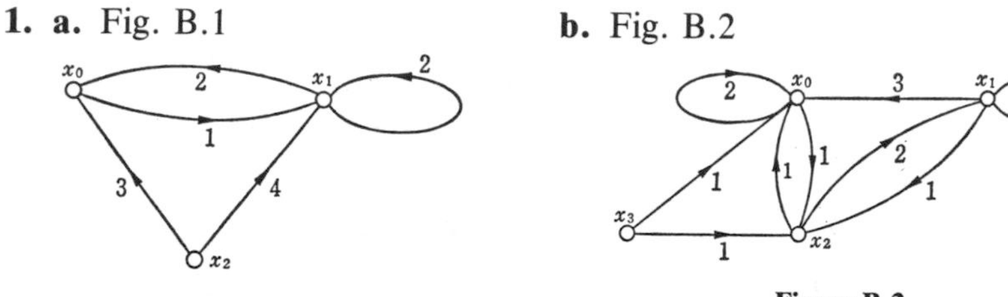

Figure B.1

Figure B.2

2. a. (a, b, c) (a, g, e) (a, c, f) (d, e, h)

3. a. (a, b, c) (c, d) (a, c, e, f) (f, g)

4. a. See Fig. B.3 and the following:

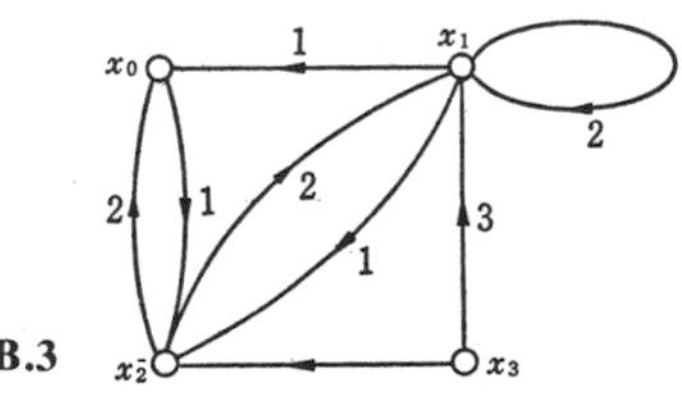

Figure B.3

Directed Loops: $C_{11} = (1)(2)$, $C_{12} = (1)(2)$, $C_{13} = (2)$, $C_{14} = (2)(1)(1)$

Directed paths: $P_1 = (3)(1)$, $P_2 = (3)(1)(2)$, $P_3 = (1)(2)$, $P_4 = (1)(2)(1)$

$$\frac{x_0}{x_3} = \frac{(3)(1) + (3)(1)(2) + (1)(2)[1 - (2)] + (1)(2)(1)}{1 - (1)(2) - (1)(2) - (2) - (2)(1)(1) + (1)(2)(2)} = -3$$

c. See Fig. B.4 and the following:

Directed Loops: $C_{11} = (1)(3)$, $C_{12} = (-2)$, $C_{13} = (2)$, $C_{14} = (1)(1)(1)$
$C_{15} = (2)(1)$, $C_{16} = (1)(1)(1)$, $C_{17} = (2)(1)(1)(1)$
$C_{18} = (1)(1)$

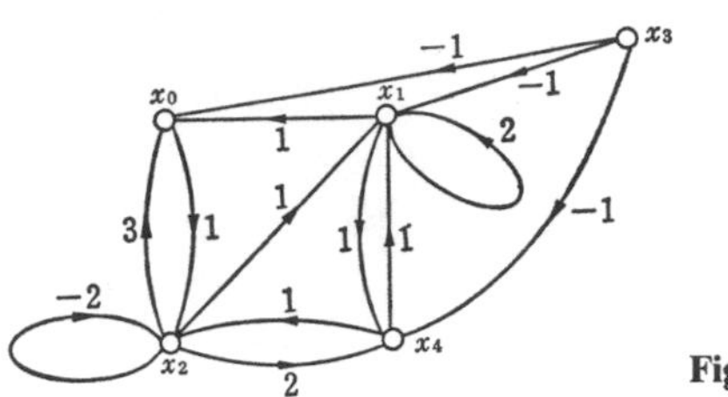

Figure B.4

Directed paths: $P_1 = (-1)$, $P_2 = (-1)(1)$, $P_3 = (-1)(1)(1)(3)$
$P_4 = (-1)(1)(1)$, $P_5 = (-1)(1)(3)$,
$P_6 = (-1)(1)(1)(1)$

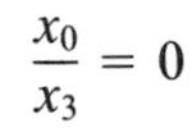

$$\frac{x_0}{x_3} = 0$$

5. a. Fig. B.5

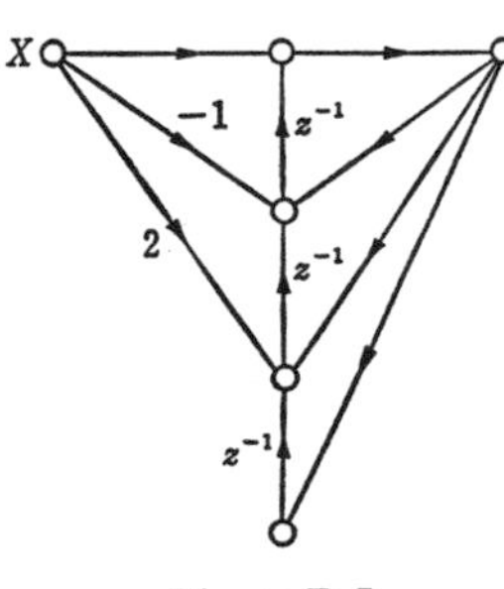

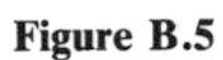

Figure B.5

c. Fig. B.6

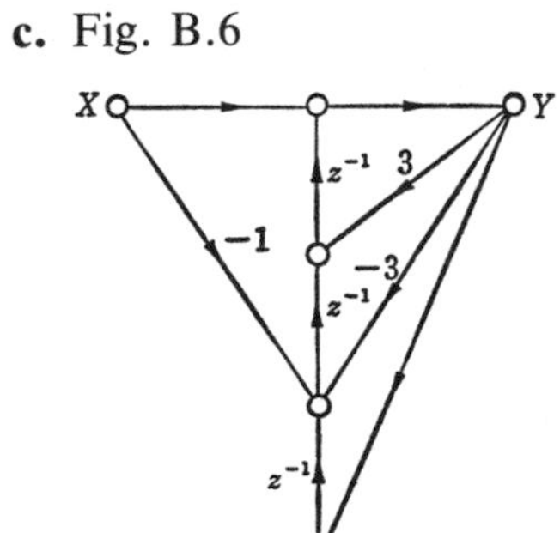

Figure B.6

6. a. Fig. B.7

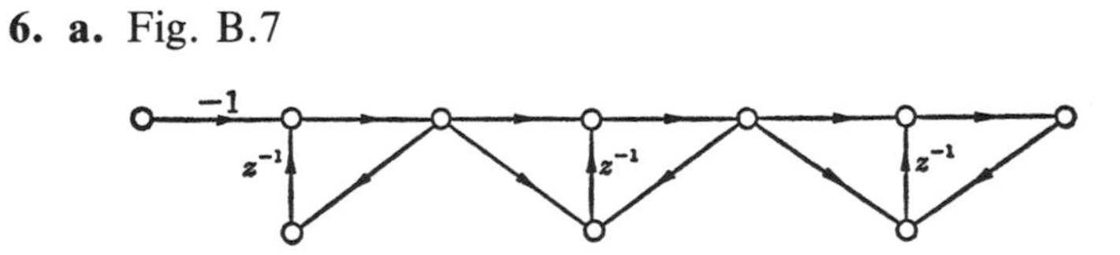

Figure B.7

c. Fig. B.8

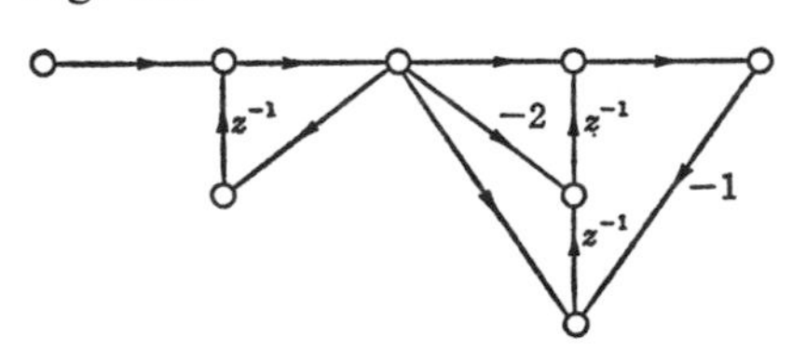

Figure B.8

CHAPTER 6

1. a. $X(0) = 8$, $X(1) = 2[1 - j(\sqrt{2} + 1)]$, $X(2) = 0$,
$X(3) = 2[1 - j(\sqrt{2} - 1)]$
$X(4) = 0$, $X(5) = 2[1 + j(\sqrt{2} - 1)]$, $X(6) = 0$,
$X(7) = 2[1 + j(\sqrt{2} + 1)]$

c. $X(0) = 0$, $X(1) = 4[1 - j(\sqrt{2} + 1)]$, $X(2) = 0$,
$X(3) = 4[1 - j(\sqrt{2} - 1)]$
$X(4) = 0$, $X(5) = 4[1 + j(\sqrt{2} - 1)]$, $X(6) = 0$,
$X(7) = 4[1 + j(\sqrt{2} + 1)]$

3. $\mathring{X}_2(2k) = \mathring{X}_1(k)$, $\mathring{X}_2(2k + 1) = 0$, $k = 0, 1, 2, \cdots, N - 1$

5. a. $\mathring{X}(0) = 15$, $\mathring{X}(1) = -4 - \sqrt{2} - j(3 + 3\sqrt{2})$, $\mathring{X}(2) = 3 + j2$,
$\mathring{X}(3) = -4 + \sqrt{2} + j(3 - 3\sqrt{2})$, $\mathring{X}(4) = 3$,
$\mathring{X}(5) = -4 + \sqrt{2} - j(3 - 3\sqrt{2})$, $\mathring{X}(6) = 3 - j2$,
$\mathring{X}(7) = -4 - \sqrt{2} + j(3 + 3\sqrt{2})$

c. $\mathring{X}(0) = 9$, $\mathring{X}(1) = -j(3 + 2\sqrt{2})$, $\mathring{X}(2) = -1$, $\mathring{X}(3) = j(3 - 2\sqrt{2})$
$\mathring{X}(4) = 1$, $\mathring{X}(5) = -j(3 - 2\sqrt{2})$, $\mathring{X}(6) = -1$, $\mathring{X}(7) = j(3 + 2\sqrt{2})$

7. a. $\mathring{x}_3(n) = \sum_{m=0}^{N-1} \mathring{x}_1(m)\mathring{x}_2(n - m)$ (See Fig. B.9)

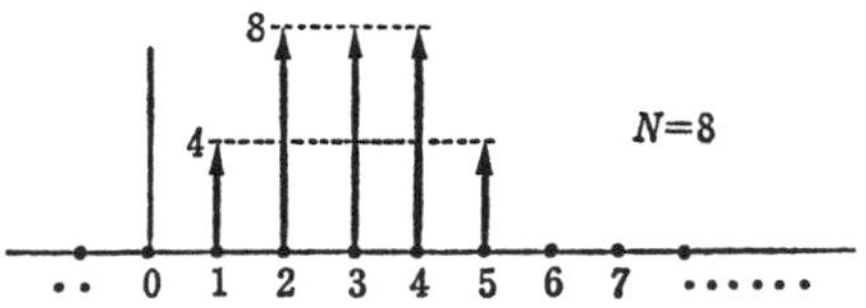

Figure B.9

8. a. $X_1(k) = \begin{cases} 8, & k = 0 \\ 0, & k = 1, 2, 3 \end{cases}$ **c.** $X_3(k) = \begin{cases} 8, & k = 2 \\ 0, & k = 0, 1, 3 \end{cases}$

e. $X_5(k) = \begin{cases} 6, & k = 0 \\ -j2, & k = 1 \\ 2, & k = 2 \\ j2, & k = 3 \end{cases}$

12. b. $x_t(n) = 0$, $n = 0, 1, 2, 3$, **d.** $x_t(n) = 12$, $n = 0, 1, 2, 3$

g. $x_t(n) = \begin{cases} 4, & n = 0, 2 \\ -4, & n = 1, 3 \end{cases}$

CHAPTER 7

3. Fig. B.10

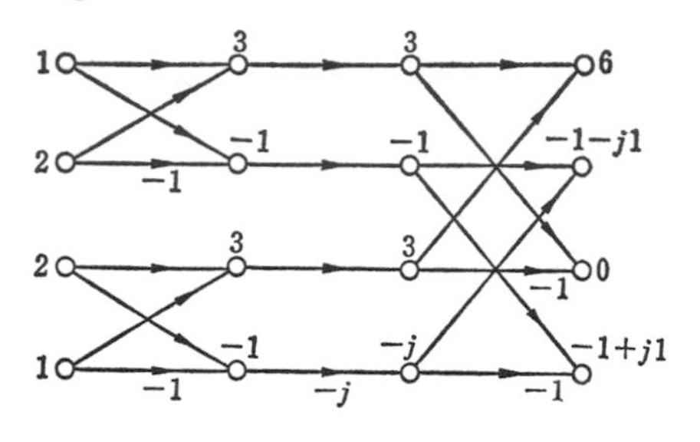

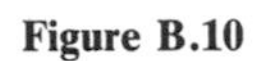
Figure B.10

5. Fig. B.11

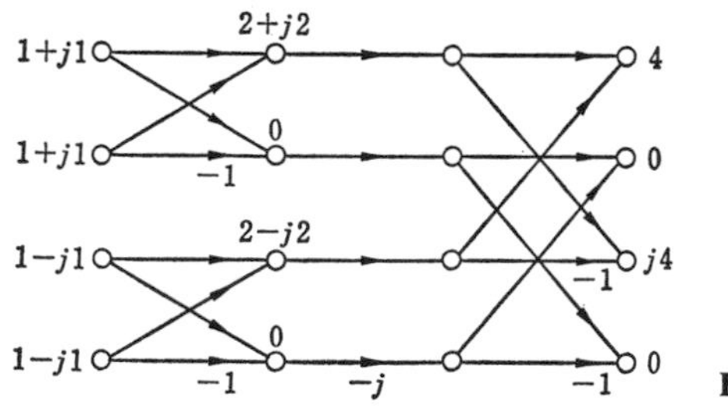

Figure B.11

9. Fig. B.12

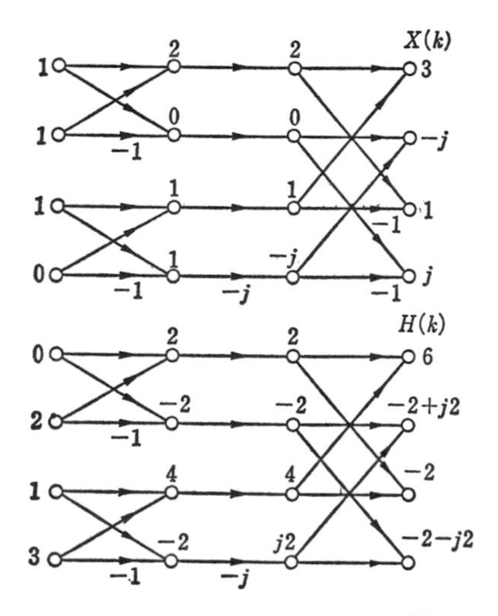

$Y(k) = X(k)\,H(k)$

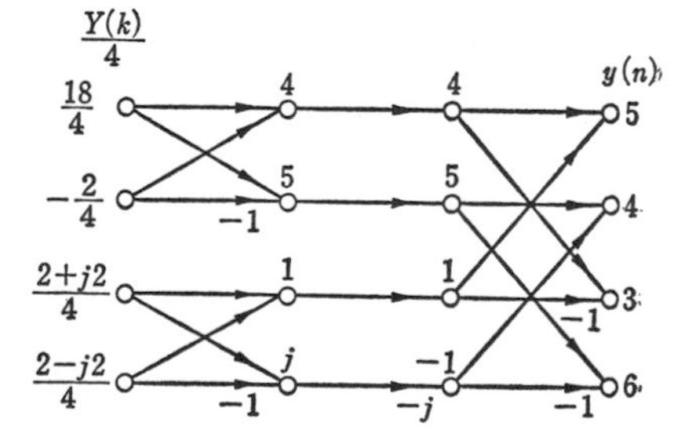

Figure B.12

CHAPTER 8

5. a.
$$\begin{aligned} H(e^{j\omega}) &= 1 + 2e^{-j\omega} + 3e^{-j2\omega} + 2e^{-j3\omega} + e^{-j4\omega} \\ &= e^{-j2\omega}[e^{j2\omega} + 2e^{j\omega} + 3 + 2e^{-j\omega} + e^{-j2\omega}] \\ &= (\cos 2\omega - j \sin 2\omega)(2 \cos 2\omega + 4 \cos \omega + 3) \end{aligned}$$

$$\underline{/H(e^{j\omega})} = \tan^{-1}\left(\frac{-\sin 2\omega}{\cos 2\omega}\right) = -2\omega$$

9. a. $H(z) = \dfrac{1}{\beta - \alpha}\left(\dfrac{1}{1 - e^{-\alpha T_0}z^{-1}}\right) + \dfrac{1}{\alpha - \beta}\left(\dfrac{1}{1 - e^{-\beta T_0}z^{-1}}\right)$

c. $H(z) = \dfrac{1}{1 - e^{-T_0}z^{-1}} + \dfrac{\frac{1}{6}(-3 + j\sqrt{3})}{1 - e^{(-1/2 - j\sqrt{3}/2)T_0}z^{-1}} + \dfrac{\frac{1}{6}(-3 - j\sqrt{3})}{1 - e^{(-1/2 + j\sqrt{3}/2)T_0}z^{-1}}$

13. a. $\dfrac{1}{11 - 10z^{-1}}$

14. a. $\dfrac{1 + z^{-1}}{21 - 19z^{-1}}$

CHAPTER 9

1. a. $t^4 + 5,\ 2t^3 + 3t$

c. $\sum_{p=0}^{N/2} a_{2p}t^{2p},\ \sum_{P=0}^{N/2} a_{2p+1}t^{2p+1}$

e. $\frac{1}{2}[e^t + e^{-t}],\ \frac{1}{2}[e^t - e^{-t}]$

2. a. $h_e(t) = \begin{cases} \frac{1}{2}(t^4 + 2t^3 + 3t + 5), & t > 0 \\ 5, & t = 0 \\ \frac{1}{2}(t^4 - 2t^3 - 3t + 5), & t < 0 \end{cases}$

c. $h_e(t) = \begin{cases} \frac{1}{2}e^t, & t > 0 \\ 1, & t = 0 \\ -\frac{1}{2}e^{-t}, & t < 0 \end{cases}$

3. $\sin^2 \omega - \cos^2 \omega + j2 \cos \omega \sin \omega$

4. a. $Q(\omega) = -\cos \omega$

5. **a.** $h_e(n) = 5n^2 + 3$ **c.** $h_e(n) = \frac{1}{2}\,(e^{j(2\pi/N)n} + e^{-j(2\pi/N)n})$

6. **a.** Fig. B.13 **c.** Fig. B.14

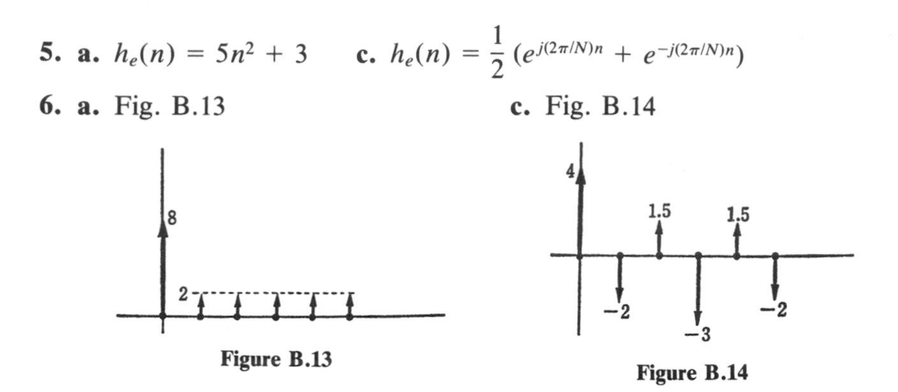

Figure B.13

Figure B.14

7. **a.** $\mathring{H}(0) = 14$, $\mathring{H}(1) = 2 - \sqrt{2} - j(2 + 3\sqrt{2})$, $\mathring{H}(2) = 4 + j2$,
$\mathring{H}(3) = 2 + \sqrt{2} + j(2 - 3\sqrt{2})$, $\mathring{H}(4) = 2$,
$\mathring{H}(5) = 2 + \sqrt{2} - j(2 - 3\sqrt{2}$, $\mathring{H}(6) = 4 - j2$,
$\mathring{H}(7) = 2 - \sqrt{2} + j(2 + 3\sqrt{2})$

8. **a.** $H(0) = 4$, $H(1) = 8 + j(4 - 4\sqrt{2})$, $H(2) = 4$,
$H(3) = 8 - j(4 + 4\sqrt{2})$, $H(4) = -12$, $H(5) = 8 + j(4 + 4\sqrt{2})$,
$H(6) = 4$, $H(7) = 8 - j(4 - 4\sqrt{2})$

CHAPTER 10

1. **a.** $X(e^{j\omega_1}, e^{j\omega_2}) = (1 + e^{-j\omega_1} + e^{-j2\omega_1})(1 + e^{-j\omega_2} + e^{-2j\omega_2})$
c. $X(e^{j\omega_1}, e^{j\omega_2}) = 2(1 + e^{-j\omega_1} + e^{-j2\omega_1} + e^{-j3\omega_1})(1 + e^{-j\omega_2} + e^{-j2\omega_2} + e^{-j3\omega_2})$

3. **a.** $X(e^{j\omega_1}, e^{j\omega_2}) = (1 + e^{-j\omega_1} + e^{-j2\omega_1} + e^{-j3\omega_1})(1 + e^{-j\omega_2} + e^{-j2\omega_2} + e^{-j3\omega_2})$

5. $X(e^{j\omega_1}, e^{j\omega_2}) = (1 + e^{-j\omega_1} + e^{-j2\omega_1} + e^{-j3\omega_1})(1 + e^{-j\omega_2} + e^{-j2\omega_2} + e^{-j3\omega_2})$
$H(e^{j\omega_1}, e^{j\omega_2}) = (e^{-j\omega_1} + 2e^{-j2\omega_1} + 3e^{-j3\omega_1})(e^{-j\omega_2} + 2e^{-j2\omega_2} + 3e^{-j3\omega_2})$
$Y(e^{j\omega_1}, e^{j\omega_2}) = X(e^{-j\omega_1}, e^{j\omega_2})H(e^{j\omega_1}, e^{j\omega_2})$

9. **a.** $X_1(k)$ is shown in Fig. B.15

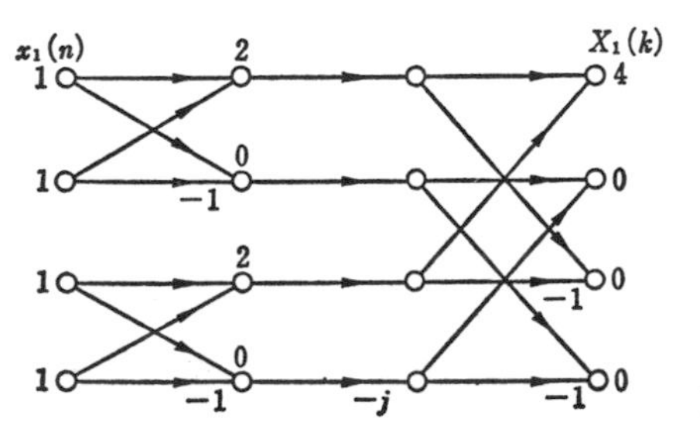

Figure B.15

$$x_2(n) = x_1(n)$$

$$X(k, r) = X_1(k)X_2(r) \begin{cases} 16, & k = r = 0 \\ 0, & \text{Otherwise} \end{cases}$$

c. $X_2(k)$ is shown in Fig. B.16.

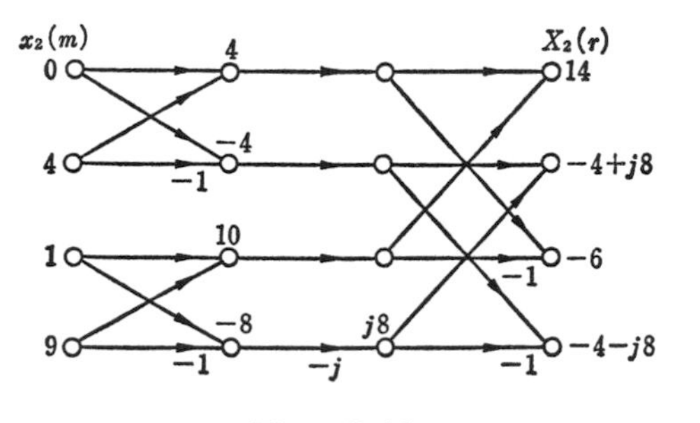

Figure B.16

Using $X_1(k)$ in part a,

$$X(k, r) = X_1(k)X_2(r)$$

$$X(k, r) = \begin{cases} 4 \cdot 14, & k = 0, r = 0 \\ 4(-4 + j8), & k = 0, r = 1 \\ 4(-6), & k = 0, r = 2 \\ 4(-4 - j8), & k = 0, r = 3 \\ 0, & \text{Otherwise} \end{cases}$$

Index

D

E

F

H

I

J

K

L

M

N

O

P

Q

S

T

U

V

Z